CW00922834

Grazing and Conservation Management

CONSERVATION BIOLOGY SERIES

Series Editors

Dr F.B. Goldsmith
Ecology and Conservation Unit, Department of Biology, University College London, Gower Street, London WC1E 6BT, UK
Tel: +44(0)171-387-7050 x2671. Fax: +44(0)171-380-7096.
email: ucbt196@ucl.ac.uk

Dr E. Duffey OBE
Chez Gouillard, 87329 Bussiere Poitevine, France

The aim of this Series is to provide major summaries of important topics in conservation. The books have the following features:

- original material
- readable and attractive format
- authoritative, comprehensive, thorough and well-referenced
- based on ecological science
- designed for specialists, students and naturalists

In the last twenty years **conservation** has been recognized as one of the most important of all human goals and activities. Since the United Nations Conference on Environment and Development in Rio in June 1992, **biodiversity** has been recognized as a major topic within nature conservation, and each participating country is to prepare its biodiversity strategy. Those scientists preparing these strategies recognize **monitoring** as an essential part of any such strategy. Chapman & Hall have been prominent in publishing key works on monitoring and biodiversity, and with this new Series aim to cover subjects such as conservation management, conservation issues, evaluation of wildlife and biodiversity.

The series contains texts that are scientific and authoritative and present the reader with precise, reliable and succinct information. Each volume is scientifically based, fully referenced and attractively illustrated. They are readable and appealing to both advanced students and active members of conservation organizations.

Further books for the Series are currently being commissioned and those wishing to contribute, or who wish to know more about the Series, are invited to contact one of the Editors.

Books already published are listed overleaf...

Already Published

1. **Monitoring Butterflies for Ecology and Conservation**
   E. Pollard and T.J. Yates (eds) (Hb 1993 o/p, Pb 1995) xiv+274pp.
   ISBN 0-412-63460

2. **Insect Conservation Biology**
   M.J. Samways (Hb 1994 o/p, Pb 1994) xv+358pp. ISBN 0-412-63450-3

3. **Monitoring for Conservation and Ecology**
   F.B. Goldsmith (ed.) (Hb/Pb 1991, Pb reprinted four times) 275pp.
   ISBN 0-412-35600-7

4. **Evaluation and Assessment for Conservation: Ecological Guidelines for Determining Priorities for Nature Conservation**
   I.F. Spellberg (Hb 1992 o/p, Pb 1994 reprinted three times) xvi+260pp.
   ISBN 0-412-44280-9

5. **Marine Protected Areas: Principles and Techniques for Management**
   S. Gubbay (ed.) (Hb 1995) xii+232pp. ISBN 0-412-59450-1

6. **Conservation and Faunal Diversity in Forested Landscapes**
   R.M. DeGraaf and R.I. Miller (eds) (Hb 1995) xxi+633pp, with colour plate section and colour foldout. ISBN 0-412-61890-7

7. **Ecology and Conservation of Amphibians**
   T.J. Beebee (Hb 1996) viii+214pp. ISBN 0-412-62410-9

8. **Conservation and the Use of Wildlife Resources**
   M. Bolton (Hb 1997) xiii+278pp, with colour plate section.
   ISBN 0-412-71350-0

9. **Conservation Management of Freshwater Habitats**
   N.C. Morgan and P.S. Maitland (Hb 1997) ISBN 0-412-59412-0

10. **Tropical Rain Forests: A Wilder Perspective**
    F.B. Goldsmith (ed.) (Hb/Pb 1998)
    ISBN Hb 0-412-81510-9, Pb 0-412-81520-6

11. **Grazing and Conservation Management**
    M.F. WallisDeVries, J.P. Bakker and S.E. Van Wieren (Hb 1998)
    ISBN 0-412-47520-0

Grazing and Conservation Management

Edited by
Michiel F. WallisDeVries
Senior Research Scientist
Department of Environmental Sciences
Wageningen Agricultural University
Wageningen, The Netherlands

Jan P. Bakker
Professor in Nature Conservation
Laboratory of Plant Ecology
University of Groningen
Haren, The Netherlands

Sipke E. Van Wieren
Senior Lecturer
Department of Environmental Sciences
Wageningen Agricultural University
Wageningen, The Netherlands

SPRINGER-SCIENCE+BUSINESS MEDIA, B.V.

A C.I.P Catalogue record for this book is available from the Library of Congress

ISBN 978-94-010-5886-5 ISBN 978-94-011-4391-2 (eBook)
DOI 10.1007/978-94-011-4391-2

Published with financial support of the Prins Bernhard Fonds

Contents

Contributors

Jan P. Bakker
Laboratory of Plant Ecology
University of Groningen
PO Box 14
9750 AA Haren
The Netherlands
e-mail j.p.bakker@biol.rug.nl

Mennobart R. Van Eerden
Ministry of Transport, Public Works and Water Management
Institute for Integral Freshwater Management and Waste Water Treatment (RIZA)
PO Box 17
8200 AA Lelystad
The Netherlands
e-mail m.veerden@riza.rws.minvenw.nl

Johan Van de Koppel
Laboratory of Plant Ecology
University of Groningen
PO Box 14
9750 AA Haren
The Netherlands

Present address:
Netherlands Institute for Ecological Research
Centre for Estuarine and Marine Research
PO Box 140
4400 AC Yerseke
The Netherlands
e-mail koppel@cemo.nioo.knaw.nl

Ger Londo
Institute for Forest Research and Nature Management
Department of Agricultural Research
PO Box 23
6700 AA Wageningen
The Netherlands

Harm Piek
Vereniging Natuurmonumenten
PO Box 9955
1243 ZS 's Graveland
The Netherlands

Richard Pott
University of Hannover
Institute of Geobotany
Nienburgerstrasse 17
D-30167 Hannover
Germany

Herbert H.T. Prins
Tropical Nature Conservation and Vertebrate Ecology Group
Department of Environmental Sciences
Wageningen Agricultural University
Bornsesteeg 69
6708 PD Wageningen
The Netherlands
e-mail herbert.prins@staf.ton.wau.nl

J. Theo Vulink
Ministry of Transport, Public Works and Water Management
Institute for Integral Freshwater Management and Waste Water Treatment (RIZA)
PO Box 17
8200 AA Lelystad
The Netherlands
e-mail t.vulink@riza.rws.minvenw.nl

Michiel F. WallisDeVries
Tropical Nature Conservation and Vertebrate Ecology Group
Department of Environmental Sciences
Wageningen Agricultural University
Bornsesteeg 69
6708 PD Wageningen
The Netherlands

Present address:
Dutch Butterfly Conservation
PO Box 506
6700 AM Wageningen
The Netherlands
e-mail vlinders@bos.nl

Sipke E. Van Wieren
Tropical Nature Conservation and Vertebrate Ecology Group
Department of Environmental Sciences
Wageningen Agricultural University
Bornsesteeg 69
6708 PD Wageningen
The Netherlands
e-mail sip.vanwieren@staf.ton.wau.nl

Preface

The use of grazing in the management of nature reserves is becoming increasingly popular. Most countries in western Europe have outstanding examples where grazing has led to high levels of biological diversity: the Camargue in France, the New Forest in England, Mols Bjerge in Denmark, Öland in Sweden, the Borkener Paradies in Germany and the Junner Koeland in The Netherlands. In many grazed reserves the grazing regime follows the pattern of traditional land use. Yet more and more grazing is applied as a tool in conservation management, a tool that can be adjusted to the needs of a particular purpose. This purpose may range from the preservation of target species and plant communities to the conservation of landscape types and restoration of ecosystems.

In The Netherlands the application of grazing as a management tool has a history of 25 to 30 years, which is longer than in most countries. The great number of grazed reserves, the diversity of settings concerned and the variety in grazing regimes imposed has generated a wealth of practical experience on the impact and the potential of grazing for nature conservation. However, both reserve managers and researchers are often confronted with the lack of a synthesis of this experience into a comprehensive framework that offers insight and guidelines for appropriate management practices. With this book we aim to fill this gap at least to some extent. We do so by reviewing the role of grazing in an ecological context and in a historical time frame. This leads to an evaluation of the potential of grazing for nature conservation and an assessment of the gaps in our current knowledge. We hope this book will provide students, researchers and reserve managers with a clearer understanding of grazing and conservation management to the ultimate benefit of nature conservation itself.

This book has a long history in the making. We are much indebted to the patience and dedication of Bob Carling and Kate Webb from Chapman & Hall, and the cooperation of John Dixon and Valerie Porter, that allowed this book to be finally published. We greatly appreciate the help and support of Rory Putman to transform the idea for this book into reality. Dick Visser once again successfully applied his invaluable skills to the drawing of clear and elegant illustrations. Maurits Gleichman and Vincent Wigbels (Biofaan) kindly supplied us with the front and back cover illustrations, respectively. Other illustrations were provided by Sip Van Wieren (Figures 3.2, 3.6, 6.9, 6.12 and 9.6), Richard Pott (Figures 4.2, 4.3, 4.6, 4.7, 4.9 and 4.11), Prof.

Schwabe-Kratochwil (Figure 4.10), Jan Bakker (Figures 5.1, 5.2 and 5.11), Kars Veling (Figure 6.6), Vincent Wigbels (Figures 7.1 and 7.11), Mennobart Van Eerden (Figure 7.15), Martijn De Jonge (Figure 7.5), Nelleke Woortman (Figure 8.1), W. Riemens (Figure 8.2), B. Muller (Figure 8.3), Jan Wamelink (Figure 8.4), Marc Pasveer (Figure 8.5) and Michiel WallisDeVries (all other photographs).

Lastly, we profoundly thank the Prins Bernhard Fonds for generous financial support that was of critical importance for the publication of this book.

Michiel WallisDeVries, Wageningen
Jan Bakker, Haren
Sipke Van Wieren, Wageningen

1

Large herbivores as key factors for nature conservation

Michiel F. WallisDeVries
Tropical Nature Conservation and Vertebrate Ecology Group, Department of Environmental Sciences, Wageningen Agricultural University, Bornsesteeg 69, 6708 PD Wageningen, The Netherlands

1.1 GRAZING AND NATURE CONSERVATION: A CONTRADICTION?

Grazing and Conservation Management – the title of this book will evoke different associations depending on one's attitude towards nature conservation. To most, grazing is linked to livestock and, therefore, the connection with nature conservation is not obvious or free from ambiguity. Thus, the first purpose of this introduction is to indicate the links between the two issues. The recent discussion in the journal *Conservation Biology* over the relationship between livestock grazing and nature conservation may serve as a good background. Noss (1994) questioned the role of grazing in the arid and semi-arid regions of the North American west:

> This we know: cattle are not native to our rangelands and they and those who manage them have caused considerable damage by performing an uncontrolled experiment over virtually the entire extent of rangelands in western North America. With this knowledge alone it is fair to conclude that range management must be drastically reformed if our conservation mission is to be fulfilled. ... In the face of uncertainty, let the burden of proof be on those who would continue grazing to show how it benefits the native ecosystem.

Noss further called for careful scientific studies to elucidate the effects of grazing on ecosystem processes and biological diversity. Fleischner (1994) and Wuerthner (1994) went on to substantiate the deleterious effects of grazing practices by reviewing the evidence. Wuerthner also argued that, as livestock production and agriculture in general are rarely economically

Grazing and Conservation Management. Edited by M.F. WallisDeVries, J.P. Bakker and S.E. Van Wieren. Published in 1998 by Kluwer Academic Publishers, Dordrecht. ISBN 0 412 47520 0.

viable in the (semi-)arid west, they should altogether cease in order to restore wildland complexes at a large scale, as outlined by Noss (1992).

This challenge against grazing practices for nature conservation was countered heavily, and sometimes furiously, by other conservation scientists (Brussard *et al.*, 1994; Curtin, 1995; Haas and Fraser, 1995; Brown and McDonald, 1995) who claimed that the reviewed evidence was severely biased, that grazing was a valuable cultural tradition, and that wise and sustainable land use should allow for an acceptable compromise between grazing and nature conservation. Brown and McDonald (1995) added an interesting argument in that many native species of the American west have evolved in the presence of large grazing mammals and, therefore, the introduction of cattle in itself is not necessarily an unnatural phenomenon in terms of ecosystem processes. This argument may not be applicable to the American west since, according to Fleischner (1994, 1996), the dominant plant species and communities of that region are not likely to have ever been subject to high levels of mammalian herbivory. Evidence presented by Truett (1996) challenges this claim by indicating that both bison (*Bison bison*) and elk (*Cervus elaphus*) were more widespread in the semi-arid southwest of the United States in recent prehistoric times than previously thought. However this may be, in a more general context it is important to view the role of herbivory over a longer, evolutionary time scale.

Clearly, the issue of grazing can and often does lead to an emotional debate among conservation scientists, let alone the other parties involved. On the one side are those who, like Noss (1992), argue that nature conservation should aim for large areas with complete indigenous animal and plant communities and as little human interference as possible. On the other side are those, such as Brussard *et al.* (1994), Curtin (1995), Haas and Fraser (1995) and Brown and McDonald (1995), who value the cultural tradition, are convinced that large wilderness areas in presently settled lands are utopic and argue that grazing and nature can be reconciled as long as humans exercise good stewardship.

In Europe the predominating standpoint is that human land use and nature conservation can fit perfectly together (e.g. Götmark, 1992); those who stress the incompatibility between the two are rarely taken seriously. However, there is general consensus that modern grazing practices are not doing any good. In the contemporary industrial era the management of livestock has become so intensive that its value for the preservation of wild plant and animal species has been virtually reduced to nil. All that remains are the benefits for a handful of species that tolerate fertilization and intensive grassland management – for example, some of the so-called meadow birds such as lapwing (*Vanellus vanellus*) and oystercatcher (*Haematopus ostralegus*) – and even these species show limits to their tolerance of intensification, as exemplified by the recent decline of the black-tailed godwit (*Limosa limosa*). Various species of wintering geese – white-fronted goose

(*Anser albifrons*), bean goose (*Anser fabialis*), greylag goose (*Anser anser*), barnacle goose (*Branta leucopsis*) – appear to tolerate intensive agricultural management, but their increase during the last two decades is also due to the cessation of hunting (Ebbinge, 1991).

In ancient traditional pastures, such as chalk grassland and montane or alpine pasture, the conservation value of grazing seems more readily apparent. The absence of chemical fertilizers, the relatively low grazing pressure and its spatial differentiation provide suitable conditions for a wide array of especially plant species and insects of open habitats (Duffey *et al.*, 1974; Bakker, 1989; Pott and Hüppe, 1991). Indeed, vegetation ecologists such as Walter (1974) and Ellenberg (1986) have lamented the loss of species after the cessation of grazing in traditional grasslands and steppe areas. The positive effects of grazing have also been stressed by the advocates of grazing in the American west. However, even in Europe, grazing effects are not always considered as beneficial. From a point of view of land use grazing is usually seen, by both foresters and ecologists, as a pest and a threat in forested regions (e.g. Walter, 1979; Ellenberg, 1986), as are the large native herbivores if these are not 'properly managed'. Indeed, European forestry only developed after livestock could be banished from the land by laws and fences and after wildlife was eradicated or controlled (Vera, 1997). This attitude persists in nature conservation practice as well as theory; conventional vegetation succession theory, based on the pioneering work of Tansley (e.g. Tansley, 1935), even excludes herbivore effects as alien disturbances. This background has led to a predominantly compartmentalized view of nature reserves: grazed open areas versus ungrazed forests. Perhaps it is not a mere coincidence that such was the landscape in the late nineteenth century when the thoughts on protecting nature took shape, inspired by the vanishing preindustrial agricultural landscape.

Can the gap between the opposing or ambivalent views towards grazing and nature conservation be bridged? The message of this book is, of course, that this is possible, but it requires a more comprehensive perspective than the ones presented above. In the heat of the debate the opposing sides tend to take positions without discussing their basic assumptions first, and thus a reasonable solution to the controversy becomes impossible. The main factors appear to be, firstly, a different view on the acceptable degree of human interference and, secondly, the automatic association of grazing with human land use. The point that needs to be made here is that there are different types of nature conservation, depending on the existing intensity of land use.

Human involvement in the landscape affects its naturalness (Anderson, 1992). This will often result in alterations of the species composition. Nevertheless, wild species of plants and animals persist even in the most human-dominated systems. To reflect the degree of human interference on the landscape several categories have been distinguished (Westhoff, 1971) which will be used throughout this book (Chapter 2):

- In the **natural** landscape, human interference is practically nil and species composition is predominantly governed by spontaneous (or natural) processes.
- In the **semi-natural** landscape, humans alter the physiognomy of the landscape but the occurring species are still, to a large extent, native and spontaneous; examples are traditional pastures, fields, coppices and heathlands.
- In the **cultivated** landscape, humans control both the physiognomy of the landscape and its species composition.

For nature conservation, the last category is of minor and the first is of major importance. However, the semi-natural landscape can be, and in fact is, another significant focus of attention because it may be the only remaining source of wild species of animals and plants in a certain region. It should be noted that the value of cultural tradition, important as it may be as a value on its own, is not an element that has any role to play in scientific discussions on conservation issues, even though, unfortunately, it frequently does – as seen in the example of the American west.

As the central topic of this book, grazing is defined as the grazing of grass-dominated vegetation by large herbivores. It is not specifically linked to livestock but can also involve wild mammals. On the other hand, even if it does concern livestock, the process of grazing may still be operating in a fashion similar to the natural process of herbivory. Therefore, although we will often discuss grazing by livestock in this book, this is by no means a necessary combination. Where it occurs, the detrimental impact of livestock is not so much due to the presence of livestock *per se*, but rather a consequence of the adopted intensity of land use. Figure 1.1 shows how the range of conservation interests encompasses a variety of natural and semi-natural grazing systems. With growing land use intensity, wild herbivores are replaced by domestic livestock, but as long as wild plant and animal species remain present, conservation interests continue to exist.

A firm answer to the leading question of this introductory section can now be given: no, grazing and nature conservation do not constitute a contradiction. The question to be answered next is why we should pay special attention to grazing when dealing with conservation.

1.2 THE KEYSTONE FUNCTION OF LARGE HERBIVORES

Grazing is a natural process directly affecting the structure and composition of plant communities (Figure 1.2). The grazing animal also exerts a direct impact by trampling and excretion of faeces and urine. In the long run basic ecosystem processes such as productivity, turnover and the distribution of nutrients may be modified. Indirectly, grazing may induce cascading effects on the entire structure and composition of an ecosystem. This is the case if

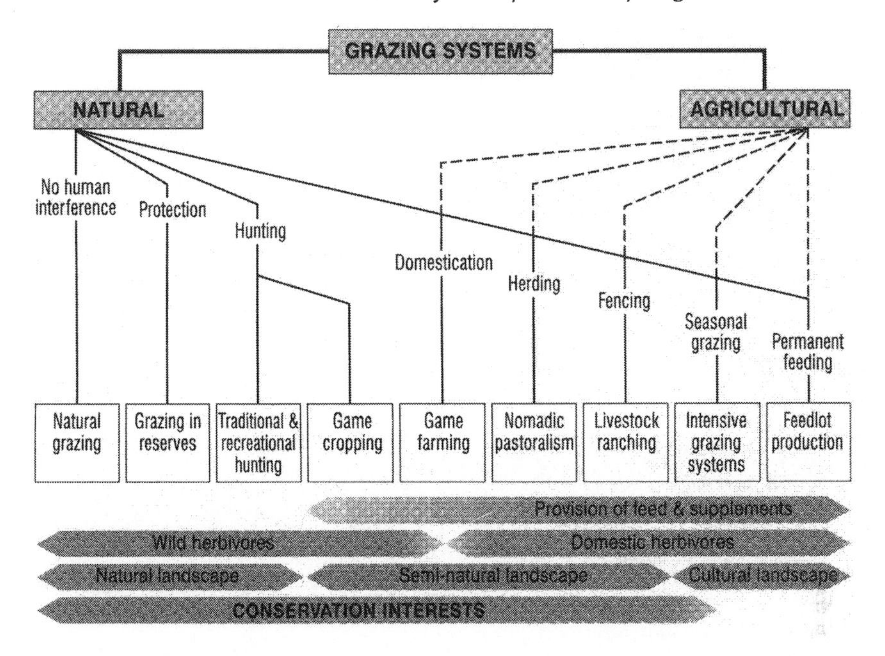

Figure 1.1 Grazing systems as a function of land use practices (after Thalen, 1987). With growing intensity of land use, wild herbivores are replaced by domestic livestock. Nature conservation interests decrease with increasing human interference but are not restricted to natural grazing systems.

direct and indirect effects provide the essential habitat conditions for other species. These impacts of grazing can be appreciated and cherished in natural ecosystems and they can be manipulated and put to use in the management of semi-natural areas. Thus, large herbivores could become key factors in conservation management by exploiting their function of keystone species.

1.2.1 What is a keystone species?

Those species whose activities have a disproportionately large impact on the patterns of occurrence, the distribution and the density of other species are termed keystone species (Paine, 1969a; Meffe and Carroll, 1994). The keystone species concept appears to be a valuable basis for the use of grazing in nature conservation but the problem is that there never has been a proper definition of what a keystone species really is, and even Paine himself, who first used the term, has only made sparse use of it – for example, Paine (1980) prefers the equivalent 'strong interactor'. Mills *et al.* (1993) proposed abandoning the concept altogether, because it is frequently used inappropriately and with widely different meanings. Instead of following this

Figure 1.2 Grazing and browsing herbivores have a large potential in shaping the structure of the vegetation.

rather fatalistic reasoning and throwing away a useful concept, it is argued here that it is more profitable to retain it but to define it properly and be more careful in its application.

Paine (1969b) introduced the term keystone species in the following passage:

> the species composition and physical appearance were greatly modified by the activities of a single native species high in the food web. These individual populations are the keystone of the community structure, and the integrity of the community and its unaltered persistence through time ... are determined by their activities and abundances.

Clearly, the essence of a keystone species is that the community structure is determined to a significant degree by its presence; conversely, its removal drastically alters the community structure. This notion can (and actually already does) occupy a central place in conservation biology. Based on this description, the following is a summary of what defines a keystone species and how it can be identified as such.

1. A keystone species markedly affects **community structure** (that is, the occurrence and relative abundance of the species belonging to the community). It is not enough to demonstrate that a species affects nutrient flows in an ecosystem, since this does not necessarily imply significant changes in community structure.

2. A keystone species can be identified by a detailed analysis of the structure of the food web. The qualification of a 'disproportionate' impact of the keystone species, used by Paine (1969b), is somewhat misleading as it does not state in relation to what quantity (numbers? biomass? nutrient flow?) it is disproportionate. However, there is no doubt that the crucial test to assess the keystone role of a species is the occurrence of a collapse or radical alteration of the community structure after its removal.
3. Although the keystone species concept was adopted initially to describe a predator-controlled community structure, many ecologists feel that it need not be limited to predators alone but that herbivores, prey species and plants may also function as keystone species (Mills *et al.*, 1993). This view is adhered to here: large herbivores should also be considered as potential keystone species.
4. The keystone species concept only applies to individual species. However, when explicitly stated, a **keystone function** could also be ascribed to a guild of species if their effects are additive instead of compensatory (Ritchie and Olff, 1998). Thus, it is conceivable that a species assemblage of large herbivores may exert a keystone function, while individual species do not.
5. In principle, only native species can be considered as keystone species. This restriction is logical when examining native communities, but when looking at the management or restoration of disturbed or fragmentary communities, it seems correct to consider the potential of introduced species as keystone species as well – that is, as long as the community is well specified and the role of the introduced species is clear on the basis of experimental evidence.
6. When a community is thoroughly influenced by human interference, discussing the keystone role of any species becomes essentially meaningless. If an introduced or even a native population of large herbivores in an area is managed to a significant extent, the only species that could possibly be performing a keystone role would be the human species, not the herbivore. Even though the herbivore's impact might be substantial, the conditions for this to happen would have been created by humans, manipulating the environment.

1.2.2 Evidence for large herbivores as keystone species

The existence of a keystone role for large herbivores in natural communities has been supported in a number of cases (Table 1.1). Only those studies have been reported here which provide some kind of comparison between community structure in the presence and in the absence of the presumed keystone species. Such comparisons have been obtained by:

- field monitoring of (changes in) communities in relation to the presence or varying density of a species;

- excluding a species by placing exclosures for a certain period;
- removal experiments including comparison with an untreated control.

Most of the evidence is derived from field monitoring and is therefore rendered somewhat circumstantial because of a lack of experimental control. On the other hand, controlled removal experiments mostly deal with a relatively short time scale and a small spatial scale. In many cases a more thorough proof for a surmised keystone function is required. Certainly when considering only mild temperate climates, a keystone function has only been documented for the non-native rabbit (but see Chapter 7 for greylag geese). It should also be noted that there is a marked lack of information on the cascading effects of keystone herbivores on the animal community. Virtually all studies have been limited to the interaction with the vegetation.

Potentially, grazers are thought to have a more significant impact on community structure and ecosystem processes than browsers, since grass-eating animals aggregate in herds and attain higher biomass concentrations than do browsers (McNaughton, 1984; Fryxell, 1991). There indeed is an accumulating body of evidence that grazers do have the potential to act as keystone species by maintaining short and open vegetation and preventing the establishment of closed woody vegetation. Grazers (or mixed feeders) for which a keystone role has been advanced include megaherbivores such as African elephant (Figure 1.3), white rhinoceros and hippopotamus, medium-sized ungulates such as bison and wildebeest, and small grazers such as rabbit (Figure 1.4) and geese (Table 1.1). It may be added that at a still smaller body size even insects can determine vegetation succession (Brown and Gange, 1992). However, in order to retain a more homogeneous functional group, throughout this book a lower threshold for 'large' will be set somewhat arbitrarily at a limit of 1.5 kg live weight: this excludes most rodents, which have widely different *r*-selected reproductive patterns (Caughley and Krebs, 1983), and cold-blooded animals, which differ in many more respects.

In at least some instances large browsing species also appear to function as keystone species, especially moose and beaver (Figure 1.5) (see Pastor and Naiman, 1992, for a modelling prediction of ecosystem changes), but possibly also impala (Table 1.1). The impact of African elephant is due in large part to their browsing activities (Van Wijngaarden, 1985; Owen-Smith, 1988).

As stated above, a detailed knowledge of the food web structure is necessary to understand truly and to ascertain correctly the keystone role of a species. This requires better insight into the impact on ecosystem processes. While the demonstration of a species' impact on nutrient flows is not sufficient proof of a keystone role, it is of great value in providing an explanatory mechanism of how the keystone function is generated. Thus, the case for a keystone role of the wildebeest is made stronger by additional studies of their impact on nutrient flows (McNaughton, 1983, 1985; Ruess and McNaughton, 1987). The same is true for the lesser snow goose (Ruess *et*

Table 1.1 Documented examples of large herbivores as keystone species: the value of the evidence provided by analysis of natural dynamics through field monitoring (FM) and exclosure experiments (EE) or removal experiments (RE) can be criticized due to the lack of experimental control in the first approach and the often limited temporal and spatial scale addressed in the other two (note the lack of evidence for repercussions on the animal community)

Species	Effects	Type of evidence	References
African elephant *Loxodonta africana*	Reduces woody cover; alters canopy structure and species composition; depresses browser densities and raises grazer densities	FM	Owen-Smith (1988) (review) Dublin *et al.* (1990)
	Shifts woodland/grassland balance to grassland	FM/EE	Van Wijngaarden (1985)
White rhinoceros *Ceratotherium simum*	Creates and maintains short sward with different species composition and higher erosion rate	EE/RE	Owen-Smith (1988)
Hippopotamus *Hippopotamus amphibius*	Creates and maintains short sward with different species composition and higher erosion rate	FM	Owen-Smith (1988) (review)
Bison *Bison bison*	Maintains grassland and prevents aspen establishment	FM	Campbell *et al.* (1994)
Moose *Alces alces*	Opens up tree canopy in aspen forest and promotes understorey; induces long-term shift to spruce forest	EE	McInnes *et al.* (1992)
Wildebeest *Connochaetes taurinus* (and other grazers?)	Maintains short grassland; reduces fire hazard; induces heavy browsing by elephants through competition for grass	FM	Dublin *et al.* (1990) Dublin (1995)
	Maintains *Andropogon-Chloris* vegetation mosaic	EE	Belsky (1986) McNaughton (1983)
Impala *Aepyceros melampus*	Prevents acacia regeneration	FM	Prins and Van der Jeugd (1993)
Beaver *Castor canadensis*	Maintains early succession riparian forest; alters hydrology, and geomorphology	FM	Naiman *et al.* (1986) Remillard *et al.* (1987)
Lesser snow goose *Chen caerulescens*	Maintains short swards; alters structure and composition of salt marsh community	EE	Bazely and Jefferies (1986)
Rabbit *Oryctolagus cuniculus*	Maintains short grassland; prevents forest regeneration and bush encroachment	FM/EE	Watt (1962,1981) Ranwell (1960) White (1961) Dobson and Crawley (1994) Van Groenendael *et al.*

Figure 1.3 As a 'bulldozer' herbivore, the elephant is a clear example of a keystone species.

Figure 1.4 Exclusion experiments reveal the effects of herbivory. In this *Agrostis/Festuca* grassland, bush encroachment is prevented by rabbit grazing.

Figure 1.5 The beaver, because of its ability to cut down adult trees and affect hydrological conditions, is a keystone species in temperate regions.

al., 1989; Hik and Jefferies, 1990; Iacobelli and Jefferies, 1991) and the moose (Pastor *et al.*, 1993; see also Molvar *et al.*, 1993).

In summary, the role of keystone species seems to apply to members from a wide spectrum of large herbivores, with grazing species as the potentially most powerful group in influencing community structure. In the course of the book we will therefore keep the emphasis on grazing species but where applicable we will address large herbivores in general, including the browsing species. Looking at entire large herbivore species assemblages might even be of crucial importance in correctly assessing their role. It may be that large herbivores only act as keystone species by complementing each other as a guild rather than as individual species. Thus, while wildebeest, in their multitude, are thought to be a keystone species in the Serengeti plains in some respects, this role might in fact be played by the combination of wildebeest, zebra (*Equus burchelli*) and Thomson's gazelle (*Gazella thomsoni*); the available evidence cannot discriminate between the two possibilities. Furthermore, competitive interactions can induce cascading effects. As hypothesized by Dublin (1995), wildebeest could increase the browsing impact of elephants by monopolizing the food resources of the grassland. This usage of complementary resources is another reason why the guild of large herbivores as a whole could perform a keystone role, even if individual species cannot. The preservation of complete species assemblages of large herbivores could thus well be a prerequisite for maintaining their impact, as argued by Prins and Olff (1998).

1.2.3 Large herbivores in conservation management: keystones and key factors

Through their impact on ecosystem processes and community structure, large herbivores can act as key factors for conservation management. In natural communities where large herbivores are found to be keystone species, conservation measures should aim to preserve this role in order to safeguard the entire system. In semi-natural communities, where humans are a permanent significant factor, the potential keystone role of large herbivores loses its meaning because they are no longer in control. The challenge is then to discover if, and how, large herbivores can be used to create or maintain desirable communities. Inspired by the animals' potential as keystone species, conservation managers could shape the conditions in which large herbivores can control and direct community structure, in analogy to natural ecosystems.

Four main avenues for the use of large herbivores as key factors in conservation management can now be distinguished:

- In more or less intact natural systems, large herbivores may be keystone species and therefore their presence can be essential for the survival of many other species, especially those from open habitats.

- Large herbivores play a great role in the preservation of semi-natural communities.
- Large herbivores are promising candidates for a main role in restoration management: as determinants of the structure of the plant community in the short run and as potential determinants of system dynamics in the long run their (re)introduction can facilitate the restoration of biologically diverse areas.
- In disturbed systems impoverished by excessive land use, artificially high densities of large herbivores may be one of the main causes of the impoverishment and ways of proper management should then be developed.

Clearly, when herbivore densities are controlled, the art of good management will be to optimize herbivore impact by regulating density and providing suitable habitat conditions.

The main problem in using large herbivores as a management tool is that it is a relatively new approach with a corresponding lack of knowledge and experience concerning the relation between cause and effect. Agricultural science and range management have yielded powerful insights in the feeding, nutrition and reproduction of livestock and their interactions with pasture growth and composition (e.g. Holechek *et al.*, 1995). Useful though these insights may be, they pertain above all to high animal density ranges and highly controlled conditions aimed at satisfying the demands of food production. They are of limited use, therefore, in dealing with the lower density ranges and the less intensive management conditions required for the preservation of wild species of plants and animals. Moreover, agricultural sciences have generated little understanding of the interrelation between large herbivores and biological diversity. This is rather the domain of the ecological disciplines (Hodgson and Illius, 1996) but these have only just started to explore the realm of systems dynamics in relation to herbivory, as exemplified by the recent modelling work of Van de Koppel *et al.* (1996). Theory and practice are still far from being reconciled. However, in the meantime practice has gone ahead on its own. Grazing is becoming a more and more popular management tool in nature conservation, especially in western Europe (Bakker, 1989; Hermy, 1989; Gordon *et al.*, 1990; Van Wieren, 1991; WallisDeVries, 1995). The need for scientific evidence to back up practical decisions is growing with the spread of this tool (Figure 1.6) (De Molenaar, 1996). The challenge we now face is: can we bridge the growing gap between management questions and scientific insights in the field of grazing and nature conservation?

1.3 SCOPE AND CONTENTS OF THIS BOOK

The two main purposes of this book readily follow from the preceding considerations. The first is to bring together both theoretical and practical evidence to increase the understanding of the impact of grazing on community

Figure 1.6 Grazing research is necessary to elucidate the impact of large herbivores on their environment.

structure and composition. The second purpose is to put this knowledge in the perspective of conservation management and thereby to facilitate the task of the people involved in the implementation of measures to preserve the wild flora and fauna for generations to come.

The book focuses on the grasslands and other characteristic open vegetation types of the temperate regions. The choice of the temperate zone is dictated by the fact that a wide array of conservation aims and methods have been developed particularly in this area, probably as a result of the relatively long history of widespread human influence in such regions in conjunction with economic wealth and the willingness to start protecting the remaining areas of natural value. The aims of conservation range from the preservation of natural ecosystems and the safeguarding of biologically diverse semi-natural systems to the restoration of areas impoverished by humans and the development of 'new nature'. In these areas large herbivores often play a significant role either as part of the ecosystem or as a tool in the management of the area.

The use of grazing as a special and powerful management tool has been recognized in many parts of western Europe, especially in The Netherlands (Chapter 2). As a result, a fairly large body of empirical evidence has accumulated from this region through studies accompanying conservation measures and studies designed to provide more general background information. The main problem is that this information has not been available in a comprehensive form but scattered over too many different journal articles, inaccessible reports and unpublished information held by conservation bodies. This book intends to put an end to such an undesirable state of affairs by drawing together contributions from conservation-oriented scientists and managers from The Netherlands and from Germany. The European perspective we present should also be relevant to other regions, as this book aims at generality by emphasizing ecological principles and management considerations.

In the following chapters there will be a focus on open vegetation, as this is the domain where large herbivores can play their greatest role. Considerable attention will also be devoted to the so-called park landscape. This picturesque combination of woodland and pasture, which will be termed **wood-pasture**, has been the model for many conservationists in Europe since its biological richness was recognized from remnants of communally grazed areas of medieval origin (Thalen, 1984; Tubbs, 1986; Pott and Hüppe, 1991). This represents an interesting shift of perspective as it transcends the conventional European compartmentalized view of nature and semi-natural reserves, mentioned earlier. It recognizes the dynamic nature of ecosystems and extends the range of valued vegetation types to include forests (see also Putman, 1986). As reviewed by Dudley (1992), the conservation status of temperate forests is abominable and it is high time that conservationists actively treat them as potentially valuable parts of the landscape instead of the usual stereotype of an ever-present invasive threat to fragile open vegetation. However, how large herbivores affect the woodland pasture remains an open question. Do they indeed act as keystone species or do they only play a role when other disturbances – be they natural or human in origin – pave their way? It would be overambitious to try to provide the final answer to this question in the following chapters, since it requires so much more knowledge of systems dynamics than we now possess. Nevertheless, this seems to be the key question underlying many problems encountered in the use of grazing in conservation management. It also provides the context in which this book should be placed.

The book consists of four main parts, each with a different emphasis. In Part One the focus is on the historical and ecological background of grazing as a management tool. Chapter 2 sketches the development in thought and practice that transformed grazing from agricultural exploitation to conservation management, initially for maintaining existing communities and later on for restoration management. As grazing is only one among many possible management practices, it is considered in the general framework of nature conservation. The next two chapters describe the natural context in which to place large herbivores and their impact on plant communities. Chapter 3 describes the origins of grassland communities on an evolutionary time scale and addresses the question of their naturalness under contemporary conditions. This perspective makes one aware of the tremendous changes that have taken place in the species composition of both plant communities and herbivore assemblages under the influence of climatic change. In doing so, it also points to the enormous difficulty as well as the challenge in providing a meaningful frame of reference for restoration management in regions with a long history of human occupation. Chapter 4 goes on to elucidate the influence of human land use and grazing by livestock in shaping the semi-natural landscape. It presents an overview of the most important plant communities of central and western Europe that depend on grazing.

Part One has thus set the stage for going into the different aspects of grazing impact on community structure, dealt with in Part Two. A plant ecological perspective is taken in Chapter 5, which treats the impact of grazing on the composition, diversity and dynamics of plant communities. This field has enjoyed the most extensive attention in grazing studies. The emerging overview is thus fairly complete – that is, with respect to short-term and medium-term changes (up to 25 years) in individual communities grazed by a single species of herbivore. The more complex long-term dynamics of a landscape mosaic with different herbivore species is the challenge for exploration in the future. Chapter 6 complements the preceding chapter with an overview of the interactions of large herbivores with the composition of the animal community. It stresses the importance of viewing the effects of herbivory as density-dependent and thereby refines the habitual conception of herbivore influence as either present or absent.

After the scientific considerations of grazing influence, Part Three continues with the application of grazing in the practice of conservation management. Chapter 7 deals with the combined effects of herbivory and cyclic changes in water level on the development of artificial wetlands. As such, it illustrates the fundamental choices that are required when opting for a systems approach to management under the constraints of a restricted reserve area. The reader is presented with fascinating large-scale developments, both spontaneous and induced by management. These changes demand adequate monitoring and experimental investigations to be understood and necessitate careful decision-making for management. Despite this considerable task, the ecosystem management approach described here will probably become increasingly important in the future of grazed nature reserves. A more general overview of the full diversity of grazing management in practice is given in Chapter 8, which concentrates on Dutch nature reserves. It provides insight into the main types of grazing management, the habitats that are covered and the results achieved. A discussion is included on current developments and the main management options that have proved successful until now.

Part Four makes up the balance of perspectives and limitations for the implementation of grazing management in nature reserves. Chapter 9 evaluates the limiting factors associated with habitat quality. It follows an animal-oriented perspective by reviewing those habitat factors that determine herbivore performance. Special attention is given to the abundance and quality of the food supply and to the role of spatial heterogeneity. The chapter concludes with general management recommendations for habitat evaluation. Chapter 10 covers the scientific foundations of grazing management by reviewing models addressing the various fields involved, from herbivore foraging and plant–herbivore interactions to decision support systems. It identifies the gaps in existing knowledge and theoretical developments that should be addressed in future research. The final chapter sums up the lessons from experience with grazing management. It evaluates the results of grazing management in practice with respect to the main management objectives and reviews the necessary conditions required for its successful application. The closing section is devoted to the identification of future areas of research to improve the value of grazing for nature conservation.

REFERENCES

Anderson, J.E. (1992) A concept for evaluating and quantifying naturalness. *Conservation Biology*, **5**, 347–352.

Bakker, J.P. (1989) *Nature Management by Grazing and Cutting*, Kluwer, Dordrecht.

Bazely, D.R. and Jefferies, R.L. (1986). Changes in the composition and standing crop of salt-marsh communities in response to the removal of a grazer. *Journal of Ecology*, **74**, 693–706.

Belsky, J.A. (1986) Population and community processes in a mosaic grassland in the Serengeti. *Journal of Ecology*, **52**, 511–544.

Brown, J.H. and McDonald, W. (1995) Livestock grazing and conservation on Southwestern rangelands. *Conservation Biology*, **9**, 1644–1647.

Brown, V.K. and Gange, A.C. (1992) Secondary plant succession: how is it modified by insect herbivory? *Vegetatio*, **101**, 3–13.

Brussard, P.F., Murphy, D.D. and Tracy, C.R. (1994) Cattle and conservation – another view. *Conservation Biology*, **8**, 919–921.

Campbell, C., Campbell, I.D., Blyth, C.B. and McAndrews, J.H. (1994) Bison extirpation may have caused aspen expansion in western Canada. *Ecography*, **17**, 360–362.

Caughley, G. and Krebs, C.J. (1983) Are big mammals simply small mammals writ large? *Oecologia*, **59**, 7–17.

Curtin, C.G. (1995) Grazing and advocacy. *Conservation Biology*, **9**, 233.

De Molenaar, J.G. (1996) *Gedomesticeerde Grote Grazers in Natuurterreinen en Bossen: een Bureaustudie – 1. De Werking van Begrazing*. IBN-rapport 231, IBN-DLO, Wageningen.

Dobson, A. and Crawley, M.J. (1994) Pathogens and the structure of plant communities. *Trends in Ecology and Evolution*, **9**, 393–398.

Dublin, H.T. (1995) Vegetation dynamics in the Serengeti-Mara ecosystem: the role of elephants, fire and other factors, in *Serengeti II – Dynamics, Management, and Conservation of an Ecosystem*, (eds A.R.E. Sinclair and P. Arcese), pp. 71–90, University of Chicago Press, Chicago.

Dublin, H.T., Sinclair, A.R.E. and McGlade, J. (1990) Elephants and fire as causes of multiple states in the Serengeti-Mara woodlands. *Journal of Animal Ecology*, **59**, 1147–1164.

Dudley, N. (1992) *Forests in Trouble: a review of the status of temperate forests worldwide*, WWF, Gland.

Duffey, E., Morris, M.G., Sheail, J. *et al.* (eds) (1974) *Grassland Ecology and Wildlife Management*, Chapman & Hall, London.

Ebbinge, B.S. (1991) The impact of hunting on mortality rates and spatial distribution of geese, wintering in the western palearctic. *Ardea*, **79**, 197–209.

Ellenberg, H. (1986) *Vegetation Mitteleuropas mit den Alpen in Ökologischer Sicht*, 4th edn, Eugen Ulmer, Stuttgart.

Fleischner, T.L. (1994) Ecological costs of livestock grazing in Western North America. *Conservation Biology*, **8**, 629–644.

Fleischner, T.L. (1996) Livestock grazing: replies to Brown and McDonald. *Conservation Biology*, **10**, 927–929.

Fryxell, J.M. (1991) Forage quality and aggregation by large herbivores. *American Naturalist*, **138**, 478–498.

Gordon, I.J., Duncan, P. Grillas, P. and Lecomte, T. (1990) The use of domestic herbivores in the conservation of the biological richness of European wetlands. *Bulletin d'Ecologie*, **21**, 49–60.

Götmark, F. (1992) Naturalness as an evaluation criterion in nature conservation: a response to Anderson. *Conservation Biology*, **6**, 455–458.

Haas, C.A. and Fraser, J.D. (1995) Grazing and advocacy. *Conservation Biology*, **9**, 234–235.

Hermy, M. (ed.) (1989) *Natuurbeheer*, Van de Wiele, Stichting Leefmilieu, Natuurreservaten en Instituut voor Natuurbehoud, Brugge, Belgium.

Hik, D.S. and Jefferies, R.L. (1990) Increases in the net above-ground primary production of a salt marsh forage grass: a test of the predictions of the herbivore-optimization model. *Journal of Ecology*, **78**, 180–195.

Hodgson, J. and Illius, A.W. (eds) (1996) *The Ecology and Management of Grazing Systems*, CAB International, Wallingford, UK.

Holechek, J.L., Pieper, R.D. and Herbel, C.H. (1995) *Range Management: Principles and Practices*, 2nd edn, Prentice Hall, Englewood Cliffs, New Jersey.

Iacobelli, A. and Jefferies, R.L. (1991) Inverse salinity gradients in coastal marshes and the death of stands of *Salix*: the effect of grubbing by geese. *Journal of Ecology*, **79**, 61–73.

McInnes, P.F., Naiman, R.J., Pastor, J. and Cohen, Y. (1992) Effects of moose browsing on vegetation and litter of the boreal forest, Isle Royale, Michigan, USA. *Ecology*, **73**, 2059–2075.

McNaughton, S.J. (1983) Serengeti grassland ecology: the role of composite environmental factors and contingency in community organization. *Ecological Monographs*, **53**, 291–320.

McNaughton, S.J. (1984) Grazing lawns: animals in herds, plant form and co-evolution. *American Naturalist*, **124**, 863–886.

McNaughton, S.J. (1985) Ecology of a grazing system: the Serengeti. *Ecological Monographs*, **55**, 259–294.

Meffe, G.K. and Carroll, C.R. (1994) (eds) *Principles in Conservation Biology*, Sinauer, Sunderland, Massachusetts.

Mills, L.S., Soulé, M.E. and Doak, D.F. (1993) The keystone-species concept in ecology and conservation. *Bioscience*, **43**, 219–224.

Molvar, E.M., Bowyer, R.T. and Van Ballenberghe, V. (1993) Moose herbivory, browse quality, and nutrient cycling in an Alaskan treeline community. *Oecologia*, **94**, 472–479.

Naiman, R.J., Melillo, J.M. and Hobbie, J.E. (1986) Ecosystem alteration of boreal forest streams by beaver (*Castor canadensis*). *Ecology*, **67**, 1254–1269.

Noss, R.F. (1992) The Wildlands project: land conservation strategy. *Wild Earth* Special Issue, 10–25.

Noss, R.F. (1994) Cows and conservation biology. *Conservation Biology*, **8**, 613–616.

Owen-Smith, R.N. (1988) *Megaherbivores – the Influence of Very Large Body Size on Ecology*, Cambridge University Press, Cambridge.

Paine, R.T. (1969a) The *Pisaster–Tegula* interaction: prey patches, predator food preference and intertidal community structure. *Ecology*, **50**, 950–961.

Paine, R.T. (1969b) A note on trophic complexity and community stability. *American Naturalist*, **103**, 91–93.

Paine, R.T. (1980) Food webs: linkage, interaction strength and community infrastructure. *Journal of Animal Ecology*, **49**, 667–685.

Pastor, J. and Naiman, R.J. (1992) Selective foraging and ecosystem processes in boreal forests. *American Naturalist*, **139**, 690–705.

Pastor, J., Dewey, B., Naiman, R.J. *et al.* (1993) Moose browsing and soil fertility in the boreal forests of Isle Royale National Park. *Ecology*, **74**, 467–480.

Pott, R. and Hüppe, J. (1991) *Die Hudelandschaften Nordwest Deutschlands*, Westfalisches Museum für Naturkunde, Münster, Germany.

Prins, H.H.T. and Olff, H. (1998) Species richness of African grazer assemblages: towards a functional explanation, in *Dynamics of Tropical Ecosystems* (eds D.N. Newberry, H.H.T. Prins and N. Brown), pp. 449–490, BES Symposium Vol. 37, Blackwell, Oxford.

Prins, H.H.T. and Van der Jeugd, H.P. (1993) Herbivore crashes and woodland structure in East Africa. *Journal of Ecology*, **81**, 305–314.

Putman, R.J. (1986) *Grazing in Temperate Ecosystems. Large Herbivores and the Ecology of the New Forest*. Croom Helm, London.

Ranwell, D.S. (1960) New Borough Warren, Anglesey. III. Changes in the vegetation in parts of the dune system after the loss of rabbits by myxomatosis. *Journal of Ecology*, **48**, 385–395.

Remillard, M.M., Griendling, G.K. and Bogucki, D.J. (1987) Disturbance by beaver (*Castor canadensis* Kuhl) and increased landscape heterogeneity. *Landscape Heterogeneity and Disturbance*, (ed. M.G. Turner), pp. 103–122, Ecological Studies 64, Springer, New York.

Ritchie, M.E. and Olff, H. (1998) Herbivore diversity and plant dynamics: compensatory and additive effects, in *Herbivores, Between Plants and Predators*, (eds H. Olff, V.K. Brown and R.H. Drent), pp. 175–204, British Ecological Society Symposium, Vol. 38, Blackwell, Oxford.

Ruess, R.W. and McNaughton, S.J. (1987) Grazing and the dynamics of nutrient and energy regulated microbial processes in the Serengeti grasslands. *Oikos*, **49**, 101–110.

Ruess, R.W., Hik, D.S. and Jefferies, R.L. (1989) The role of lesser snow geese as nitrogen processors in a sub-arctic salt marsh. *Oecologia*, **79**, 23–29.

Tansley, A.G. (1935) The use and abuse of vegetational concepts and terms. *Ecology*, **16**, 284–307.

Thalen, D.C.P. (1984) Large mammals as tools in the conservation of diverse habitats. *Acta Zoologica Fennica*, **172**, 159–163.

Thalen, D.C.P. (1987) Begrazing in een Nederlands perspectief, in *Begrazing in de Natuur*, (eds S. de Bie, W. Joenje and S.E. van Wieren), pp. 3–14, Pudoc, Wageningen, The Netherlands.

Truett, J. (1996) Bison and elk in the American Southwest: in search of the pristine. *Environmental Management*, **20**, 195–206.

Tubbs, C.R. (1986) *The New Forest: A Natural History*, Collins, London.

Van de Koppel, J., Huisman, J., Van der Wal, R. and Olff, H. (1996) Patterns of herbivory along a productivity gradient: an empirical and theoretical investigation. *Ecology*, **77**, 736–745.

Van Groenendael, J., Boot, R., Van Dorp, D. and Rijntjes, J. (1982) Vestiging van meidoornstruwelen in duingrasland. *De Levende Natuur*, **84**, 11–18.

Van Wieren, S.E. (1991) The management of populations of large mammals, in *The Scientific Management of Temperate Communities for Conservation*, (eds I.F. Spellenberg, F.B. Goldsmith and M.G. Morris), pp. 103–127, 31st British Ecological Society Symposium, Blackwell, Oxford.

Van Wijngaarden, W. (1985) *Elephants – Trees – Grass – Grazers: relationships between climate, soils, vegetation and large herbivores in a semi-arid savanna ecosystem (Tsavo, Kenya)*. Doctoral thesis, Wageningen Agricultural University. ITC Publication No. 4, Enschede, The Netherlands.

Vera, F.W.M. (1997) Metaforen voor de Wildernis: Eik, Hazelaar, Rund en Paard. Doctoral thesis, Wageningen Agricultural University, Wageningen, The Netherlands.

WallisDeVries, M.F. (1995) Large herbivores and the design of large-scale nature reserves in Western Europe. *Conservation Biology*, **9**, 25–33.

Walter, H. (1974) *Die Vegetation Osteuropas, Nord- und Zentralasiens*, Gustav Fisher, Stuttgart.

Walter, H. (1979) *Vegetation of the Earth and Ecological Systems of the Geo-biosphere*, 2nd edn, Springer, New York.

Watt, A.S. (1962) The effect of excluding rabbits from Grassland A (Xerobrometum) in Breckland 1936–1960. *Journal of Ecology*, **50**, 181–198.

Watt, A.S. (1981). Further observations on effects of excluding rabbits from Grassland A in East Anglian Breckland: the pattern of change and factors affecting it (1936–1973). *Journal of Ecology*, **69**, 509–539.

Westhoff, V. (1971) The dynamic structure of plant communities in relation to the objectives of conservation, in *The Scientific Management of Animal and Plant Communities for Conservation* (eds E. Duffey and A.S. Watt), pp. 3–14, 11th British Ecological Society Symposium, Blackwell, Oxford.

White, D.J.B. (1961) Some observations on the vegetation of Blakeney Point, Norfolk, following the disappearance of rabbits in 1954. *Journal of Ecology*, **49**, 113–118.

Wuerthner, G. (1994) Subdivisions versus agriculture. *Conservation Biology*, **8**, 905–908.

Part One

Historical and Ecological Background

2

Grazing for conservation management in historical perspective

Jan P. Bakker[1] and Ger Londo[2]
[1]Laboratory of Plant Ecology, University of Groningen, PO Box 14, 9750 AA Haren, The Netherlands
[2]Institute for Forest Research and Nature Management, Department of Agricultural Research, PO Box 23, 6700 AA Wageningen, The Netherlands

2.1 INTRODUCTION

In the last decades the importance of grazing as a management tool for nature conservation has increased considerably. This chapter will clarify the context in which grazing takes place. It will consider the increasing impact from large indigenous herbivores to livestock introduced by humans. For thousands of years grazing was superimposed on abiotic conditions, which have been strongly influenced by human impact. This includes the change from large-scale common grazing areas towards a landscape divided into private properties. Grazing as a conservation tool has to cope with such patterns in the landscape. We will review the changes from agricultural exploitation via maintenance management towards restoration management. The questions about the ecological frame of reference for the goal of restoration will be discussed. Finally various approaches to current grazing management will be treated.

During the development of northwestern Europe we may discern three periods: the natural period, the semi-natural period and the cultural period. The **natural period** is characterized by the dominance of communities, landscapes and processes without any noticeable human influence (Londo, 1997). The major patterns in the landscape were largely determined by climatic and geological factors; they were inserted upon the geological matrix (Van Wirdum *et al.*, 1992). There was grazing and browsing by indigenous herbivores. Hence, the natural landscape can be defined by the species assemblage of the original indigenous flora, vegetation and fauna (Westhoff, 1971, 1983).

Grazing and Conservation Management. Edited by M.F. WallisDeVries, J.P. Bakker and S.E. Van Wieren. Published in 1998 by Kluwer Academic Publishers, Dordrecht. ISBN 0 412 47520 0.

The first agricultural invasion took place about 7000 BP followed by a second one around 4600 BP. These people grew arable crops in a shifting cultivation system after the clearance of primeval forest. For the greater part, indigenous large herbivores were gradually replaced by livestock (Chapter 3). In medieval times degradation and destruction of primeval forests continued and large oligotrophic bogs, mesotrophic fens and eutrophic reedbeds (Succow, 1988) were drained, reclaimed and even completely removed for fuel (Louwe Kooijmans, 1974, 1985). Not only did the natural communities disappear but so did certain natural landscape-building processes, through the exclusion of the influence of the sea and rivers and through the regulation of hydrological conditions (Baerselman and Vera, 1995). Although the open landscape was new, the majority of the species that invaded grasslands and heathlands were already present as elements in the understorey of open forest, in small glades, fringes along streams, fens and bogs, and in larger open areas along the coast. Hence, the definition of the semi-natural landscape includes the original flora and fauna but also a transformation of the original vegetation by humans (Westhoff, 1971, 1983; see also Dierschke, 1984; Sjörs, 1986). These are the landscapes of the **semi-natural** period (Londo, 1990, 1997).

The character of human impact also changed. At the end of the natural period and in the beginning of the semi-natural period, humans only influenced the biotic component by cutting trees and grazing livestock. Abiotic conditions were only influenced indirectly by, for instance, trampling and nutrient transport, and directly by superficial ploughing. Large areas of semi-natural landscapes, like heathland and grassland on unfertile soils used for common grazing, were not enclosed into private fields (Figure 2.1). We refer to it as the communal semi-natural landscape. These often large-scale common grazing grounds were predominantly found on the drier, not yet reclaimed sandy parts. Here the geological matrix remained more or less intact. As human impact was stable during many centuries, it became superimposed by a historical matrix (Van Wirdum *et al.*, 1992). In the semi-natural period, regulation of hydrological conditions by drains and ditches in wet parts enabled direct influences on the abiotic conditions by reclamation, deep ploughing and soil levelling. These activities, as well as the division of the landscape into private properties, resulted in the enclosed semi-natural landscape, where fields became delimited by ditches, hedgerows and hedgebanks. The geological matrix became severely perturbed.

The transition from the semi-natural to the cultural period in north-western Europe was caused by the introduction of organic manure or waste from large cities and artificial inorganic fertilizer. The large-scale reclamation and subsequent fertilization of common grassland and heathland occurred after 1920 when mechanization in agriculture started. It resulted in the development of the cultivated landscape, in which not only the vegetation but also the flora and fauna became heavily influenced by humans

Figure 2.1 Traditional forms of land use are still applied on heathlands as a tool in conservation management.

(Westhoff, 1949, 1983). Indigenous species were eradicated by herbicides and non-indigenous species were introduced. These landscapes represent the **cultural period** (Londo, 1997) in which we are living now.

The present cultivated landscape still harbours semi-natural remnants, most of which are now designated nature reserves, and are often degraded. Natural landscapes, i.e. not influenced by humans, no longer exist in northwestern Europe. This has consequences for nature conservation with respect to the ecological frame of reference (section 2.7). Although near-natural landscapes can still be found in parts of the dunes, salt marshes, bogs, and old-growth forests, eutrophication and acidification take place everywhere by atmospheric deposition. However, these diffuse sources of human impact need not always frustrate the idea of restoring natural landscapes or ecosystems.

Nature conservation nowadays aims at endangered near-natural and semi-natural landscapes. Apart from nature conservation, these landscapes may also fulfil amenity goals. We have to address the question of what nature conservation means in these landscapes and to what extent grazing can be involved as a tool for conservation management.

2.2 AGRICULTURAL EXPLOITATION IN SEMI-NATURAL LANDSCAPES

The aforementioned processes can partly be illustrated by the changes along a small river valley, Drentsche A, on the Pleistocene sand plateau in the northeastern part of The Netherlands (Figure 2.2). In the river valley itself degraded remnants of primeval forest and riparian grassland still occurred around 1650. On the plateau, arable fields were found near the villages and

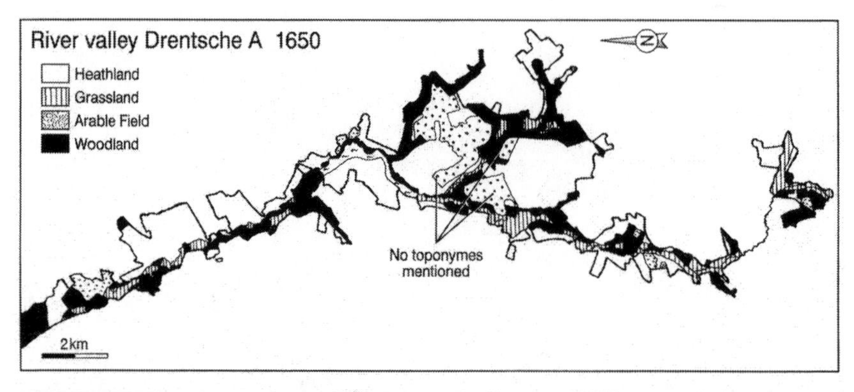

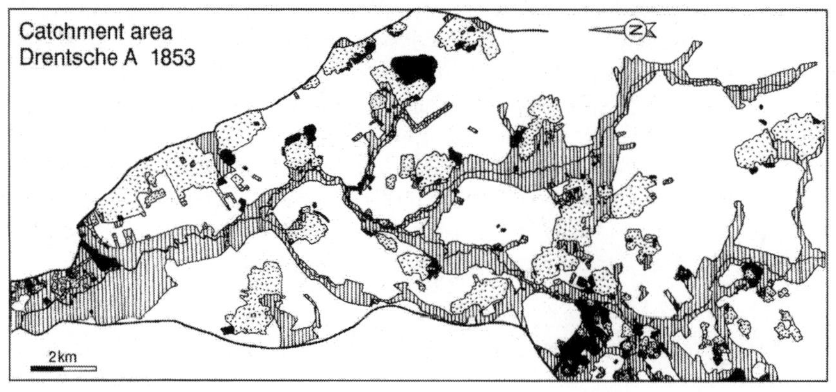

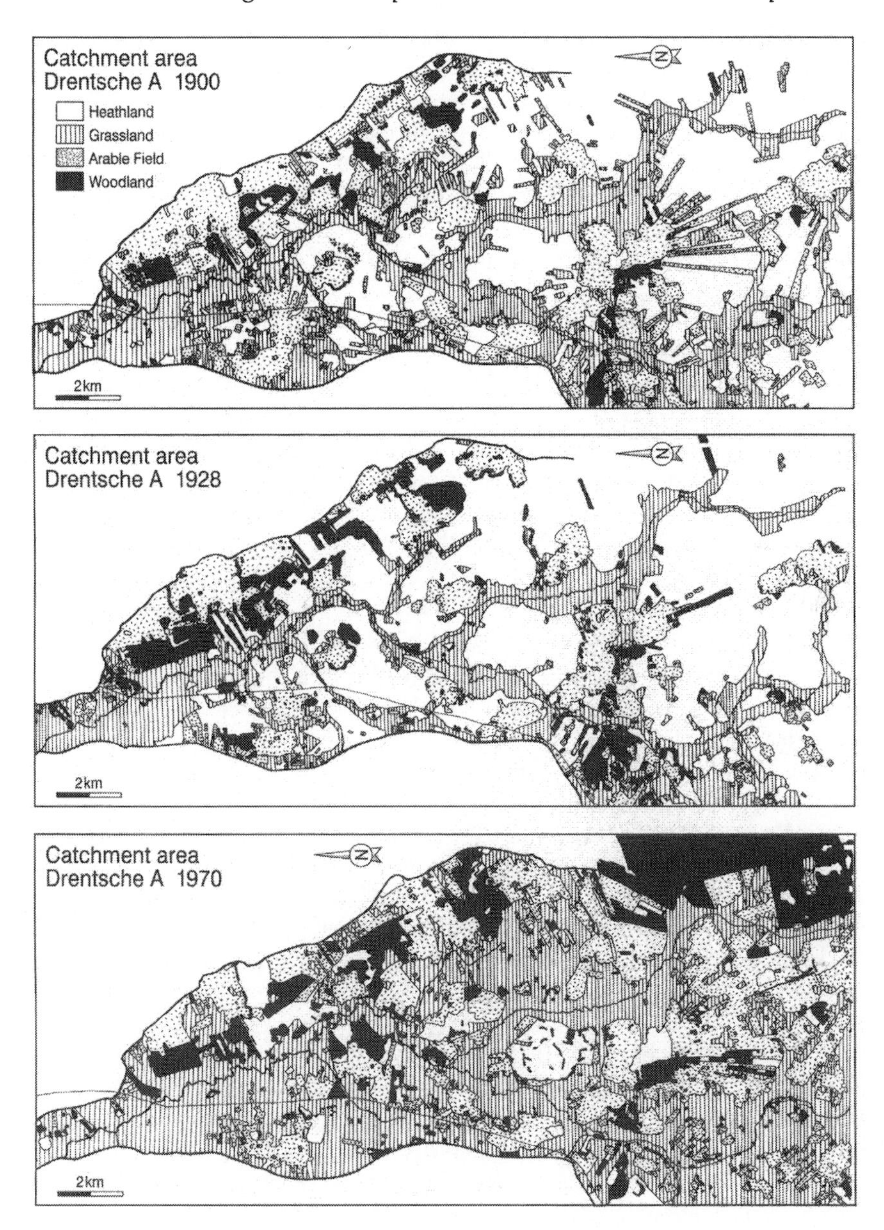

Figure 2.2 Changes in location of main landscape elements (heathland, grassland, arable field, woodland) in the Drentsche A catchment area in the period 1650–1970. (After Bakker, 1989.)

large tracts of heathland between villages. Except for the disappearance of forest from the valley in favour of the extension of grasslands, the semi-natural landscape of a catchment area could still be discerned in the Napoleonic period, on the eve of the intensification of agricultural practices. In 1970 the cultivated landscape of intensified grasslands and arable fields dominated. Woodlands had been planted for timber production and recreational purposes. Some semi-natural communities were left in small heathlands and grassland areas along the river itself. The latter in particular were an important reason for establishing the Drentsche A river valley as a nature reserve in the 1960s.

Haymaking or grazing took place for centuries in semi-natural landscapes and they harbour characteristic plant and animal communities. The idea of uninterrupted occupation and hence agricultural exploitation of village territories is not only based on the comparison of a large number of different villages on the Drenthe Plateau. It is also substantiated by excavations in Peelo, which revealed continuous occupation from 2400 BP until the present day (Bardet *et al.*, 1983; Waterbolk, 1984). The development of the Drenthe Plateau is not an isolated phenomenon. Behre (1976) and Waterbolk (1979) suggested that, with the exception of the fringes, there has been continuous occupation of all the uplands of the northern part of The Netherlands and the adjacent northwestern part of Germany.

Apart from the supposedly stable long-term agricultural exploitation, these landscapes featured gradients in human impact. Such gradients around settlements can be expected as a result of economical costs of carriage of agricultural products. Von Thünen (1842) described a theoretical agricultural system on boulder clay soils in the eastern part of Germany, where concentric circles (German: *Thünische Ringe*) existed, with the village at the centre. The circles, from the centre outwards, included horticulture and cattle on the farmstead fields, coppice for the supply of (fire)wood, arable fields, cattle or sheep-grazing for manure production, and woodland for hunting (see also Bouwer, 1985; Bieleman, 1987). Soil chemistry analyses in various parts of The Netherlands have shown lower phosphate and potassium conditions when the fields were further away from the farmstead (Vermeulen, 1954; Draaisma, 1958; Koopmans, 1960; Pieters, 1971). These gradients disappeared after the re-allotment programmes because they involved the removal of farmsteads into the open rural landscape.

Generally it can be concluded that human impact lasted for centuries in semi-natural landscapes, but that this impact was far from constant on the individual field level. Wells *et al.* (1976) reached similar conclusions for a chalk grassland complex in the United Kingdom. The traditional view that most areas of present-day chalk grasslands were used as sheepwalk for many centuries was not supported by studies of land use history, vegetation and soils on the Porton Ranges. Written evidence and the plough marks visible on air photographs indicate that over 75% and probably more of the Ranges were cultivated.

Examples of long-lasting land use in semi-natural landscapes are also known, however, such as the Haselüner Kuhweide (Tüxen, 1974) and Borkener Paradies (Burrichter *et al.*, 1980), now nature reserves in northern Germany but formerly used for common grazing since medieval times. These wood-pasture landscapes will be discussed further in Chapter 4. Outside nature reserves extensive areas of semi-natural landscapes in current agricultural systems amount to 56 million ha in Europe. So-called low-intensity farming systems still feature in 82% of the agricultural area in Spain, 61% in Greece, 60% in Portugal, 35% in Ireland, 31% in Italy, 25% in France, 23% in Hungary, 14% in Poland, and 11% in the United Kingdom (Bignal & McCracken, 1996). Large transhumance livestock movements took place in Greece and Spain until recently (Beaufoy *et al.*, 1994), in Germany until the 1950s (Hornberger, 1959), and until the early 1900s in the limestone region of The Netherlands, and adjacent areas of Belgium and Germany (Hillegers, 1993).

2.3 FROM AGRICULTURALLY EXPLOITED LANDSCAPES TO NATURE CONSERVATION

The intensification of agricultural practices during the twentieth century was possible by drainage, reclamation and the application of artificial fertilizers. It resulted in an enormous degradation of nature in The Netherlands (Stumpel, 1981; Westhoff and Weeda, 1984; Weeda, 1985; Weinreich and Musters, 1989; Bink *et al.*, 1994), and a loss of species (Weeda *et al.*, 1990; Siebel *et al.*, 1992). Most groups of organisms show a decline during the last few decades; only birds and mammals show a larger increase than decrease (Table 2.1).

The reduction of species diversity through the application of fertilizers has been described for the United Kingdom (Williams, 1978), Germany (Klapp,

Table 2.1 Recent changes in The Netherlands of the number of species in different taxonomic groups (from Weinrich and Musters, 1989)

Taxonomic group	*N* species	% Extinct locally	% Declined	% Increased
Mushrooms	314	–	54	36
Vascular plants	1436	2.3	34	8
Butterflies	70	10	69	–
Carabidae	343	together *c.* 90		–
Freshwater fish	46	–	64	7–30
Amphibia	14	–	100	–
Reptiles	7	–	30	30
Breeding birds	180	1	30	36
Wintering birds	76	–	20	36
Mammals	62	5	26	62

1965), Belgium (Van Hecke *et al.*, 1981) and The Netherlands (Oomes and Mooi, 1981; Elberse *et al.*, 1983; Willems, 1983). The degradation of flora and vegetation in natural and semi-natural landscapes has become a matter of great concern. The problem affects the whole of Western Europe as well as many other parts of the temperate world, but The Netherlands appears to occupy the top position. This may be attributed to the high population density and the advanced level of agriculture and technological development resulting in expanding urban and industrial areas, connected by a vast network of roads. The rural areas outside the population concentrations still have a predominantly agricultural land use. The intensity of agricultural exploitation, however, leaves little room for species diversity.

A backlash against the degradation of flora and vegetation has developed. From the beginning of the twentieth century onwards, areas have been acquired by private organizations for landscape and nature conservation purposes, more recently also by the State. There are about 1000 nature reserves in The Netherlands and in size they are equally divided between those larger than 1000 ha and those of less than 1000 ha. This means that many nature reserves represent small fragments of areas with conservation interest. This is also true for Belgium, Denmark, Ireland and Greece. All other European countries feature a relatively large proportion of reserves larger than 10 000 ha (WallisDeVries, 1995).

Since characteristic landscapes and communities disappeared as a result of increasing human impact, the main conservation practices (probably as a reaction) initially implied complete non-interference, effectuated by fencing-in nature reserves. During the Second World War both British (Anonymous, 1943; Westhoff, 1945; Gabrielson, 1947) and Dutch conservation pioneers (Westhoff, 1945; Westhoff and Van Dijk, 1952) developed the idea of the necessity of intervention in semi-natural landscapes to preserve communities of nature conservation interest. Reduction of species diversity in grasslands without human interference is well known from the comparison of managed and unmanaged grassland (UK: Wells, 1980; Germany: Schiefer, 1981; Switzerland: Zoller *et al.*, 1984; Sweden: Persson, 1984; Tyler, 1984; The Netherlands: Willems, 1983; During and Willems, 1984, 1986). The organizations owning nature reserves were, therefore, transformed into authorities in charge of active management.

Since most semi-natural grasslands and heathlands are marginal from an agricultural point of view, these areas tend to be the first to be neglected or abandoned. This was common practice earlier, but recently it has been enforced by the European Union agricultural policy; this facilitated highly productive farms and closed down less productive ones, on which so-called marginal agriculture or low-intensity farming was practised (Beaufoy *et al.*, 1994; Bignal and McCracken, 1996). Nature management through the continuation of existing agricultural exploitation in semi-natural landscapes is the most logical option to maintain characteristic and often species-rich grassland and heathland communities. During the last 10–20 years the scope of nature management has been extended to the restoration of formerly species-rich plant communities that have been degraded by abandonment or the application of fertilizers.

2.4 WHICH LANDSCAPES TO MANAGE FOR CONSERVATION INTEREST?

The majority of grazed landscapes are grasslands or short herbaceous vegetation. These open communities cannot simply be called amenity grasslands. The Natural Environmental Research Council in the United Kingdom defined amenity grassland as 'all grassland with recreational, functional or aesthetic value and of which agricultural productivity is not the primary aim' (NERC, 1977, in Rorison, 1980). Vast areas of amenity grasslands are managed intensively and can be considered as cultivated landscapes, e.g. lawns, urban parks and open spaces, urban road verges, golf fairways. Semi-natural amenity grasslands include rural road verges, golf courses, common land, National Trust land and nature reserves. Only the latter two are interesting for our purpose, because they will be managed primarily for nature conservation purposes. The low-intensity farming grasslands form part of the semi-natural amenity grasslands and they are restricted to nature reserves. Abandoned grasslands occur in agricultural areas when farmers lose interest, or in nature reserves where a shortage of money prevents the application of envisaged nature management practices.

Low-intensity farming is not an absolute concept. It is characterized by relatively low yields and a poor quality of forage compared with intensive agriculture. In general, low-intensity farming is practised on grasslands in remote valleys and hilly and rough mountain areas, but also in lowland areas where sufficient drainage is too expensive to become economically profitable. The character of these grasslands, therefore, depends to a significant extent on abiotic environmental factors: their locations are too dry, too wet, or too steep. But they can also simply be situated too far from the farm or too difficult to reach and exploit because of obstacles, especially in small-scale landscapes (hedgerows, hedgebanks, ditches and small woods).

The differences between countries should be taken into account for a local definition of low-intensity farming areas. Relatively low yields in some countries are regarded as being relatively high in other countries with an overall lower agricultural production. In other words, low-intensity farming areas taken out of the agricultural system in, for instance, The Netherlands or Denmark are still exploited agriculturally in, for example, the United Kingdom or France. That is the reason why even artificially fertilized grasslands can be taken out of the agricultural system in The Netherlands with the aim of restoring species-rich grassland communities or heathland – plant communities still widespread in other countries. Striking in this respect are the attempts to restore former hayfield communities in The Netherlands, while comparable remaining plant communities in northeastern Poland are being degraded due to recent introduction of artificial fertilizers (Bakker and De Vries, 1982). This, in fact, has a double negative influence. The productivity of the grasslands adjacent to the farm increases to such an extent that the farmer can produce the same amount from a smaller area than he did before the introduction of fertilizers, while the second effect is the abandoning of vast grassland areas situated further away.

One should not misunderstand the above. An interchange of plant communities might be suggested between The Netherlands and Poland but this is impossible since grassland communities on comparable abiotic substrate show local variation due to geographical differences. The above reasoning is a plea to maintain still valuable semi-natural grassland and heathland communities (Londo, 1990; Bignal and McCracken, 1996) including their geographical differences. It has become clear that it takes a great deal of time and money to restore former species-rich communities once they have been degraded, if restoration is possible at all (Bakker, 1989).

2.5 FROM MAINTENANCE MANAGEMENT TOWARDS RESTORATION MANAGEMENT

To achieve the aim of nature conservation in semi-natural grasslands or heathlands in nature reserves (in brief: high species and community diversity including rare and endangered characteristic species), several notions have been used with respect to nature management practices. In the United Kingdom, Wells (1980) distinguished between reclamation management, performed only once, and regular maintenance management. When grasslands or heathlands are taken out of the agricultural system, coarse grasses, sedges and scrub take over. When such an abandoned area is required for some form of amenity use, it first has to be reclaimed before regular or maintenance management can proceed. For instance, a great deal of scrub removal and burning of grass litter took place in many public open spaces that had been neglected during the Second World War, before the land was designated for amenity purposes.

In The Netherlands two different situations have to be faced. In the first, conservation interest is already high; management practices aimed at maintaining such a level of interest of an area have been called '**nature management *sensu stricto***' (Londo and Van Wirdum, 1994; Londo, 1997) and are similar to maintenance management (Figure 2.3). This implies the continuation of existing management practices – such as coppicing, haymaking, cutting sods, and livestock grazing – with a certain regularity. Except for the cutting of sods all these management practices include affecting the structure of the vegetation by harvesting it, whether for agricultural exploitation as in former times, or for present-day nature conservation.

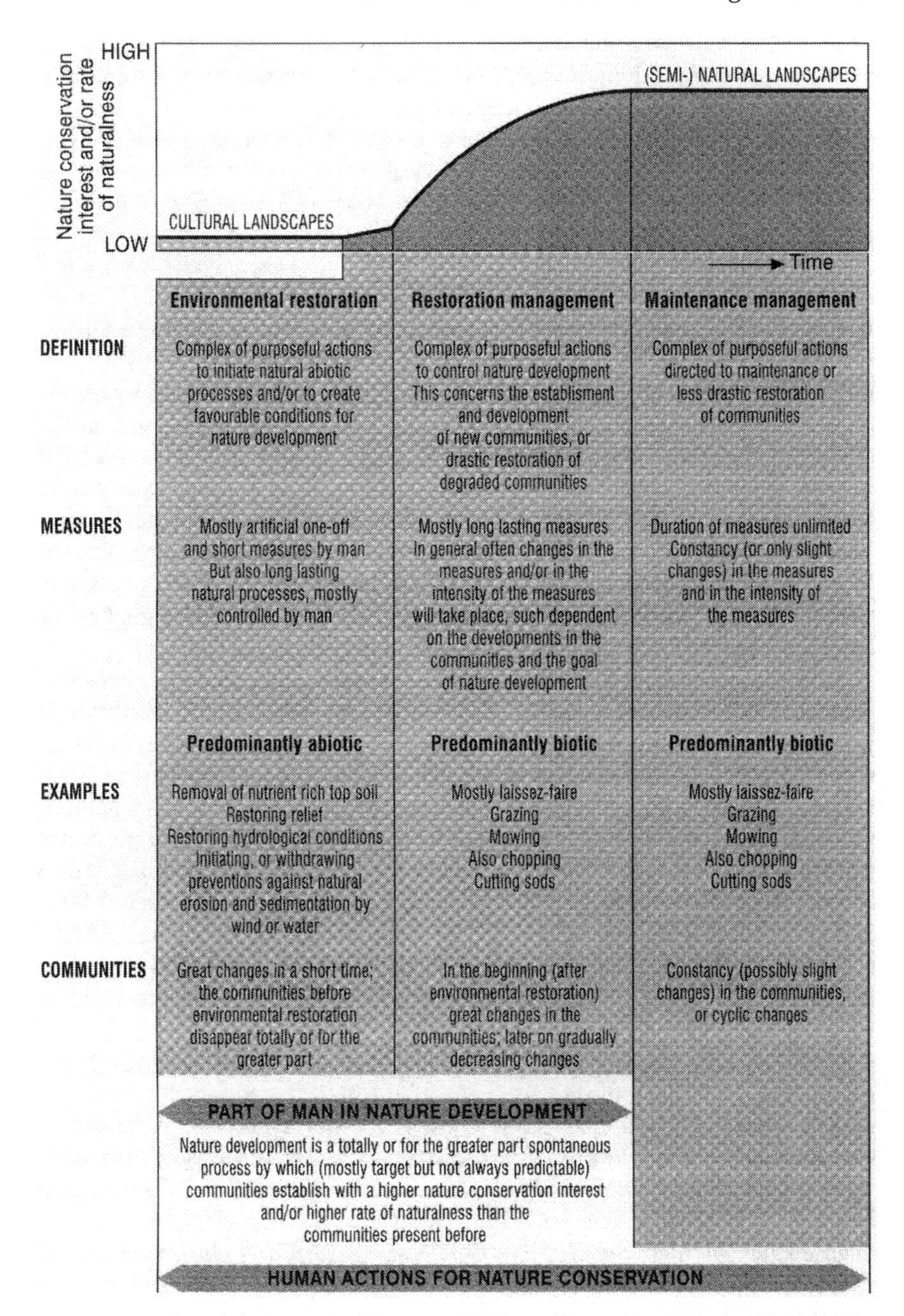

Figure 2.3 Definitions, measures, examples and communities in relation to human activities for nature conservation in the framework of cultural and (semi-)natural landscapes. (After Londo, 1997.)

In the second situation the nature conservation interest is low, mostly because of former intensive agriculture. New nature has to develop, so in The Netherlands this process is called **nature development**. In this (for the greater part spontaneous) process, two phases in human activities are distinguished (Figure 2.3). The first phase of 'environmental restoration' is only necessary when the abiotic environment has been degraded – for example, by lowering the groundwater table or by levelling down the original relief. Environmental restoration is necessary to restore the geological matrix (as far as possible). This may include removal of the eutrophic topsoil by cutting sods or the whole agriculturally affected topsoil up to 50 cm, restoration of the relief, and raising the groundwater table. In this case, environmental conditions for the target plant communities are created by affecting the abiotic conditions. The quality of the new nature will depend on the quality of environmental conditions. If the abiotic conditions have not been degraded, environmental restoration is not necessary and the second phase, 'restoration management', can be implemented directly (Figure 2.3). This includes a change in the existing management practices – for example, cessation of fertilizer application, followed by haymaking or grazing at a low stocking rate. It creates the environmental conditions for the target plant communities by affecting existing plant communities.

When environmental restoration has taken place new communities will develop. Which management measures are required depends on the goal (near-natural or semi-natural communities/landscapes). For the development of open or half-open semi-natural landscapes in which grasslands and/or heathlands play an important role, the appropriate management is grazing. There is experimental research on environmental restoration and restoration management in small-scale plots in the United Kingdom (Gibson *et al.*, 1987; Mountford *et al.*, 1996), France (Grévilliot and Muller, 1996), Denmark (Bülow-Olsen, 1980; Buttenschøn and Buttenschøn, 1982), Germany (Schreiber and Schiefer, 1985; Kiehl *et al.*, 1996) and The Netherlands (Bokdam and Gleichman, 1989; Olff and Bakker, 1991; Bobbink and Willems, 1993; Oomes *et al.*, 1996), to study the effects of grazing, burning and cutting regimes. Questions of burning or grazing versus cutting, and return of the cuttings versus removal, can thus be evaluated for large-scale practical management. However, in the absence of the necessary knowledge and conceptual framework at the landscape level, management is sometimes carried out by trial and error, and intuitively, rather than by rational decisions (Snaydon, 1980). With or without a scientific basis, the maintenance of high species diversity in the sward and the management costs of cutting give rise to major dilemmas (Green, 1980). Restoration management practices outside The Netherlands seem to be restricted to the reclamation of abandoned grasslands (Wells, 1980) but do not form an important part of the activities of authorities in charge of the management. Recently, however, some restoration projects have been carried out in, amongst other countries, Belgium (Leroy, 1991), the United Kingdom (Jordan *et al.*, 1987; Buckley, 1989) and the United States (Cairns, 1986;

Jordan and Packard, 1989). In the latter two countries, environmental restoration and restoration management are referred to as 'biological habitat reconstruction' and 'restoration, reclamation and regeneration of degraded and destroyed ecosystems'.

In The Netherlands, on the contrary, environmental restoration and restoration management practices are usually performed in nature reserves. A minor part consists of managing abandoned grassland and heathlands; the major part consists of reducing the residual effects of fertilizers. The tradition of common grazing almost disappeared with the decline of heathland area in The Netherlands but controlled haymaking was started long ago in nature reserves. As in other countries, the costs of cutting became very high. Another disadvantage of mowing is the uniformity of its impact, whereas grazing at low stocking rates may enhance a greater community diversity.

Insights into the potential of grazing were gained during, for example, an excursion of several members of The Netherlands Research Institute for Nature Management (Van Leeuwen *et al.*, 1971) to the Baltic islands of Gotland and Öland in Sweden. In particular, the very species-rich *ängar* attracted attention. In April/May, remaining leaves and sticks were raked (which facilitated later cutting by scythe) and burnt. The ashes were spread over the grassland as a fertilizer. Cutting took place in July to early August. The aftermath was grazed from late August until September. Every third or fourth year, trees were lopped as winter fodder for livestock. During the winter period trees were thinned for firewood.

Such exploitation for many centuries must be the explanation for this species-rich system (Ekstam *et al.*, 1988). The species-rich grasslands had been exploited by grazing at low stocking rate and the combination of these practices resulted in a small-scale differentiation into grassland, scrub and woodland in a wood-pasture landscape (Chapter 4). Descriptions of plant communities in wood-pastures were given by Westhoff *et al.* (1983). During recent decades the *ängar* were more or less abandoned (Persson, 1984) and the lands were grazed during a longer period of the year, which may have overestimated the influence of grazing in the former exploitation. Because of their high biological diversity the establishment of wood-pastures subsequently became an important goal for grazing management in nature reserves in The Netherlands.

2.6 GRAZING AS A MANAGEMENT TOOL IN NATURE CONSERVATION

2.6.1 Aims of grazing management

As mentioned earlier, Westhoff (1945) developed ideas about grazing as a management tool in semi-natural landscapes:

> We should dare to carry out the experiment of introducing moose and aurochs into some of our largest nature reserves; for lack of aurochs we might take cattle. Of course, firstly, careful research is necessary to

estimate the appropriate stocking rate for the plant communities. The vegetation would benefit from it, in particular the natural sward, as grasses depend on ruminants to the same extent as ruminants on grasses.

As the tradition of common grazing had almost disappeared in The Netherlands, there were initial objections against grazing on heathlands (De Boer, 1949; Van Bemmel and De Smidt, 1955). Nevertheless, referring to examples abroad, grazing as a management tool for the maintenance or restoration of characteristic semi-natural landscapes was introduced in the early 1970s, in particular by members of The Netherlands Research Institute for Nature Management. Its aim was threefold (Oosterveld, 1975, 1977; Thalen, 1984):

- **Management and maintenance of open space** in existing nature reserves. A few areas were found at that time where the existing grazing management already fulfilled to a certain extent the aim of maintaining characteristic plant communities of semi-natural landscapes. The introduction of grazing was aimed particularly at replacing cutting in nature reserves where it became too costly and impractical. This, in fact, is maintenance management. It can also be applied as reclamation management where maintenance management has ceased some time ago and open spaces have succumbed to encroachment by bushes. Large herbivores can be introduced to stop or remove bush encroachment, or the bush can be cut so that the herbivores can maintain the newly created open space.
- **Management of low-intensity farming areas** to be removed from agricultural exploitation and integrated into existing nature reserves. If the aim with low-intensity farming areas after abandonment is not woodland, the introduction of grazing may be useful. Often these areas can be integrated with existing natures reserves and hence contribute to the removal of sharp boundaries in the enclosed landscape. If the grazing management is extensive (i.e. when production is higher than utilization of the forage by grazing and trampling), it is expected that a pattern will develop with heavily grazed short patches and lightly grazed taller areas in which scrub or trees may establish. Such a pattern should initiate the formation of the intended wood-pasture, which will only emerge after a very long time.
- **Management of newly embanked areas** which are no longer inundated. In several coastal areas of The Netherlands, large tracts are no longer affected by tidal influences. Some of these areas are designed to develop as nature reserves. As most of these areas are virtually flat, it was expected that extensive monotonous stands of a few dominant plant species would establish. The introduction of grazing was expected to play a role in the diversification of these areas. Hence comparisons were made between various management practices like abandoning, grazing and cutting.

2.6.2 Guiding principles for grazing management

There were some basic guiding principles to accomplish the above three aims of grazing management (Oosterveld, 1975, 1977):

- **Grazing should be practised at the lowest possible stocking rate.** Taking into account the cessation of fertilizer or manure application, grazing should be extensive, i.e. the utilization of the forage (removal by grazing and trampling) should be lower than forage production. Without presenting data on production, it was advocated as a rule of thumb that stocking rates should not exceed one cow or horse per 3 ha, or one sheep or goat per hectare.
- **Grazing should continue for as long in the season as possible.** During spring and summer the herbivores are supposed to forage on the green parts of grasses and herbs. Only when they stay out in the field during autumn and winter can they be expected to forage on less palatable material, such as coarse grasses and browse and bark from shrubs and trees. Therefore the influence of large herbivores is assumed to be more profound if their year-round presence is possible.
- **Grazing management should use breeds that need little care.** Livestock breeds for high production of milk, meat and wool (e.g. Holstein-Friesian cattle, Texel sheep) are supposedly unable to maintain themselves year-round in the field, because of the decreasing forage quality in the off-season and the ensuing weight loss. It is therefore assumed that unimproved, so-called primitive breeds are a necessary prerequisite for such a grazing programme. Although this assumption lacks the support of solid evidence (Chapter 9), traditional Dutch breeds of Schoonebeker, Drenthe Heath, Kempen Heath and later Mergelland sheep, Dutch Landrace goats, and Lakenvelder and Witrik cattle were introduced. Later their breeding programme became a goal in itself (Clason, 1980; Fokkinga and Felius, 1995). Apart from Dutch breeds, so-called winter-resistant breeds from abroad were introduced: Icelandic ponies, Shetland ponies, Swedish Mountain (Fjäll) cattle and, later, Polish Konik horses, Scandinavian Fjord horses, Russian Przewalski horses and, from Scotland, Highland and Galloway cattle and Soay sheep. There has been a growing awareness of the importance of conserving livestock breed diversity for future agricultural purposes (Hall and Bradley, 1995) but the conservation of breeds and the conservation of nature may lead to conflicting management goals.
- **Grazing should be practised over as large an area as possible.** The assumption is that small fenced-in areas will develop a homogeneous stand of vegetation when grazed: they will show only edge effects, and not the intended variation in canopy structure as in nature reserves. In experimental paddocks, herbivores are removed once they have grazed the canopy to a certain height or leaf area index. Moreover, if in small areas the number of herbivores is decreased for some reason, the grazing

capacity of the herbivores will be severely decreased, whereas in a large area with over one hundred herbivores there will be hardly any difference. In practice it was suggested that grazing should be introduced into areas over 10 ha.

2.6.3 Application of grazing management

Some examples of how management goals are oriented will illustrate the outcome of discussions on options in which grazing plays a role. It needs to be taken into account that in many cases the management practices have started only recently or are soon to be implemented; thus the results cannot (yet) be evaluated properly.

In small-scale enclosed landscapes, grazing is managed to keep the sward short and is often used as a replacement for cutting regimes. The management goal is the species composition of the vegetation, and also the habitat for breeding meadow birds, winterstaging waterfowl and entomofauna (Figure 2.4).

Isolated and often small existing nature reserves, harbouring remnants of heathland, are combined with formerly intensively exploited agricultural land after removal of the eutrophicated topsoil. Large herbivores are used to maintain the semi-natural heathland and create a mosaic with developing forest. The management is oriented towards a transformation of an enclosed semi-natural and cultivated landscape into a large semi-natural common landscape (Figure 2.5).

Figure 2.4 On wet, peaty soils summer grazing is practised for the benefit of meadow birds.

Figure 2.5 One of the first year-round grazing experiments in The Netherlands was with Icelandic ponies on heathland and former arable land (Baronie Cranendonck).

Large-scale heathlands and forests, which were managed separately until recently, are now interconnected by the removal of fences and the construction of 'cerviducts' over which animals can cross motorways. The management is heavily biased towards the populations of large herbivores and their impact on the landscape (Worm and Van Wieren, 1996).

Recently ideas have been put into practice to restore a more natural landscape by local removal of summer-dikes in order to give access to the dynamics of river water. The exploitation of brick-making clay in the river forelands should create a pattern of open water alternating with more sandy levees. Superimposed on the dynamics of the river, large herbivores are supposed to create variation in the patterns of developing reeds, scrub and forest. The management is oriented towards natural landscape-building processes.

2.7 NATURE CONSERVATION AND THE ECOLOGICAL FRAME OF REFERENCE

2.7.1 The goal of nature conservation

It becomes apparent that nature conservation may be aimed at the maintenance of near-natural or semi-natural landscapes which still have enough potential. In such cases the best way of achieving this goal is to change as little as possible, i.e. to practise maintenance management. The ecological frame of reference is merely the existing situation.

It is important to realize that areas that have formerly been considered as untouched or hardly influenced by humans usually do in fact betray considerable human impact. *Quercus petraea* woodlands on acid soils on steep, rocky slopes in high-rainfall areas of western Britain and Ireland have long been assumed to be either 'wildwood' or closely related to the original 'wildwood'. Yet the evidence suggests that stands have been heavily influenced by coppicing, clear felling, burning and grazing during the last 500–700 years (Birks, 1996). The structure and composition of an old-growth (presumably pristine) forest in Sweden have been shown to be a result of changes in management, including fire and grazing, only a few centuries ago (Bradshaw and Hannon, 1992). Bogs declared as nature reserves because of their presumed naturalness do not always reveal a natural hydroseral succession but show local human interference over the last two centuries (Birks, 1996). Apparently palaeo-ecological data are needed to support ideas on naturalness. Human memory can sometimes be very short.

Even the Darien of Panama in Central America, a remote area rich in endemic species and assumed to have been one of the few remaining undisturbed neotropical wildernesses, has been subject to at least 4000 years of human activities, including agriculture (Bush and Colinvaux, 1994). In the Serengeti–Mara region in East Africa (Tanzania), following the great rinderpest epidemic of 1890, both human and animal populations are thought to have been reduced to very low numbers. Explorers and hunters of the early 1900s encountered a Serengeti–Mara characterized by broad, open expanses of grassland including occasional *Acacia* trees. Over the next 30–50 years the prevailing conditions enabled the establishment of dense woodland and thickets. The Serengeti–Mara region was set aside for protection in the 1930s. Ironically, the colonial administrators worked diligently to protect these 'pristine' woodlands, but did not recognize that this area had been open grassland less than 50 years earlier (Dublin, 1995). These examples stress that the ecological frame of reference for nature conservation should be used carefully, and with historical knowledge. They also stress the need for protected areas as ecological baselines (Arcese and Sinclair, 1997).

As soon as it has been decided that the nature conservation interest of a cultivated landscape or a degraded semi-natural landscape is to be enhanced, the goal of conservation and the ecological frame of reference need to be defined. After the realization, following the Second World War, that semi-natural landscapes have been influenced by human impact, these often enclosed landscapes became the main ecological frame of reference until the 1980s. It was the source of inspiration for Oosterveld (1975) in formulating aims and methods for the introduction of the grazing programme discussed in the previous section. An important point in considering grazing as a tool was that it implied that nature conservation, in particular of the vegetation, is the aim and the grazing animal is the means, replacing cutting machinery (Thalen, 1984b; Bakker, 1987). Restoration of semi-natural landscapes of the second half of the nineteenth century was the objective, including haymaking and some grazing in the enclosed landscape, and grazing in the common landscape (Figure 2.6).

Increasingly it is being appreciated that, for instance, deep drainage in the modern landscape imposes severe restrictions on the possibilities for restoration. Deep drainage has resulted in an irreversible desiccation of the soil, with subsequent mineralization of organic soils, and acidification as a result of the replacement of deep calcium-rich seepage water by shallow calcium-poor seepage water (Grootjans *et al.*, 1996). This process of eutrophication and acidification is enhanced by the present atmospheric deposition. It limits the restoration of semi-natural and near-natural landscapes. For successful restoration management a strong decrease of eutrophication and desiccation is a prerequisite.

2.7.2 A new ecological frame of reference

During the past decade more attention has been paid to the development of (potential) near-natural landscapes with a high degree of self-regulation, and hence low costs for regular management. Two major events had an enormous impact on these ideas in The Netherlands. During two subsequent winter periods, in 1972 and 1973, extensive areas in planted woodland were destroyed by storms. In particular, monotonous *Pinus sylvestris* plantations

Figure 2.6 Wood-pasture landscape in Öland, Sweden. This species-rich landscape type became a new frame of reference for management in The Netherlands during the 1980s.

on dry sandy soil suffered severely: 6000 ha were windblown (Van Nispen tot Sevenaer, 1975). According to local forest law, the gaps thus created should be replanted as soon as possible. Suddenly nature conservationists saw the opportunity to create more diverse deciduous woodlands that included large herbivores (Van De Veen and Van Wieren, 1980; Thalen, 1981; Van De Veen, 1985). It resulted in a large-scale experiment on forest grazing by Scottish Highland cattle in *Pinus sylvestris* stands with an understorey of *Deschampsia flexuosa* in part of a complex of nature reserves in the southern fringe of the Veluwe area (Figure 2.7) (Van Wieren, 1985, 1988).

The second impetus came from the spontaneous development of a large marsh area in the polder Zuidelijk Flevoland, which was embanked in 1967. Its northeastern, low-lying corner was predestined to become an industrial area, but the economy was less prosperous than expected and the area was abandoned for 20 years. As it remained wetland, a peerless marsh area called the Oostvaardersplassen emerged. It harbours breeding great cormorant (*Phalacrocorax carbo*), spoonbill (*Platalea leucorodia*), great white egret (*Egretta albus*), greylag goose (*Anser anser*) and migrating osprey (*Pandion haliaëtus*) and white-tailed eagle (*Haliaëtus albicilla*). Apart from the surrounding artificial dike, its development was initially governed by natural landscape-building processes, including hydrological processes and the impact of herbivores. These herbivores were not large mammals but flocks of greylag geese. Later, large herbivores were introduced into the drier surroundings of the marsh in order to influence the vegetation (Vera, 1988) (Chapter 7).

Figure 2.7 Forest grazing with Highland cattle on poor sandy soils.

The view about the introduction of large herbivores as a means ('grazing machinery') or as a goal in itself, i.e. as part of the ecosystem, was elucidated by Van Wieren (1991). In small nature reserves livestock are brought in with the aim of managing species of plants and animals that depend on a short sward (entomofauna, small mammals, birds) in often enclosed semi-natural landscapes. The large herbivores are simply tools and the management of their populations is not important: they are covered by veterinary law and need supplementary food in poor seasons and veterinary care.

In large nature reserves (generally exceeding 10 000 ha) mosaics of grassland, heathland, scrub and woodland often occur. Management of surplus vegetation may be necessary by preventing scrub encroachment or by temporary exclosures for the establishment of new cohorts of trees. Apart from livestock, natural herbivores such as roe deer and red deer may play a role in structuring the vegetation. The populations of all herbivores, whether domesticated or wild, are regulated by humans. The New Forest in southern England is an example of such a large-scale semi-natural common landscape including some natural components.

In very large reserves (more than 100 000 ha) the climax vegetation of temperate regions may be preserved, including the accompanying large herbivore fauna – roe deer (*Capreolus capreolus*), red deer (*Cervus elaphus*), wild boar (*Sus scrofa*), European bison (*Bison bonasus*) and moose (*Alces alces*). The niches of the extinct aurochs (*Bos primigenius*) and tarpan or European wild horse (*Equus ferus ferus*) might be taken by Heck cattle and Konik horses, respectively, (Chapter 7) but their impact on the vegetation is poorly known. Large predators should also be present. Human impact should be restricted to a minimum in such natural landscapes.

The newly created natural landscapes will harbour forest, with open areas where abiotic conditions (too saline, too wet or too dry) prevent the development of forest. Large herbivores are supposed to contribute further to create and maintain open places by their grazing and browsing activities, as they might have done before human impact became important in northwestern Europe around 5000 BP. Therefore the ecological frame of reference was put at that period, the Atlanticum (8000–5000 BP). As argued in Chapter 3, however, the activities of large herbivores were probably not of major importance at that time, and hence were hardly able to create and maintain glades in the forest.

Wood-pastures instead of forests have been present at least since 8000 BP according to Vera (1997), who claims that pollen diagrams of forests from the Atlanticum period are similar to those of recent wood-pastures and that, therefore, wood-pastures have been present for millennia. He reasons that palynologists have always unjustly taken it for granted that primeval forests were dense, because they did not include the impact of indigenous large herbivores. The herbivores are thought to be responsible for cyclic processes, including phases of grassland, thorny scrub and woodland. *Corylus avel-*

lana, *Quercus robur* and *Q. petraea* in particular, being light-demanding species, should have benefited from these cyclic processes, otherwise their large proportion in pollen diagrams in the presence of more competitive tree species could not be explained. Although thorny shrub species, grasses and herbs are found in present wood-pastures, they are supposed to be virtually absent from pollen diagrams because of limited pollen dispersal in the dense thorny scrub, and also because the herbs and grasses would have been prevented from flowering profusely by intensive grazing. Under heavy grazing, however, one may expect that the number of herbivores will be limited by the standing crop in the winter period. This should enable a portion of the grasses and herbs to flower during spring and summer. Moreover, if herbivore density was as high as claimed by Vera, more remnants should have been found from hunting bags when humans appeared on the scene. The problem arises that, on drier soils, conditions for the preservation of bone material are poor. Remains are only available from clay and peat soil, which may bias the conclusions on palaeo-environments. The available data suggest that in central Europe an agricultural system spread from 7500 BP onwards; the main source of meat was livestock, and only locally was 25% or more of the meat taken from wild animals (Benecke, 1994). In the period 5000–4000 BP, some remains were found in The Netherlands indicating the presence of red deer and roe deer, with very little from moose and aurochs, but none from tarpan. In contrast, remains of fur animals – mainly otter (*Lutra lutra*) and beaver (*Castor fiber*) – and of birds and domestic mammals (mainly cattle) were found frequently (Zeiler, 1997). In Sjælland, Denmark, the period 8000–5000 BP is well documented with remnants. The initial mammalian fauna was highly diverse, with faunal elements from both forest and open habitats. During the period under study, aurochs and moose disappeared; the body size of herbivores such as red deer, roe deer and wild boar decreased – which may indicate poorer forage quality. The carbon-isotope ratio in bone collagen changed, indicating a shift in the ratio in the food eaten by the herbivores, and suggesting a transition from grassland to forest forage. Pollen analysis indicates a transition from *Corylus avellana*/*Pinus sylvestris* to *Tilia cordata* forest (Noe-Nygaard, 1995). The herbivores seem to have been unsuccessful in maintaining a wood-pasture in this region. We may conclude that there is no consensus on the controversy of dense forest versus wood-pasture in relation to indigenous large herbivores. Wood-pastures might have emerged at the transition of hunting/gathering to agricultural culture, and hence not earlier than at the transition of Neolithic and Bronze Ages.

Van de Veen and Lardinois (1991) put great emphasis on the impact of human hunters in the past. They argue that the ecological frame of reference for maximal unfolding of natural processes in western European ecosystems is before the introduction of long-distance weapons such as the throwing spear and bow and arrows which killed so many large herbivores. As we now live in a relatively warm period, we may have to look for our ecological frame of reference in the previous warm interglacial period, i.e. the Eemien about 100 000 years ago (Van de Veen and Lardinois, 1991).

In order to have some actual picture of ecological frames of reference, Dutch nature conservationists nowadays tend to refer to the relatively intact marshes in former Eastern Germany and in Poland (Wassen, 1990; Wassen *et al.*, 1992). Examples for river foreland systems are derived from the rivers Rhine, Danube and Loire in central and southern Europe (De Bruin *et al.*, 1987). The famous example of the old-growth forest of Bialowieza in Poland is often referred to (Koop, 1993; Meeuwissen, 1993). For relatively open wood-pasture, the New Forest in England and the Borkener Paradies in Germany serve as examples (Van de Veen and Lardinois, 1991). Relatively recently developed woodlands on abandoned 'old fields' in the United States may give some impression of the processes to be expected for the development of woodlands on marginal agricultural land to be taken out of agricultural exploitation in the framework of European Union policy.

A new perspective was developed by the 'vision on nature development' (Baerselman and Vera, 1995). The main item of this vision was to return to natural landscapes and communities, including the natural landscape-building processes and the role of large herbivores superimposed on them. Water was regarded as a major factor in the dynamics of former natural landscapes, as we have seen before in this chapter. Restoration of former hydrological conditions is, therefore, an important issue. This can be attained by raising the groundwater table, inundation of polders, replacement of mineral-poor/eutrophic groundwater by mineral-rich/mesotrophic groundwater, and removal of summer-dikes along rivers and even along the coast.

2.7.3 Comparison of ecological frames of reference

A good impression of the different approaches to nature conservation described above might be gained from the experiences of an outsider to the European situation, such as those reported by Wedin (1992). He described how the North American tallgrass prairie was a vast natural system shaped by climate, topography, fire and herbivores – until 200 years ago. He remarked that both chalk grasslands in Europe without grazing and tallgrass prairie without fire in Northern America can be displaced by woody vegetation within a few decades, which suggests that the tallgrass prairie could be a semi-natural landscape instead of a natural landscape. Indeed the tallgrass prairie appears to be a replacement community that has arisen since the last glaciation and had a significant human contribution in its origins and maintenance (Axelrod, 1985). Unfortunately, in the Midwest, bison (*Bison bison*) were extinct half a century before early ecologists began to describe the prairie in detail, and therefore there is no good record of the original grazing regimes. Nowadays the prairie persists as small isolated fragments in an intensively exploited agricultural landscape, i.e. like islands, subjected to eutrophication and the loss of fire and grazing.

The problems of fragmentation, eutrophication and abandonment as threats to biological diversity are similar to those in northwestern Europe, as we have seen in the preceding sections. An important threat to the diversity of grasslands on both continents has been the loss of the ecological factors that originally maintained these communities. American managers of tallgrass prairie have realized that fire must be restored in order to maintain prairie remnants. The pre-settlement fire regime (if it can be reconstructed) may not be sufficient to counter the increased productivity of the present prairie remnants by atmospheric deposition. Reintroduction of bison may also raise problems by increasing the turnover rate of nitrogen (Hobbs *et al.*, 1991). Although haymaking is not as 'natural' as bison grazing, it might be more appropriate from an ecological point of view. This brings Wedin (1992) to the lamentation that the heavy North American emphasis on 'naturalness' affects the setting of conservation goals, the determination of what is worthy of protection and restoration, and the decision of what management tools are appropriate or inappropriate. He was struck by the observation that conservation management in Europe considers both ecosystem function (with respect to, for example, hydrological regimes) and ecosystem structure (with respect to population biology and species composition of threatened communities). Wedin concludes:

> The lesson of European grassland conservation is that unless North Americans begin to address these two problems together, we may not find a solution to the problems of conserving our threatened grassland biodiversity.

Although the idea of using livestock as a conservation management tool (Bokdam and WallisDeVries, 1992; Hobbs and Huenneke, 1992; Fleischner, 1994; WallisDeVries, 1995) is not unknown in North America, it is not judged to be an appropriate tool. This is mainly because the exclusion of livestock seems to result in more successful ecosystem protection (Fleischner, 1994). Fleischner's review evoked a discussion about grazing management (Chapter 1) which is certainly not new to conservationists in northwestern Europe. He tended towards the conclusion that, because of over-exploitation by livestock, a management option could be the complete removal of livestock. Others replied that the cessation of grazing can also lead to habitat degradation (e.g. Curtin, 1995 and quoting Milchunas and Lauenroth, 1993).

A tendency to move from intensive grazing towards no grazing at all is also seen in northwestern Europe. A good example is given by the mainland salt marshes along the North German coast. These artificial salt marshes were heavily grazed by sheep until recently, which had resulted in monotonous 'greens'. Now ideas about the function of salt marshes, in particular artificial marshes, are changing, as summarized by Kiehl and Stock (1994), which implies that there are all sorts of management options. Two main options for the future development of salt marshes as nature reserves are possible. The first is to abandon all human interference and leave the marshes to entirely natural development – this is the 'wilderness' concept. In this concept the aim and practice of management are identical: namely, the unimpeded outcome of natural development. The outcome of this process with respect to target communities is less important. This is largely the present approach in Germany. Large areas are incorporated into the National Park Schleswig-Holsteinisches Wattenmeer. By definition, there is no longer any place for livestock, which will eventually result in the cessation of grazing impact by large herbivores. The second option aims at target communities with high species richness by maintaining or creating variation in abiotic conditions – this is the 'biodiversity' concept, in which the choice for management practices is derived from the management aims. It is the more common policy for the salt marshes of Denmark and The Netherlands (Bakker *et al.*, 1997).

The different options for nature conservation are all practised at the same time in various parts of The Netherlands. In fact this policy is a continuation of the characteristically pluriform nature conservation in The Netherlands (Van der Windt, 1995).

This chapter can be summarized by highlighting the shift in ideas on nature conservation, with the role of grazing as one of the central issues. Grazing in nature reserves has been practised increasingly since the early 1970s. The main emphasis was at first on botanical management (and that of the dependent meadow birds) of small nature reserves featuring enclosed semi-natural landscapes with herbivores acting as cutting machinery. A new management vision evolved from the early 1980s onwards with an emphasis on extensive, unenclosed nature reserves featuring large-scale semi-natural and near-natural landscapes, with landscape-building processes involving a minimum of human interference, and large herbivore populations as an integral part of the ecosystem. This shift in emphasis appears to imply a transition from pattern-oriented management towards process-oriented management (Londo and Van Wirdum, 1994), or from vegetation-oriented towards ecosystem-oriented.

REFERENCES

Anonymous (1943) *Nature Conservation and Nature Reserves*. Report of the British Ecological Society.

Arcese, P. and Sinclair, A.R.E. (1997) The role of protected areas as ecological baselines. *Journal of Wildlife Management*, 63, 587-602.

Axelrod, D.I. (1985) Rise of the grassland biome, central North America. *The Botanical Review*, 51, 163-201.

Baerselman, F. and Vera, F.W.M. (1995) *Nature Development. An exploratory study for the construction of ecological networks*, Ministry of Agriculture, Nature and Fisheries, Den Haag.

Bakker, J.P. (1987) Diversiteit in de vegetatie door begrazing, in *Begrazing in de Natuur* (eds S. De Bie, W. Joenje and S.E. Van Wieren), pp. 150–164, Pudoc, Wageningen.

Bakker, J.P. (1989) *Nature Management by Grazing and Cutting*, Kluwer, Dordrecht.

Bakker, J.P. and De Vries, Y. (1982) Enige indrukken van beekdalvegetaties in Noordoost-Polen. *Natura*, **79**, 381–393.

Bakker, J.P., Esselink, P. and Dijkema, K.S. (1997) Salt-marsh management for nature conservation, the value of long-term experiments. *Wadden Sea News Letter*, **1997-1**, 19–24.

Bardet, A.C., Kooi, P.B., Waterbolk, H.T. and Wieringa, J. (1983) Peelo, historisch geografisch en archeologisch onderzoek naar de ouderdom van een Drents dorp. *Mededelingen Koninklijke Nederlandse Akademie van Wetenschappen afdeling Letterkunde, Nieuwe Reeks*, **46**, 1–25.

Beaufoy, G., Baldock, G. and Clark, J. (1994) *The Nature of Farming: Low-intensity Farming Systems in Nine European Countries*, Institute for European Environmental Policy, London.

Behre, K.E. (1976) Beginn und Form der Plaggenwirtschaft in Nordwestdeutschland nach Pollenanalytischen Untersuchungen in Ostfriesland. *Neue Ausgrabungen und Forschungen in Niedersachsen*, **10**, 197–224.

Benecke, N. (1994) *Archäozoologische Studien zur Entwicklung der Haustierhaltung in Mitteleuropa und Südskandinavien von den Anfängen bis zum ausgehenden Mittelalter*, Akademie Verlag, Berlin.

Bieleman, J. (1987) *Boeren op het Drentse zand 1600–1910. Een nieuwe visie op de oude landbouw*, Hes, Utrecht.

Bignal, E.M. and McCracken, D.I. (1996) Low-intensity farming systems in the conservation of the countryside. *Journal of Applied Ecology*, **33**, 413–424.

Bink, R.J., Bal, D., Van den Berk, V.M. and Draaijer, L.J. (1994) *Toestand van de natuur 2*, Report IKC-NBLF, Wageningen.

Birks, H.J.B. (1996) Contributions of Quaternary palaeoecology to nature conservation. *Journal of Vegetation Science*, **7**, 89–98.

Bobbink, R. and Willems, J.H. (1993) Restoration management of abandoned chalk grassland in The Netherlands. *Biodiversity and Conservation*, **2**, 616–626.

Bokdam, J. and Gleichman, J.M. (1989) De invloed van runderbegrazing op de ontwikkeling van Struikheide en Bochtige smele. *De Levende Natuur*, **90**, 6–14.

Bokdam, J. and WallisDeVries, M.F. (1992) Forage quality as a limiting factor for cattle grazing in isolated Dutch nature reserves. *Conservation Biology*, **6**, 399–408.

Bouwer, K. (1985) De ontwikkeling van het cultuurlandschap, in *Geschiedenis van Drenthe*. (eds J. Heringa, D.P. Blok, M.G. Buist and H.T. Waterbolk), pp. 91–140, Boom, Meppel.

Bradshaw, R. and Hannon, G. (1992) Climatic change, human influence and disturbance regime in the control of vegetation dynamics within Fiby Forest, Sweden. *Journal of Ecology*, **80**, 625–632.

Buckley, G.P. (ed.) (1989) *Biological Habitat Reconstruction*, Belhaven Press, New York.

Bülow-Olsen, A. (1980) Changes in the species composition in an area dominated by *Deschampsia flexuosa* as a result of cattle grazing. *Biological Conservation*, **18**, 257–270.

Burrichter, E., Pott, R., Raus, T. and Wittig, R. (1980) Die Hudelandschaft 'Borkener Paradies' im Emstal bei Meppen. *Abhandlungen aus dem Landesmuseum für Naturkunde zu Münster in Westfalen*, **42**(4), 1–69.

Bush, M.B. and Colinvaux, P.A. (1994) Tropical forest disturbance: palaeoecological records from Darien, Panama. *Ecology*, **75**, 1761–1768.

Buttenschøn, J. and Buttenschøn, R.M. (1982) Grazing experiments with cattle and sheep on nutrient poor, acidic grassland and heath. I. Vegetation development. *Natura Jutlandica*, **21**, 1–18.

Cairns, J. (1986) Restoration, reclamation, and regeneration of degraded or destroyed ecosystems, in *Conservation Biology*, (ed. M.E. Soulé), pp. 465–484, Sinauer Associates, Sunderland, Mass.

Clason, A.T. (1980) *Zeldzame Huisdierrassen*, Thieme, Zutphen.

Curtin, C.G. (1995) Grazing and advocacy. *Conservation Biology*, **9**, 233.

De Bruin, D., Hamhuis, D., Van Nieuwenhuijze, L. *et al.* (1987) *Ooievaar. De toekomst van een rivierengebied*. Stichting Gelderse Milieufederatie, Arnhem.

Dierschke, H. (1984) Natürlichkeitsgrade von Pflanzengesellschaften unter besonderer Berücksichtigung der Vegetation Mitteleuropas. *Phytocoenologia*, **12**, 173–184.

Draaisma, A. (1958) Het productieniveau onderzoek. I. Teelt en bemesting op bouwland in de praktijk. *Verslagen Landbouwkundige Onderzoeken*, **64.9**, 116–120.

Dublin, H.T. (1995) Vegetation dynamics in the Serengeti–Mara ecosystem: the role of elephants, fire, and other factors, in *Serengeti II. Dynamics, Management, and Conservation of an Ecosystem* (eds A.R.E. Sinclair and P. Arcese), pp. 71–90, University of Chicago Press, Chicago.

During, J.H. and Willems, J.H. (1984) Diversity models applied to a chalk grassland. *Vegetatio*, **57**, 103–114.

During, H.J. and Willems, J.H. (1986) The impoverishment of the bryophyte and lichen flora of the Dutch chalk grasslands in the thirty years 1953–1983. *Biological Conservation*, **36**, 143–158.

Ekstam, U., Aronsson, M. and Forshed, N. (1988) *Ängar. Om naturliga slåttermarker i odlingslandskapet*, LTs förlag, Stockholm.

Elberse, W.Th., Van den Bergh, J.P. and Dirven, J.G.P. (1983) Effects of use and mineral supply on the botanical composition and yield of old grassland on heavy clay soil. *Netherlands Journal of Agricultural Sciences*, **31**, 62–88.

Fleischner, T.L. (1994) Ecological costs of livestock grazing in Western North America. *Conservation Biology*, **8**, 629–644.

Fokkinga, A. and Felius, M. (1995) *Een land vol vee*, Misset, Doetinchem.

Gabrielson, I.N. (1947) Management of nature reserves on the basis of modern scientific knowledge, in *Proceedings and Papers of the 6th Technological Meeting IUCN*, pp. 27–35, IUCN, London.

Gibson, C.W.D., Watt, T.A. and Brown, V.K. (1987) The use of sheep grazing to recreate species-rich grassland from abandoned arable land. *Biological Conservation* **42**, 165–183.

Green, B.H. (1980) Management of extensive amenity grassland by mowing, in *Amenity Grassland, an Ecological Perspective*, (eds I.H. Rorison and R. Hunt), pp. 155–161, Wiley, Chichester.

Grévilliot, F. and Muller. S. (1996) Etude de l'impact des changements des pratiques agricoles sur la biodiversité végétale dans les prairies inondables du Val de Meuse: présentation méthodologique et premiers résultats. *Acta Botanica Gallica*, **143**, 317–338.

Grootjans, A.P., Van Wirdum, G., Kemmers, R. and Van Diggelen, R. (1996) Ecohydrology in The Netherlands: principles of an application-driven interdiscipline. *Acta Botanica Neerlandica*, **45**, 491–516.

Hall, S.J.G. and Bradley, D.G. (1995) Conserving livestock breed biodiversity. *Trends in Ecology and Evolution*, **10**, 267–270.

Hillegers, H.P.M. (1993) Heerdgang in Zuidelijk Limburg. *Publicaties Natuurhistorisch Genootschap Limburg*, **Reeks XL-1**, 1–160.

Hobbs, N.T., Schimmel, D.S., Owensby, C.E. and Ojima, D.S. (1991) Fire and grazing in the tallgrass prairie: contingent effects on nitrogen budgets. *Ecology*, **72**, 1374–1382.

Hobbs, R.J. and Huenneke, L.F. (1992) Disturbance, diversity and invasion: implications for conservation. *Conservation Biology*, **6**, 324–337.

Hornberger, T. (1959) *Die kulturgeographische Bedeutung der Wanderschäferei in Süddeutschland*, Selbstverlag der Bundesanstalt für Landeskunde, Remagen/Rhein.

Jordan, W.R., Gilpin, M.E. and Aber, J.D. (eds) (1987) *Restoration Ecology. A synthetic approach to ecological research*, Cambridge University Press, Cambridge.

Jordan, W.R. and Packard, S. (1989) Just a few oddball species: restoration practice and ecological theory, in *Biological Habitat Reconstruction*, (ed. G.P. Buckley), pp. 18–26, Belhaven Press, New York.

Kiehl, K. and Stock, M. (1994) Natur- oder Kulturlandschaft? Wattenmeersalzwiesen zwischen den Anspruchen von Naturschutz, Küstenschutz und Landwirtschaft, in *Warnsignale aus dem Wattenmeer*, (eds J.L. Lozán, E. Rachor, K. Reise *et al.*), pp. 190–196, Blackwell, Berlin.

Kiehl, K., Eischeid, I., Gettner, S. and Walter, J. (1996) Impact of different sheep grazing intensities on salt marsh vegetation in northern Germany. *Journal of Vegetation Science*, **7**, 99–106.

Klapp, E. (1965) *Grünlandvegetation und ihre Standort*, Parey, Berlin.

Koop, H. (1993) Ecosysteemvisie Bos: natuurbosreferenties. *Nederlands Bosbouwkundig Tijdschrift*, **65**, 227–236.

Koopmans, J. (1960) Het productie niveau onderzoek. II. De bemesting van grasland in de praktijk. *Verslagen Landbouwkundig Onderzoek*, **66**(5), 35–39.

Leroy, P. (1991) Natuurontwikkelingsplan Vlaanderen. *Landschap*, **8**, 135–140.

Levitt, J. (1972) *Responses of Plants to Environmental Stresses*, Academic Press, New York.

Londo, G. (1990) Conservation and management of semi-natural grasslands in northwestern Europe, in *Vegetation and Flora of Temperate Zones*, (eds U. Bohn and R. Neuhäusl), pp. 69–77, SPB Academic Publishing, Den Haag.

Londo, G. (1997) *Natuurontwikkeling*, Backhuys Publishers, Leiden.

Londo, G. and Van Wirdum, G. (1994) Natuurlijkheidsgraden en natuurontwikkeling. *De Levende Natuur*, **95**, 10–16.

Louwe Kooijmans, L.P. (1974) The Rhine/Meuse delta. PhD thesis, University of Leiden.

Louwe Kooijmans, L.P. (1985) *Sporen in het Land. De Nederlandse delta in de prehistorie*, Meulenhoff Informatief, Amsterdam.

Meeuwissen, T.W.M. (1993) Ecosysteemvisie Bos: het bestaande bos. Welke natuur willen we waar en hoeveel? *Nederlands Bosbouwkundig Tijdschrift*, **65**, 237–247.

Milchunas, D.G. and Lauenroth, W.K. (1993) Quantitative effects of grazing on vegetation and soils over a global range of environments. *Ecological Monographs* **63**, 327–366.

Mountford, J.O., Lakhani, K.H. and Holland, R.J. (1996) Reversion of grassland vegetation following the cessation of fertilizer application. *Journal of Vegetation Science*, 7, 219–228.

Noe-Nygaard, N. (1995) Ecological, sedimentary, and geochemical evolution of the late-glacial to postglacial Åmose lacustrine basin, Denmark. *Fossils and Strata*, **37**, 1–436.

Olff, H. and Bakker, J.P. (1991) Long-term dynamics of standing crop, vegetation composition and species richness after the cessation of fertilizer application to hay-fields. *Journal of Applied Ecology*, **28**, 1040–1052.

Oomes, M.J.M. and Mooi, H. (1981) The effect of cutting and fertilizing on the floristic composition and production of an *Arrhenatherion elatioris* grassland. *Vegetatio*, **47**, 233–239.

Oomes, M.J.M., Olff, H. and Altena, H.J. (1996) Effects of vegetation management and raising the water table on nutrient dynamics and vegetation change in a wet grassland. *Journal of Applied Ecology*, **33**, 576–588.

Oosterveld, P. (1975) Beheer en ontwikkeling van natuurreservaten door begrazing. *Natuur en Landschap*, **29**, 161–171.

Oosterveld, P. (1977) Beheer en ontwikkeling van natuurreservaten door begrazing. I, II, III. *Bosbouwvoorlichting*, **16**, 18–21, 66–68, 94–98.

Persson, S. (1984) Vegetation development after the exclusion of grazing cattle in a meadow area in the south of Sweden. *Vegetatio*, **55**, 65–92.

Pieters, J.H. (1971) *Een Bodemvruchtbaarheidsonderzoek op Grasland in de Friese Wouden*. Report 12 Institute for Soil Fertility, Haren.

Rorison, I.H. (1980) The current challenge for research and development, in *Amenity Grassland, an Ecological Perspective*, (eds I.H. Rorison and R. Hunt), pp. 3–10, Wiley, Chichester.

Schiefer, J. (1981) Bracheversuche in Baden-Württenberg. Vegetations- und Standortsentwicklung auf 16 verschiedenen Versuchsflächen mit unterschiedlichen Behandlungen. *Beiheifte Veröffentlichungen Naturschutz und Landschaftsplege, Baden Württemberg*, **22**, 1–325.

Schreiber, K.F. and Schiefer, J. (1985) Vegetations- und Stoffdynamik in Grünlandbrachen – 10 Jahre Bracheversuche in Baden Württenberg. *Münstersche Geographische Arbeiten*, **20**, 111–153.

Siebel, H.N., Aptroot, A., Dirkse, G.M. *et al.* (1992) Rode lijst van in Nederland verdwenen en bedreigde mossen en korstmossen. *Gorteria*, **18**, 1–20.

Sjörs, H. (1986) On the gradient from near-natural to man-made. *Transacta of the Botanical Society of Edinburgh*, 150th Anniversary Supplement, 77–84.

Snaydon, R.W. (1980) Ecological aspects of management – a perspective, in *Amenity Grassland, an Ecological Perspective*, (eds I.H. Rorison and R. Hunt), pp. 219–231, Wiley, Chichester.

Stumpel, A.H.P. (1981) Threats to and conservation of reptiles and amphibians in the Netherlands, in *Proceedings European Herpetological Symposium, Oxford*, (ed. J. Coborn), pp. 97–100, Cotswold Wildlife Park, Burford.

Succow, M. (1988) *Landschaftsökologische Mooreskunde*, Fischer Verlag, Jena.

Thalen, D.C.P. (1981) Grote grazers en snelle snoeiers, het beheer van vegetaties, in *Verslag van het Veluwe-Symposium 09.05.80*, pp. 21–39, Gelderse Milieufederatie, Arnhem.

Thalen, D.C.P. (1984) Large animals as tools in the conservation of diverse habitats. *Acta Zoologica Fennica*, **172**, 159–163.

Thalen, D.C.P. (1987) Begrazing in een Nederlands perspectief, in *Begrazing in de natuur* (eds S. De Bie, W. Joenje and S.E. Van Wieren), pp. 3–14, Pudoc, Wageningen.

Tüxen, R. (1974) Die Haselüner Kuhweide. Die Pflanzengesellschaften einer mittelalterlichen Gemeindeweide. *Mitteilungen Floristisch-Soziologische Arbeitsgemeinschaft Neue Fassung*, **17**, 69–102.

Tyler, C. (1984) Calcareous fens in South Sweden. Previous use, effects of management and management recommendations. *Biological Conservation*, **30**, 69–89.

Van Bemmel, A.C.V. and De Smidt, J.T. (1955) Heideschapen. *Natuur en Landschap*, **9**, 97–105.

Van de Veen, H.E. (1985) Natuurontwikkelingsbeleid en bosbegrazing. *Landschap*, **2**, 14–28.

Van de Veen, H.E. and Van Wieren, S.E. (1980) Van Grote Grazers, Kieskeurige Fijnproevers en Opportunistische Gelegenheidsvreters; over het gebruik van grote herbivoren bij de ontwikkeling en duurzame instandhouding van natuurwaarde. Internal report 80/11, Instituut voor Milieuvraagstukken, Free University Amsterdam.

Van de Veen, H.E. and Lardinois, P. (1991) *De Veluwe Natuurlijk!* Schuyt & Co., Haarlem.

Van der Windt, H.J. (1995) *En dan: wat is natuur nog in dit land? Natuurbescherming in Nederland 1880–1990*, Boom, Amsterdam/Meppel.

Van Hecke, P., Impens, I. and Behaeghe, T.J. (1981) Temporal variation of species composition and species diversity in permanent grassland plots with different fertilizer treatments. *Vegetatio*, **47**, 221–232.

Van Leeuwen, C.G., Londo, G. and Van Wijngaarden, A. (1971) Verslag van een Studiereis naar Gotland en Öland in 1971. Internal report (mimeo), Research Institute for Nature Management, Leersum.

Van Nispen tot Sevenaer, W.J.C.M. (1975) De gevolgen van de stormrampen voor de bosbouw in Nederland. *Nederlands Bosbouwkundig Tijdschrift*, **47**, 41–52.

Van Wieren, S.E. (1985) *Begrazingsproef met Schotse Hoogland-runderen in het natuurgebied de Imbos*. Tweede voortgangsrapport, Instituut voor Milieuvraagstukken, Free University Amsterdam.

Van Wieren, S.E. (1988) *Runderen in het bos: begrazingsproef met Schotse Hooglandrunderen in het natuurgebied de Imbos*. Report 88/3, Instituut voor Milieuvraagstukken, Free University Amsterdam.

Van Wieren, S.E. (1991) The management of populations of large mammals, in *The Scientific Management of Temperate Communities for Conservation*, (eds I.F. Spellerberg, F.B. Goldsmith and M.G. Morris), pp. 103–127, Blackwell, Oxford.

Van Wirdum, G., Den Held, A.J. and Schmitz, M. (1992) Terrestrializing fen vegetation in former turbaries in the Netherlands, in *Fens and Bogs in the Netherlands: vegetation, history, nutrient dynamics and conservation*, (ed. J.T.A. Verhoeven), pp. 323–360, Kluwer Academic, Dordrecht.

Vera, F.W.M. (1988) *De Oostvaardersplassen: van spontane natuuruitbarsting tot gerichte natuurontwikkeling*, IVN and Grasduinen Oberon, Amsterdam.

Vera, F.W.M. (1997) Metaforen voor de Wildernis: eik, hazelaar, rund en paard. PhD thesis, Wageningen Agricultural University.

Vermeulen, F.H.B. (1954) Invloed van de afstand tot de boerderij op de vruchtbaarheid van graslanden. *Landbouwvoorlichting*, **11**, 424–426.

Von Thünen, J.H. (1842) *Der isolierten Staat im Beziehung auf Landwirtschaft und Nationalökonomie*, Fischer, Stuttgart.

WallisDeVries, M.F. (1995) Large herbivores and the design of large-scale nature reserves in Western Europe. *Conservation Biology*, **9**, 25–33.

Wassen, M.J. (1990) Water flow as a major landscape ecological factor in fen development. PhD thesis, University of Utrecht.

Wassen, M.J., Barendregt, A., Palczynski, A. *et al.* (1992) Hydro-ecological analysis of the Biebrza mire (Poland). *Wetlands Ecology and Management*, **2**, 119–134.

Waterbolk, H.T. (1979) Siedlungskontinuität im Küstengebiet der Nordsee zwischen Rhein und Elbe. *Probleme der Küstenforschung im südlichen Nordseegebiet*, **13**, 1–21.

Waterbolk, H.T. (1984) Gebruik van het landschap in de Romeinse tijd en de Vroege Middeleeuwen, in *Het Drentse landschap*, (eds J. Abrahamse, S. Bottema, M.H. Buruma *et al.*), pp. 40–47, Walburg Pers, Zutphen.

Wedin, D.A. (1992) Biodiversity conservation in Europe and North America: grasslands, a common challenge. *Restoration and Management Notes*, **10**, 137–143.

Weeda, E.J. (1985) Veranderingen in het voorkomen van vaatplanten in Nederland, in *Atlas van de Nederlandse flora, deel 2*, (eds J. Mennema, A.J. Quené-Boterenbrood and C.L. Plate), pp. 9–47, Scheltema and Holkema, Utrecht.

Weeda, E.J., Van der Meijden, R. and Bakker, P.A. (1990) Floron-Rode Lijst 1990. Rode Lijst van de in Nederland verdwenen en bedreigde planten. (Pterydophyta en Spermatophyta) over de periode 1.1.1980–1.1.1990. *Gorteria*, **16**, 2–26.

Weinreich, J.A. and Musters, C.J.M. (1989) *Toestand van de Natuur. Veranderingen in de Nederlandse natuur*. SDU Uitgeverij, Den Haag.

Wells, T.C.E. (1980) Management options for lowland grassland, in *Amenity Grassland, an Ecological Perspective*, (eds I.H. Rorison and R. Hunt), pp. 175–195, Wiley, Chichester.

Wells, T.C.E., Sheail, J., Ball, D.F. and Ward, L.K. (1976) Ecological studies on the Porton Ranges: relationships between vegetation, soils and land-use history. *Journal of Ecology*, **64**, 589–626.

Westhoff, V. (1945) *Biologische Problemen der Natuurbescherming*. Inleiding gehouden voor het N.J.N. congres op 15-8-1945 te Drachten.

Westhoff, V. (1949) Schaakspel met de natuur. *Natuur en Landschap*, **3**, 54–62.

Westhoff, V. (1971) The dynamic structure of plant communities in relation to the objectives of conservation, in *The Scientific Management of Animal and Plant Communities for Conservation*, (eds E. Duffey and A.S. Watt), pp. 3–14, 11th Symposium British Ecological Society, Blackwell, Oxford.

Westhoff, V. (1983) Man's attitude towards vegetation, in *Man's Impact on Vegetation*, (eds W. Holzner, M.J.A. Werger and I. Ikusima), pp. 7–24, Junk, Den Haag.

Westhoff, V. and Van Dijk, J. (1952) Experimenteel successie-onderzoek in natuurreservaten, in het bijzonder in het Korenburgerveen bij Winterswijk. *De Levende Natuur*, 55, 5–16.

Westhoff, V. and Weeda, E. (1984) De achteruitgang van de Nederlandse flora sinds het begin van deze eeuw. *Natuur en Milieu*, **8**, 8–17.

Westhoff, V., Schaminée, J. and Sykora, K.V. (1983) Aufzeichnungen zur Vegetation der schwedischen Inseln Öland, Gotland und Stora Karlsö. *Tüxenia*, 3, 179–198.

Willems, J.H. (1983) Species composition and above ground phytomass in chalk grassland with different management. *Vegetatio*, **52**, 171–180.

Williams, E.D. (1978) *The Botanical Composition of the Park Grass Plots at Rothamsted 1856–1976*. Rothamsted Experimental Station, Harpenden.

Worm, P.B. and Van Wieren, S.E. (1996) Reactie van edelherten op veranderend beheer van de Vereniging Natuurmonumenten. *De Levende Natuur*, **97**, 27–32.

Zeiler, J.T. (1997) Hunting, fowling and stock-breeding at Neolithic sites in the western and central Netherlands. PhD thesis, University of Groningen.

Zoller, H., Bischof, N., Erhardt, A. and Kienzle, U. (1984) Biocoenosen von Grenzertragflächen und Brachland in den Berggebieten der Schweiz, Hinweize zur Sukzession zum Naturschutzwert und zur Pflege. *Phytocoenologia*, **12**, 373–394.

–3

Origins and development of grassland communities in northwestern Europe

Herbert H.T. Prins
Tropical Nature Conservation and Vertebrate Ecology Group, Department of Environmental Sciences, Wageningen Agricultural University, Bornsesteeg 69, 6708 PD Wageningen, The Netherlands

3.1 INTRODUCTION

Most of the countryside of northwestern Europe is characterized by an absence of forest. Indeed, forest covers only about 25% of France, 27% of Germany, 10% of The Netherlands and 8% of England and Wales; in western Europe only 1% is considered to be 'old-growth' forest (Dudley, 1992). This quintessence was captured by many seventeenth century painters, who emphasized the sky with its clouds over near-treeless landscapes. To many a citizen of today, heaths, downs, limestone grasslands and other open vegetation types are viewed as original, natural and ancient. Yet many of these vegetation types are artificial and, as such, are as unnatural as most forests of northwestern Europe.

This chapter will elucidate the origins of open vegetation dominated by grasses, and will pay particular attention to the effect of human influences on vegetation composition and development, and to the impact of large herbivores on the vegetation. Grass and humanity – there is more to their interrelation than is often surmised when one is sitting on a well kept lawn, since grasses form the mainstay of domesticated herbivores. Grasses form the production basis for milk, meat and wool and grasses were the fuel that made the first semi-intensive agriculture possible through ploughing by oxen and horse. Indeed, through human manipulation of vegetation over the last millennia, Europe's present landscape, including most of its nature reserves, came into being. The question of which of today's open vegetation types are least disturbed and most akin to natural vegetation, and which ones are to be viewed as culturally derived, has important implica-

Grazing and Conservation Management. Edited by M.F. WallisDeVries, J.P. Bakker and S.E. Van Wieren. Published in 1998 by Kluwer Academic Publishers, Dordrecht. ISBN 0 412 47520 0.

tions for the management of these vegetation types and, thus, for the management of nature reserves.

3.2 OPEN VEGETATION BEFORE THE PLEISTOCENE

The basis for understanding the present open vegetation in northwestern Europe lies in understanding the Pleistocene. The study of vegetation development during interglacial periods allows human influence to be teased apart from natural development during the Holocene, in which we live. This section briefly describes vegetation development prior to the Pleistocene.

Grasses (Gramineae, about 8000 species) and sedges (Cyperaceae, about 4000 species) belong to the Order Cyperales which, together with other monocotyledons, probably derived from dicotyledons during the beginning of the Cretaceous – 135 million years before present (BP) (Cronquist, 1968; Von Denffer *et al.*, 1976; Doyle, 1978). There is good evidence that the first angiosperms lived in locally disturbed habitats, such as stream margins, away from closed forests dominated by conifers and ferns (Doyle, 1978) and that until the end of the Cretaceous (65 million years ago), monocotyledons grew mainly in swamps (Von Denffer *et al.*, 1976). During this epoch there is very little evidence for open vegetation, and the Earth was covered to a large extent by forests dominated by woody species (Cronquist, 1968; Kurtén, 1972; Von Denffer *et al.*, 1976; Webb, 1977, 1978). Evidence for this stems both from the types of plants that have been found and from the absence of remains of grazing animals. During the early Tertiary (Eocene, 45 million years BP) the average temperature in northwestern Europe was about 22°C and the area was covered with forests resembling present tropical monsoon forests. Yet on Spitsbergen and Greenland the first representatives of the species that would later dominate Europe were already established: representatives of *Castanea*, *Juglans*, *Fagus*, *Quercus*, *Acer*, *Betula*, *Populus*, *Salix*, *Tilia*, *Ulmus* and *Alnus* were growing, as well as *Picea* and *Pinus* (Mai, 1965; Hess *et al.*, 1967). During the Miocene (20 million years BP) the average temperature dropped to about 20°C and Europe was covered by evergreen subtropical forests. These forests comprised about one-third a temperate element (131 higher plant species) and two-thirds a subtropical element (226 species) although the mosses that occurred all have a tropical distribution at present (Hess *et al.*, 1967; Schmidt, 1969). During the following epoch the general cooling continued, and during the Pliocene (2 million years BP) the average annual temperature dropped to 15–18°C and in Switzerland even to 10–17°C (Schmidt, 1969). At that time European forests were characterized by temperate species but included species that until the present only survived outside Europe in East Asia or North America (*Ginkgo*, *Liriodendron*, *Magnolia*, and *Metasequoia*) (Hess *et al.*, 1967; Walter, 1968; Schmidt, 1969).

As well as cooling during this period, the climate became much drier because new mountain ranges arose with rain-shadows in their lee. This was mainly caused by orogenesis of the Alps, the Pyrenees and mountain ranges on the Balkan, and elsewhere the Himalayas, the Andes and the Rocky Mountains (Kurtén, 1972; Webb, 1977, 1978). It should be realized that at that time montane plant species had disappeared from Europe because the mountain ranges that had formed during the Hercynian orogeny (about 300 million years BP: Gotthard-Aar Massif, Black Forest, Vosges, Ardennes, Massif Armoricain) had eroded long before the Pliocene (Hess *et al.*, 1967). This desiccation did not result in the disappearance of the European forests – in contrast to North and South America and Central Asia, where steppe formations developed (Webb, 1977, 1978), although in Europe and North Asia many plant species became extinct during the cooling and desiccation in the course of the Miocene and later. During the Pliocene the occurrence of the first modern plant species can be demonstrated (Reid and Chandler, 1915): in Reuverian deposits from Reuver, Swalmen and Brunssum 214 species were identified, of which 98 are still extant and of which 22 occur at present in Europe:

> *Alisma plantago-aquatica*, *Carpinus betulus*, *Cicuta virosa*, *Corylus avellana*, *Ilex aquifolium*, *Larix decidua*, *Lycopus europaeus*, *Malus sylvestris*, *Myriophyllum verticillatum*, *Najas marina*, *Picea abies*, *Prunus spinosa*, *Quercus robur*, *Ranunculus nemorosus*, *Sagittaria sagittifolia*, *Scirpus lacustris* ssp. *glaucus*, *S. lacustris* ssp. *lacustris*, *S. mucronatus*, *Sparganium erectum*, *Trapa natans*, *Urtica dioica*, *Urtica urens*.

The newly developed mountain ranges in Europe derived their typical cold-adapted mountain flora mainly from two sources: the Mediterranean and the high Asian mountains (Table 3.1). Many herbs that came to dominate the vegetation were derived from woody species, and Cronquist (1968) stressed the point that herbaceous groups of angiosperms relate not directly to each other, but separately to the 'woody' core of the division (see also Doyle, 1978). The result of both evolution and temperature changes was that the flora of Europe became gradually more modern, and during the Pleistocene nearly all species were apparently identical to those of the present (Table 3.2).

In North America, in the rain-shadow of the Rocky Mountains, open grassland communities developed but not (at that stage) the associated fauna. The first herbivorous ungulates date back to the Palaeocene (65 million years BP) but these were apparently still browsers rather than grazers. It appears that the first mammals adapted to feeding on grasses developed in Asia, where the first finds of hypsodontic mammals date back to the Oligocene (40 million years BP). Hypsodontism refers to the type of cheek-teeth: high-crowned with clear enamel ridges well suited for grinding the siliceous plant material that is typical in grasses. These mammals penetrated North America, but even during the Miocene no steppes had developed. The open landscape was savanna-like, interspersed with gallery forests. However, tough grasses such as *Stipa* date back to this epoch and typical

Table 3.1 Plant genera from which some species have colonized the Alps (Hess *et al.*, 1967)

Species from Asian mountains	Species from the Mediterranean
Aconithum	*Allium* *
Androsace	*Anthyllis*
Artemisia *	*Biscutella*
Astragalus	*Campanula*
Crepis	*Cerinthe*
Delphinium	*Colchicum*
Gentiana	*Crocus*
Oxythrophis	*Dianthus*
Pedicularis	*Globularia*
Primula	*Gypsophila*
Rhododendron	*Helianthemum*
	Herniaria
	Linaria
	Narcissus
	Ononis
	Saponaria
	Sesleria
	Silene

* Species from these genera are also considered to have originated in the steppe area of eastern Europe (Szafer, 1966, cited in Godwin, 1975).

Table 3.2 The modernization of the European flora since the Miocene (after Szafer, 1946, cited in Godwin, 1975)

Epoch and fossil deposit arranged by geological age	*N* species investigated	% Species local at present
Miocene (*c.* 20 million BP)		
Salzhausen	22	4.5
Herzogenrath	39	12.8
Wieliczka	27	18.5
Niederlausitz	62	22.6
Sosnice	46	31.1
Pliocene (*c.* 2 million BP)		
Pont-du-Gail	36	25.0
Frankfurt	89	29.2
Kroscienko	113	33.8
Reuver	116	35.0
Willershausen	44	41.2
Castle Eden	41	61.0
Pleistocene (*c.* 1 million BP)		
Tegelen	85	75.2
Schwanheim	47	78.9
Vogelheim	55	94.6
Cromer	132	98.4

grazing ruminants had developed by the end of the Miocene (Webb, 1977, 1978). Modern associations between grasslands and herbivores could now start to evolve, and during the following Pliocene period herd-forming grazing ungulates reached their highest species diversity (Kurtén, 1972; Webb, 1977, 1978) to be followed by a continuing loss of diversity during the Pleistocene. However, during this long time span from the Cretaceous until the Pliocene, there is no evidence for grasslands in northwestern Europe. Most of Europe was covered in forest, with the exception of the peaks of the new mountain ranges.

3.3 OPEN VEGETATION DURING THE PLEISTOCENE

The Pleistocene period was characterized by the occurrence of widespread glaciations in the northern hemisphere. During these periods woody vegetation disappeared to a large extent (Table 3.3; Figure 3.1). It should be noted at this point that the reconstruction of prehistoric vegetation is largely based on fossil remains. One of the main sources of information about the vegetation at that time is through the study of fossil pollen. From these pollen assemblages it is clear that woody species gave way to Gramineae and Ericales (Table 3.4) although it should be realized that the pollen of many herb species is strongly underrepresented because it is not dispersed by wind. The use of fossil plant remains, whether they are macroscopic plant parts or whether they are pollen, to gain an insight into palaeovegetation is fraught with difficulties but the basic methods appear valid (Janssen, 1979).

The glacial periods during the Pleistocene alternated with interglacial periods. During glacial periods the average temperature dropped to about 0°C or even lower (Atkinson *et al.*, 1987) and plant species that were adapted to cold and dry conditions dominated the vegetation. During the interglacials, temperatures increased to present values, or became even higher, and present-day species dominated the vegetation. The first cool period at the junction of the Pliocene and the Pleistocene was the Pre-Tiglian (Table 3.4), which can perhaps be characterized as a glacial period (Zagwijn, 1960).

Table 3.3 Decline of woody species in western Europe (after Reid and Chandler, 1933; Godwin, 1975)

Epoch	Location	% Woody species
Lower Eocene	London Clay	97
Upper Eocene	Hordle	85
Middle Oligocene	Bembridge	57
Late Pliocene	Reuver	57
Early Pleistocene	Tegelen	28
Middle Pleistocene	Cromer	22

Table 3.4 Dominance of graminean species in the Pleistocene (after Zagwijn, 1960; data from the following locations: Meinweg, Bouwberg, Waubach, Herkenbosch, Susteren and Zeeuws-Vlaanderen)

	Percentage of pollen sum		
Epoch and deposit	Woody species	Ericales	Gramineae
Pliocene			
Susterian	85	10	5
Brunssumian	85	10	5
Reuverian	85	10	5
Pleistocene			
Pre-Tiglian	50	10	40
Tiglian[a]	70	15	15
Eburonian	45	15	40
Menapian	20	20	60
Late Pleistocene	20	1	80

[a] The Tiglian was a comparatively warm period.

3.3.1 Interglacial periods

After the Pre-Tiglian, the Tiglian (about 2 million years ago) in northwestern Europe was again a warm period, which can be considered as the first interglacial. Elephants, rhinoceroses, a wild pig species and possibly even tapirs (see Table 3.6) lived in a landscape that was dominated by riverine

Figure 3.1 The Mongolian steppe. Natural grasslands such as these dominated Europe during the Ice Age.

woodlands and swamps floristically similar to those that nowadays occur in the Caucasus (Van den Hoek Ostende, 1990). Most modern plant species that lived during the Tiglian are now affiliated with marshy plant communities and open water; a few species are affiliated with forest communities, but only very few with present-day grassland communities (Table 3.5). Grassland communities as they occur at present were still very rare in Europe, though steppe formation worldwide had its widest distribution during the Pliocene and grazing mammal species reached peak variety at that time (Kurtén, 1972; Webb, 1977, 1978).

Table 3.5 Vegetation during Early and Middle Pleistocene, deduced from flora of Tegelen (Dutch Limberg) and Cromer, respectively

Vegetation formation and class	*N* character species (Pleistocene)	
	Early[a]	Middle[b]
Open water		
Potametea	11	14
Total	11 (= 20%)	14 (= 11%)
Marshes		
Phragmitetea	8	7
Bidentetea tripartiti	7	–
Parvocaricetea	1	4
Littoreletea	–	2
Isotea–Nanojuncetea	–	2
Oxycocco-Sphagenetea	–	4
Total	16 (= 29%)	26 (= 21%)
Ruderal vegetation		
Chenopodietea	4	9
Artemesietea vulgaris	3	13
Plantaginetea majoris	3	6
Secalietea	2	2
Total	12 (= 22%)	30 (= 24%)
Forest		
Querco-Fagetea	4	16
Rhamno-Prunetea	3	6
Quercetea robori-petraeae	1	4
Trifolio-Geranietea sanguinei	1	3
Alnetea glutinosae	–	8
Franguletea	–	4
Epibolietea angustifolii	–	1
Total	9 (= 16%)	42 (= 34%)

Table 3.5 (*continued*)

Vegetation formation and class	N character species Pleistocene)	
	Early[a]	Middle[b]
Grassland		
Koelerio-Corynephoretea	3	2
Molinio-Arrhenatheretea	3	9
Festuco-Brometea	1	–
Total	7 (= 13%)	11 (= 9%)

[a] Of 55 species that still occur in northwestern Europe, the present-day associations (classified after Westhoff and Den Held, 1969) give some indication that grasslands could have already occurred in the Early Pleistocene to some extent. If these formations occurred, then their species composition was so different from the modern ones that it is difficult to suggest what the vegetation looked like. Many other forest species occurred that are now extinct, such as *Alnus viridis*, *Acer opulifolium*, *Acer limburgensis*, *Magnolia* spec., *Phellodendron* spec., *Pterocarya limburgensis*.

[b] During a Middle Pleistocene interglacial (Cromer flora based on Reid and Chandler, 1933), the flora was already more modern, with 123 recognizable species, but the distribution over the different formations did not differ (χ^2 = 7.867, df = 4; NS) from that of Tegelen.

With the alternations of glacials with interglacials during the Pleistocene, the proportion of plant species living in Europe became increasingly similar to the present, and during the Middle Pleistocene the number of species that can be used for a vegetation reconstruction on the basis of character species for present-day plant communities increases (Table 3.5). It appears safest not to try to identify plant communities from the fossil material from before the end of the Pleistocene, but to limit the reconstruction to formations (that is, physiognomic units such as forest, grassland, bog) (Janssen, 1979; see also Delcourt and Delcourt, 1991). Quite a lot of information is available for the Cromerian (Voigtstedtian) interglacial (about 1 million years ago) but from the available data there is very little indication of the occurrence of grasslands during that interglacial period (Table 3.5). Also fossils of mammals, such as straight-tusked elephant (or forest elephant, *Elephas antiquus*, also known as *Palaeoloxodon antiquus*), extinct deer (*Cervus accronatus*), wild boar (*Sus scrofa*) and extinct rhinoceroses (*Dicerorhinus kirchbergensis* and *D. etruscus*) (Stuart, 1991; Kahlke, 1994) indicate the presence of forests. The hippo species at that time, *Hippopotamus* cf. *major*, could indicate riverine floodplains but the ecology of this species may have been different from the present hippo species: its occurrence appears to be associated with coniferous forest (*Pinus*, *Picea* and *Abies*) mixed with alder and a rich marsh community without much evidence for real grasslands (Stuart and Gibbard, 1986).

The last major interglacial was the Eemian (Ipswichian), about 130 000 to 110 000 years ago. This period is quite well understood. At that time the straight-tusked elephant occurred in northwestern Europe again, together with fallow deer (*Dama dama*), red deer (*Cervus elaphus*), the ancestor of the wisent (that is, Steppe bison, *Bison priscus*) and a rhinoceros (*Dicerorhinus kirchbergensis*) akin to the Sumatran rhino (Table 3.6). The Eemian interglacial was warmer than today and the period was characterized by an early development of climax forest, where *Quercus* had its culmination before *Corylus*. The period ended with dominance by spruce and *Pinus*. Also the beaver (*Castor fiber*) lived here during this period. When the climate became colder during the Early Weichselian, the vegetation changed from birch and pine forest to one characterized by *Betula nana*, *Juniperus* and herbaceous vegetation. This sequence is a near mirror-image of the vegetation development during the reforestation in the Holocene (Aaris-Sørensen *et al.*, 1990). The Eemian interglacial is perhaps the best example of the potential appearance of northwestern Europe today had there been no human interference.

Are there reliable indications of the occurrence of open grasslands at that time? A species of particular interest is the hippo (*Hippopotamus amphibius*), of which fossils have been found that date from the Eemian interglacial (Table 3.6) (Stuart and Gibbard, 1986). Hippo was typically found together with fallow deer, steppe rhino (*Dicerorhinus hemitoechos*), spotted hyena (*Crocuta crocuta*) and the straight-tusked (forest) elephant, while the vegetation could be characterized as mixed oak forest (*Quercus*, *Pinus*, *Acer*, *Alnus* and *Corylus*) or as temperate forest with hornbeam (*Carpinus*). Quite often hippo fossils have been found in association with giant deer (*Megaceros giganteus*), aurochs (*Bos primigenius*) and red deer. Investigations on pollen spectra from hippo bones reveal a preponderance of tree pollen but with indications of herb-dominated riverine floodplain plants (grasses, sedges, *Plantago*, Compositae and a few aquatics) (Stuart and Gibbard, 1986). Judging from the present-day effect of hippo on African vegetation, it has been concluded that during the warmest period of the Eemian large herbivores were maintaining, and possibly even initiating, herb-dominated areas on the floodplains (Turner, 1975; Stuart and Gibbard, 1986). However, it is not clear whether these riverine grasslands have to be viewed as grazing swards or as *Phragmites* reedlands, since the pollen of Gramineae includes those of *Phragmites* (Turner, 1975). The presence of fallow deer would suggest real grazing swards and thus the occurrence of grasslands in the strict sense of the word during the Eemian. These grasslands then occurred within the mixed forest, and could have been used by aurochs, red deer, giant deer, steppe rhino and hippo. It thus appears justified to conclude that most of northwestern Europe was covered by closed forests and swamps and by riverine grassy floodplains during previous interglacial periods (Kurtén, 1972).

Table 3.6 Large mammals occurring in Europe since the Pliocene (based on Anderson, 1984; but for Proboscids on Mol and Van Essen, 1992)

Species	Pliocene	Pleistocene Early	Pleistocene Middle	Pleistocene Late	Holocene
Primates					
Macaca florentina[T]	+	–	–	–	–
Wolves					
Cuon (alpinus) (dhole)	–	–	+	+	–
Canis (lupus) (wolf)	–	+	+	+	+
Bears					
Ursus etruscus[T]	+	–	–	–	–
Ursus spelaeus (cave bear)[1]	–	–	+	+	–
Ursus arctos (brown bear)	–	–	+	+	+
Ursus maritimus (polar bear)	–	–	–	+	+
Large cats					
Dinofelis (sabertooth)	+	+	–	–	–
Machairodus (sabertooth)	+	+	–	–	–
Meganteron (sabertooth)	+	+	–	–	–
Aciconyx pardinensis (giant cheetah)	–	+	–	–	–
Aciconyx jubatus (cheetah)	–	–	–	+	–
Homotherium latidus (scimitar cat)	–	–	–	+	–
Panthera schreuderea[T,2]	+	–	–	–	–
Panthera leo (lion) (incl. cave lion ssp. *spelaea)*	–	–	–	+	+[3]
Panthera pardus (leopard)	–	–	–	+	–
Hyenas					
Euryboas (hunting hyena)	+	+	–	–	–
Chasmaporthes (hunting hyena)	–	+	–	–	–
Crocuta crocuta (spotted hyena) (incl. cave hyena ssp. *spelaea)*	–	–	–	+[4]	–
Hyaena perrieri[T]	+	–	–	–	–
Hyaena brevirostris (short-faced hyena)	–	+	+	–	–
Hyaena hyaena (striped hyena)	–	–	–	+	–
Beavers					
Trogontherium[T] (Eurasian giant beaver)	+	+	+	+	–
Castor fiber (beaver)	–	–	–	?	+
Proboscids					
Deinotherium (deinothere)	+	–	–	–	–
Anancus arvernensis (straight-tusked gomphotere)	+	+	–	–	–
Mammut borsoni (Borson's mastodont)	+	+	–	–	–
Elephas namadicus[5]	+	+	+	+	–
Mammuthus meriodionalis[T] (southern mammoth)	+	+	–	–	–

Table 3.6 (*continued*)

Species	Pliocene	Pleistocene			Holocene
		Early	Middle	Late	
Mammuthus throgonterii[6] (steppe mammoth)	–	–	+	–	–
Mammuthus primigenius (woolly mammoth)	–	–	–	+	+[7]
Rhinoceroses					
Diceros bicornis (black rhino)	+	–	–	–	–
Dicerorhinus etruscus[T]	+	+	–	–	–
Dicerorhinus kirberchensis (cf. Sumatran rhino)	?	+	+[8]	–	–
Dicerorhinus hemitoechus (steppe rhino)	+	+	+	+	–
Elasmotherium (giant rhino)	–	+	+	+	–
Coelodonta antiquitatus (woolly rhino)	–	–	+	+[9]	–
Tapirs					
Tapirus spp.[10] (tapir species)	(+)[T,11]	+	+	–	–
Horses					
Hipparion spp. (three-toed horse species)	+	–	–	–	–
Equus spp.[12] (horse species)	+[T]	+	+	+	+
Hippopotamus					
Hippopotamus cf. *major* (extinct hippo)	–	–	–	+[13]	–
Hippopotamus amphibius (hippo)	–	–	+[14]	+[14]	–
Pigs					
Sus spp. (boar – different species)	+[T,15]	+	+	–	–
Sus scrofa (wild boar)	–	+	+[16]	+	+
Deer					
Eucladocerus (bush-antlered deer)	+[T,17]	+	+	–	–
Libralces gallicus (extinct moose)	–	+	–	–	–
Alces latifrons (giant moose)	–	–	+	–	–
Alces alces (moose)	–	–	–	+	+
Capreolus capreolus (roe deer)	–	–	+	+	+
Cervus rhenanus[T]	+	–	–	–	–
Cervus elaphus (red deer)	–	+	+	+	+
Dama dama (fallow deer)	–	+[18]	+[18]	–	–
Megalocerus giganteus (giant deer)	–	–	+	+	+[19]
Rangifer tarandus (reindeer)	–	–	+	+	+
Bovids					
Soergelia (primitive musk ox)	–	–	+	–	–
Praeovibos (extinct musk ox)	–	–	+	–	–
Ovibos moschatus (musk ox)	–	+	+	+	+[20]
Gazella spp. (gazelle species)	+	–	–	–	–

Table 3.6 (*continued*)

Species	Pliocene	Pleistocene			Holocene
		Early	Middle	Late	
Rupicapra rupicapra (chamois)	–	–	–	+	+
Hippotragus (cf. sable and roan antelope)	+	–	–	–	–
Saiga tatarica (saiga antelope)	–	+	+	+	–
Leptobos[T,21]	+	–	–	–	–
Bison priscus (steppe bison)	–	–	+	+[22]	–
Bison bonasus (wisent)	–	–	–	–	+[23]
Bos primigenius (aurochs)	–	–	–	+[24]	+
Capra spp. (goat and ibex)	–	+	+	+	+
Ovis (sheep)	–	+	?[25]	?[25]	–
Large carnivores (excl. bears)	6	8	3	8	2
Large herbivores (excl. bears)	19	21	25	21	14

[T] Late Pliocene mammals from Tegelen (The Netherlands), after Van den Hoek Ostende (1990).
[1] The cave bear was a strict vegetarian (Anderson, 1984) and became extinct about 15 000–11 000 BP (Stuart, 1991).
[2] *Panthera schreuderea* may be identical to *P. schaubi* and/or *P. toscana*.
[3] Lion became extinct about 11 000 BP (Stuart, 1991).
[4] Spotted hyena apparently became extinct *c.* 20 000 BP in England and 14 000 BP in southern France (Stuart, 1991).
[5] *Elephas namadicus* is also known as *Palaeoloxodon antiquus* (Mol and Van Essen, 1992); it was a browsing species living in broad-leaved forests, and continued to live during the Weichselian (about 100 000 BP in Italy (Stuart, 1991).
[6] Also known as *Mammuthus armeniacus* (Mol and Van Essen, 1992).
[7] Extinct in Europe 10 000 BP (Lister, 1991).
[8] The rhinoceros *Dicerorhinus hemitoechus* was more a grazer, *D. kirchbergensis* more a browser; these species became extinct quite early in northwestern Europe but continued to live in the Levant till about 20 000 BP (Stuart, 1991).
[9] Woolly rhinoceros was a grazer; it was extinct 12 000 BP (Stuart, 1991).
[10] Tapir systematics needs revision and is not clear (Anderson, 1984).
[11] The dating of *Tapirus arvernensis* is not certain (Van den Hoek Ostende, 1990).
[12] Horse systematics in state of confusion (Anderson, 1984) but, for the Late Glacial, I only find references to *Equus ferus*. There is, *inter alia*, no evidence that it survived in the wild state in Britain.
[13] *Hippopotamus* cf. *grandis* is known from the Cromer Forest Bed series (Stuart and Gibbard, 1986); it might have been a different species than *H. amphibius*.
[14] Hippopotamus lived in Europe during interglacials only (Anderson, 1984); Stuart (1991) considers *H. antiquus* a subspecies of *H. amphibius*.
[15] The boar species reported for Tegelen is *Sus strozzii*.
[16] Wild boar had a refugium in Italy during the Last Glacial (Stuart, 1991).
[17] The bush-antlered deer of Tegelen is reported to be *E. teguliensis*.
[18] Fallow deer lived in northwestern Europe during the Eemian interglacial (Anderson, 1984) but it occurred during the Weichselian in Italy (Stuart, 1991).
[19] Giant deer survived until about 10 500 BP (Stuart, 1991).
[20] Musk ox survived until the Early Dryas (about 12 000 BP) in southern Germany and 15 000 BP in southern France, and became extinct on Taymir Peninsula 3000 BP (Stuart, 1991).
[21] *Leptobos* was ancestral to both *Bison* and *Bos*.
[22] Steppe bison survived until about 12 000 BP in southern France but appears to have been absent from northwestern Europe during the Late Glacial; before that time it was apparently common (Stuart, 1991).
[23] Wisent appears to be a recent immigrant from North America (Anderson, 1984).
[24] Aurochs is first recorded from the Holstein interglacial.
[25] I have not been able to find records of sheep or ibex from the Middle or Late Pleistocene.

Climatic fluctuations were a recurrent phenomenon during the Pleistocene, and several interstadials took place during which forest vegetation returned after a glacial period. However, the pattern of forest development over time varied: different patterns of tree species immigration rates resulted in slight differences in species assemblages during the different warm periods (Delcourt and Delcourt, 1991). This makes it difficult to use previous warm periods as exact models of natural vegetation development during the Holocene – that is, the period in which we now live. On the other hand, Bennet *et al.* (1991) showed that in Europe the beginning of a warm period is generally characterized by the occurrence of *Betula* and *Pinus*, followed by *Ulmus* and *Quercus*, with a final stage characterized by *Carpinus* and *Abies*. With the onset of colder conditions, populations of most tree species became extinct again, with *Pinus* and *Betula* replacing the other species; *Salix* populations may have continued to exist north of the Alps during a glacial period. Bennet *et al.* (1991) stressed the point that (with the possible exception of *Betula* and *Pinus*) there was no southward migration of trees ac the end of a warm period. Instead, all evidence points to relict populations of broad-leaved trees and apparently species such as *Taxus* and *Picea* survived glacial periods in the mountains of the Balkans, Italy and Greece, from where recolonization of northwestern Europe took place during warm climatic conditions. Frequently, interstadial vegetation was characterized by steppe–tundra, not forest, and the next section focuses on this steppe–tundra.

3.3.2 Glacial periods

During glacial periods it appears that although the average annual temperature dropped considerably, the summer temperatures were moderate though still below 10°C (Atkinson *et al.*, 1987). Low winter temperatures (about –25°C; Atkinson *et al.*, 1987) prevented tree growth (Bell, 1969). The July wind directions were from east to northeast (not from the southwest as is the present normal pattern), which may have resulted in lower precipitation than today. The resulting vegetation in northwestern Europe can best be characterized as steppe–tundra, i.e. a vegetation that had an affiliation both with modern tundra vegetation and with the steppe. Typical species indicating an affiliation with steppe vegetation are (Bell, 1969; Godwin, 1975):

> *Ajuga reptans*, *Androsace septrentionales*, *Aphanes arvenis*, *Artemisa* spp., *Blysmus* (*Scirpus*) *rufus*, *Carex arenarea*, *Coryspermum* spec., *Demasonium alisma*, *Diplotaxis tenuifolia*, *Festuca rubra*, *Filipendula ulmaria*, *Groenlandia densa*, *Helianthemum canum*, *Linum perenne*, *Lycopus europaeus*, *Medicago sativa falcata*, *Onobrychis viciifolia*, *Potentilla fruticosa*, *Ranunclus sardous*, *Rumex maritimus*, *Rhinanthus* spp., *Stipa* spp.

The large herbivore assemblage from this environment is discussed below.

During the last glaciation (Weichselian), the steppe–tundra covered southern England and Ireland, southwestern France, The Netherlands, northern Germany to central Poland and from there to Siberia and Beringia. Real steppes appear to have covered most of the loess soils. They occurred in northern France, central and southern Germany, the Rhone area in France, northern Italy, and Hungary; from there the steppe stretched to the Black Sea and the Caspian Sea. A park-tundra, physiognomically perhaps like the open savannas of East Africa, covered Spain and Portugal, most of Italy and the Balkan Peninsula; it also covered the area north of the Caspian Sea and from there north of the Mongolian Plateau further to the east. Coniferous forests covered Corsica and Sardinia, Sicily and southern Italy, Greece, and the southern shores of the Black Sea and the Caspian Sea (Kurtén, 1972).

On the basis of radiocarbon-dated finds of woolly mammoth (*Mammuthus primigenius*), Aaris-Sørensen *et al.* (1990) concluded that an ice-free period occurred during the Weichselian (Devensian) glacial period between 45 000 and 20 000 years ago (see also Lister, 1991). Mammoth disappeared from northwestern Europe when the ice advanced: the vegetation became very sparse and plant cover discontinuous, while soil surface layers were unstable (Aaris-Sørensen, 1990). Large herbivores, apparently, could no longer find food here and disappeared from the record, though a few woolly mammoth finds date from this period (Lister, 1991). The ice-free period, however, was characterized by a steppe-like biome: the steppe–tundra. Because this steppe–tundra, which stretched from northwestern Europe to Beringia, was characterized everywhere by the mammoth, this vegetation type has been termed 'mammoth steppe' by Guthrie (1990). It was dominated by sedges and grasses, and only a few shrubs and herbs (Aaris-Sørensen *et al.*, 1990). Mammoth typically fed on grasses and sedges, but herbs and even mosses and ferns were consumed and occasionally browse from shrubs and trees (Lister, 1991; see also Guthrie, 1990). These steppe–tundras, steppes and park-tundras were ideally suited for large herbivores. Apart from species that are now extinct, such as mammoth, predators that are now considered typical of Africa – such as spotted hyena, lion (*Panthera leo*) and leopard (*Panthera pardus*) (Guthrie, 1990; see the rock paintings discovered in France in 1994) – preyed upon the large herbivores (Table 3.6). Besides the woolly mammoth, the herbivore species of the Weichselian mammoth steppe were giant deer, reindeer (*Rangifer tarandus*), saiga antelope (*Saiga tatarica*), musk ox (*Ovibos moschatus*), woolly rhinoceros (*Coelodonta antiquitatis*) and steppe bison (Figure 3.2; Table 3.6) (Aaris-Sørensen, 1999). Typical also were wild horse (*Equus ferus*) (Figure 3.3) and wild sheep (*Ovis* spec.) (Klein, 1974) but also steppe rhinoceros (*Dicerorhinus hemitoechus*), common hamster (*Cricetus cricetus*) and steppe pika (*Ochotona pusilata*) (Roebroeks, 1990) – all indicative of steppe-like vegetation. It should be stressed that this steppe-like landscape was not identical to present-day steppes of Eurasia: the mammoth steppe has no parallel today.

With the continuing cycle of glacial periods alternating with interglacial periods, and with the subsequent alternation of open vegetation and forests, the number of herd-forming herbivore species dwindled not only in Europe

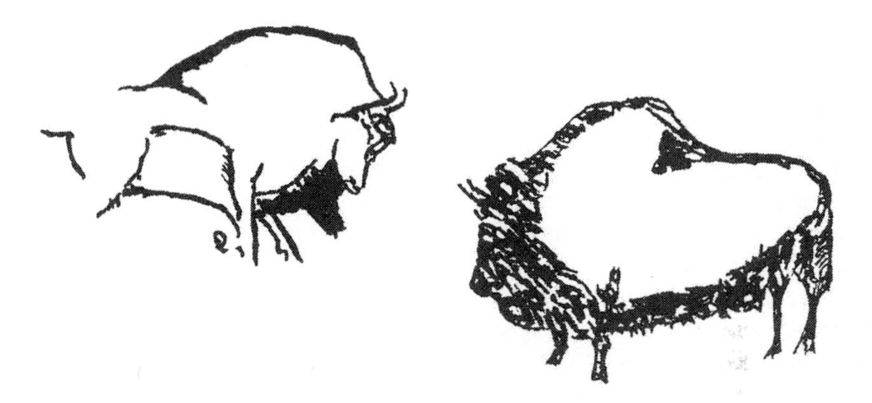

Figure 3.2 Steppe bison are frequently portrayed in Paleolithic art. These examples are from Front-de-Gaume (left) and Le Portel (right) (redrawn after Guthrie, 1990).

Figure 3.3 In the late Pleistocene, many large herbivores became extinct. The Przewalski horse (or *takhi*) escaped this fate but nearly became extinct in historical times. It is now being reintroduced into Mongolia.

but also in North America (Webb, 1977) and South America (Webb, 1978). In fact, at the end of the Pliocene 30 species of this type of herbivore, with a social organization typical of grazers (Estes, 1991), lived in Europe but only six remained there at the very end of the Pleistocene (Good, 1964) (Table 3.6). Thus, it is likely that the potential for keeping the vegetation open through grazing diminished during the Pleistocene. Other herbivores, more typical for a shrub vegetation, were moose (*Alces alces*), red deer, giant deer and wild goat (*Capra* spec.) (Klein, 1974). The main effect of the remaining wild herbivores, with a predominance of browse in their diet, will have been on structuring the species composition of woody species. Indeed, moose and hare (*Lepus* spec.) can have quite a considerable impact on nutrient cycling and on woody species composition in the forest-steppe of southern Russia (Zlotin and Khodashova, 1980). This has also been described for moose in boreal forest (Pastor *et al.*, 1993) and a significant herbivore influence on woodland structure has been observed in African savanna systems (Prins and Van der Jeugd, 1993).

Structuring of the woody species composition does not imply that the balance cover between woody species and the herbaceous layer was influenced towards a higher percentage cover of herbs and grasses or that browsers would have facilitated the development of grassland vegetation. Perhaps with the extinction of the 'bulldozer herbivores' (elephant, mammoth) this was no longer possible in northwestern Europe.

3.4 THE TRANSITION FROM PLEISTOCENE TO HOLOCENE

After the last retreat of the ice from northwestern Europe (Late Weichselian period, 13 000 to 10 000 years ago), woolly mammoth again immigrated to England, Denmark and southern Sweden about 13 000 years ago together with the associated mammoth-steppe 'disharmonious' faunal assemblage (Aaris-Sørensen *et al.*, 1990; Lister, 1991). One thousand years later mammoth became extinct in Europe – in France, England (Lister, 1991), Denmark and southern Sweden (Aaris-Sørensen *et al.*, 1990) – but survived another few thousand years in Siberia.

The Late Weichselian steppe–tundra was characterized by a mixture of floral elements from several modern biomes: tundra, steppe and even mediterranean vegetation (Table 3.7 and above). Likewise, the faunal assemblage is considered to be disharmonious (Table 3.8). Disharmonious species assemblages are characterized by the coexistence of species that today are allopatric and presumably ecologically incompatible (Graham and Lundelius, 1984). Typical examples of such species combinations are spotted hyena–red deer, spotted hyena–lynx (*Lynx lynx*)–brown bear (*Ursus arctos*)–hippopotamus, spotted hyena–reindeer, or spotted hyena–wolverine (*Gulo gulo*); all these combinations have been found in France (Graham and Lundelius, 1984) (Table 3.8). The disharmony may have caused a disequilibrium between species which was created by the disruption of coevolutionary interactions between plants and animals at the end of the Pleistocene, and this may have been a cause of many extinctions of large herbivores at the beginning of the Holocene (Graham and Lundelius, 1984). Late Pleistocene extinction of large mammals is certainly real (Table 3.6) (Graham and Lundelius, 1984). Much of the so-called unusually high mortality of the Pleistocene is best regarded as a natural consequence of high faunal turnover caused by major oscillations during the Pleistocene and the Holocene in climate and environmental heterogeneity (Gingerich, 1984). There is no convincing evidence for overkill of large herbivore populations by early humans (Anderson, 1984; Graham and Lundelius, 1984). When the climate changed during the transition from the Pleistocene to the Holocene 10 000 years ago, different species showed individual readjustments to environmental change. This implies that community structure would not have been stable over long periods of time; thus, modern communities cannot be used as direct analogues in the reconstruction of Pleistocene environments or Pleistocene communities.

The improvement of the climate after 13 000 BP was deduced originally from the palynological record. It was assumed that the occurrence of plant species was a good reflection of the climate and from the palynological record it appeared that the transition was rather slow. However, based on *Coleoptera* data, it appears there was a rapid climatic amelioration around 13 000 BP. This leads to the conclusion (Tipping, 1991) that:

> The lag between climate and vegetation in the [Late Weichselian] makes estimates from palynological data of the rates of thermal improvement unwise ... The detection of climatic deterioration from palynological data might have a firmer basis, provided the deterioration is sufficiently pronounced to have disrupted the vegetation.

From these observations it can be concluded that the vegetation and the associated large mammal fauna were not in equilibrium with climatic conditions. There was an important time lag of about 1000 years in the vegetation response to the climatic improvement in, for example, southern Sweden; this confirms the idea that plant species cannot be used directly as palaeoclimatic indicators at that time (Berglund *et al.*, 1984) (see Figure 3.4). Insects, because of their quick colonization of newly available habitats, are much more reliable (Coope, 1994). Also, *Coleoptera* assemblages are remarkably 'harmonious' during the last million years or so, indicating that their physiological adaptation to climatic factors, for example, has not changed: a time lag between *Coleoptera* assemblage structure and climate has not been observed (Coope, 1987).

During the Pleniglacial (22 000–18 000 BP) the average temperature in July was about 10°C and in January about –16°C. When the ice retreated the coldest winter months even became as cold –20 to –25°C and the climate was

Table 3.7 Steppe–tundra species assemblage during the Older Dryas (*c.* 13 000 BP): the assemblage was different from modern tundra and modern steppe; trees were almost absent, and broad-leaved trees were only present in northern Yugoslavia (area 7)

	Area												
Vegetation	1	2	3	4	5	6	7	8	9	10	11	12	13
Grasses and herbs													
Poaceae	+	+	–	+	+	+	–	+	u	+	+	–	–
Cyperaceae	+	+	–	+	+	+	–	+	u	+	+	–	–
Compositae	–	+	+	+	+	–	–	–	u	–	–	–	–
Artemisia	–	+	+	+	+	–	–	–	u	–	+	+	+
Centaurea	–	+	–	–	–	–	–	–	u	–	+	–	+
Chenopodiaceae	–	+	+	+	–	–	–	+	u	–	+	–	–
Rumex	–	+	–	–	–	–	–	–	u	–	+	–	–
Plantaginaceae[1]	–	+	+	–	–	–	–	–	u	–	+	–	+
Caryophyllaceae[2]	–	–	–	–	–	–	–	–	u	–	–	–	+
Ranunculaceae	–	+	+	–	+	–	–	–	u	–	–	–	–
Thalictrum	–	+	+	–	–	–	–	–	u	–	+	–	–
Saxifragaceae													
Saxifraga	–	+	+	–	–	–	–	–	u	+	+	–	–
Parnassia	–	–	–	–	–	–	–	–	u	+	–	–	–
Cystaceae[3]	–	+	+	–	+	–	–	–	u	–	+	–	+
Plumbaginaceae[4]	–	–	–	+	–	–	–	–	u	–	+	+	–
Rosaceae	–	+	–	–	–	–	–	–	u	–	–	–	–
Sanguisorba	–	–	–	–	–	–	–	–	u	–	–	+	–
Filipendula	–	+	–	–	+	–	–	–	u	–	–	–	–
Polygonaceae[5]	–	+	–	+	+	–	–	–	u	–	–	–	–
Rubiaceae	–	+	+	–	–	–	–	–	u	–	–	–	–
Umbelliferae	–	+	+	–	–	–	–	–	u	–	–	–	–
Spore plants													
Selaginella	–	+	+	+	–	+	–	–	u	+	+	–	+
Botrychium	–	+	+	–	–	–	–	–	u	–	+	–	–
Shrubs and dwarf trees													
Hippophaë	+	–	+	+	–	+	–	–	u	+	+	–	+
Salix[6]	+	+	+	+	+	+	–	+	u	+	+	–	–
Betula nana	+	+	–	+	+	–	+	+	u	–	+	–	–
Ephedra[7]	+	+	+	+	–	+	–	+	u	–	–	–	–
Juniperus	–	+	+	+	–	+	–	+	u	+	+	–	–
Pinus	–	+	–	+	+	+	–	+	u	+	–	–	–
Trees													
Larix	–	–	–	–	–	–	–	–	u	+	–	–	–
Corylus	–	–	–	–	–	–	+	+	u	–	–	–	–
Populus	–	+	–	–	–	–	–	–	u	–	–	–	–
Broad-leaved[8]	–	–	–	–	–	–	+	–	u	–	–	–	–

Table 3.7 (*continued*)

u = unvegetated
[1] *Plantago*
[2] *Gypsophilla repens*
[3] *Helianthemum*
[4] *Armeria*
[5] *Polygonum*
[6] *Salix polaris* or *S. repens*
[7] *Ephedra* cf. *dystachia*
[8] includes besides *Quercus* also *Acer*, *Carpinus*, *Fraxinus*, *Tilia* and *Ulmus*.
Area 1: Western Alps (Wegmüller, 1972)
Area 2: Northern Alps in Switzerland (Welten, 1972)
Area 3: Bayerische Voralpen (Schmeidl, 1972)
Area 4: Northern Alps, near Salzburg (Klaus, 1972)
Area 5: Lower Austria (Perschke, 1972)
Area 6: Basin of Vienna, Austria (Klaus, 1972)
Area 7: Southeastern Alps and northern Jugoslavia (Serceli, 1972)
Area 8: Karnische Alpsa, Austria (Fritz, 1972)
Area 9: Dachstein, central Austria (Krasl, 1972)]
Area 10: Western Carpathians (Ralska-Jasiewiczowa, 1972)
Area 11: Lowlands of northern and central Europe (Firbas, 1949)
Area 12: Northwestern Germany (Behre, 1967)
Area 13: The Netherlands (Waterbolk, 1954)

much more continental. The temperature range between the warmest and coldest month was about 30–35°C in the period between 14 500 until just before 13 000 BP, compared with 14°C at present (Atkinson *et al.*, 1987). The evidence now points to a very fast climatic improvement around 13 000 BP (with a rise in temperature of 7–8°C in summer and 25°C in winter), a slow cooling between 12 500 and 10 500 BP (with summer temperature of about 10°C and winter temperatures of about –15 to –20°C during the Younger Dryas around 10 500 BP) and again a fast warming around 10 000 BP (Atkinson *et al.*, 1987) (Figure 3.4). There were short-term climatic deteriorations in the period from 13 000 to 10 000 BP during the Older Dryas, Dryas 2, and the Younger Dryas (Figure 3.5) (Cordy, 1991). Recently it was shown that the final transition of a glacial climate to the present temperate climate took only 50 years at the end of the Younger Dryas (Dansgaard *et al.*, 1989). Indeed, plants typical of tundra conditions were growing while the July temperature had already increased to 18°C and the January temperature to –5°C. This time lag underscores the point made about plant and animals assemblages that were not harmonious, or in equilibrium with the prevailing climatic conditions at the end of the Late Glacial. In the face of such a major change it cannot be expected that regional floras and mammalian faunas (in contrast to insects) were indicative of the climate after such a sudden transition (Coope, 1994): colonization rates would have precluded this.

Whether the disharmony of faunal assemblages was a fact and not an artefact from admixture of different layers or the lack of a fine enough stratigraphy is not very easy to clarify. The detailed Cordy (1991) study of

Table 3.8 Presence of Late Glacial mammal species indicative of a 'disharmonious' mammal assemblage (A–C) and of mammal species from the Boreal period indicative of a 'harmonious' assemblage (D)

	Late Glacial			Boreal
Species	A	B	C	D
Sorex araneus (common shrew)	+	–	–	–
Desmana moschata (desman)	–	–	+	–
Erinaceus europaeus (hedgehog)	–	–	–	+
Spermophilus spp. (suslik, groundsquirrel)	+	–	+	–
Sciurus vulgaris (squirrel)	–	–	–	+
Castor fiber (beaver)	+	+	+	–
Clethrionomys glareolus (bank vole)	+	–	–	–
Arvicola terrestris (water vole)	+	+	–	–
Microtus agrestis (field vole)	+	–	+	–
Microtus oeconomus (northern vole)	+	+	–	–
Microtus gregalis (tundra vole)	+	+	–	–
Apodemus sylvaticus (woodmouse)	+	–	–	–
Dicrostonyx torquatus (Arctic lemming)	+	+	–	–
Lemnus lemnus (Norway lemming)	+	+	–	–
Lepus spp. (hare)	+	+	+	+
Ochotona pusilla (steppe pika)	+	+	–	–
Canis lupus (wolf)	+	+	+	+
Alopex lagopus (Arctic fox)	+	+	–	–
Vulpes vulpes (red fox)	+	+	–	+
Ursus arctos (brown bear)	+	+	+	+
Ursus spelaeus (cave bear)	+	–	–	–
Crocuta crocuta (spotted hyena)	+	–	–	–
Gulo gulo (wolverine)	+	–	+	–
Martes martes (pine marten)	–	–	–	+
Mustela putorius (polecat)	–	–	–	+
Meles meles (badger)	–	–	–	+
Lutra lutra (otter)	–	–	–	+
Panthera leo (lion)	+	–	+[1]	–
Felis sylvestris (wild cat)	+	–	–	+
Lynx lynx (lynx)	+	+	–	+
Mammuthus primigenius (woolly mammoth)	+	+	+	–
Coelodonta antiquitatis (woolly rhinoceros)	+	–	–	–
Sus scrofa (wild boar)	–	–	–	+
Equus ferus (wild horse)	+	++	+	–
Megalocerus giganteus (giant deer)	+	–	+	–
Alces alces (moose)	+	–	+	+
Capreolus capreolus (roe deer)	–	–	–	+
Rangifer tarandus (reindeer)	+	+	+	–
Cervus elaphus (red deer)	–	++	–	+
Saiga tatarica (saiga antelope)	–	+	–	–
Ovibos moschatus (musk ox)	+	–	–	–
Bos primigenius (aurochs)	+[2]	+	+	+
Bison bonasus (wisent)	–	–	+	–

Table 3.8 *(continued)*

[1] This record is from the Late Dryas (about 10 000 BP), The Netherlands (Housley, 1991).
[2] It is likely that this was aurochs and not wisent, since the latter is considered a post-glacial immigrant from North America (Anderson, 1984).
A. 18 000 to 14 000 BP, Wales (David, 1991).
B. 13 000 to 11 000 BP (Bølling to Younger Dryas), Somerset (Currant, 1991); the fauna was dominated by red deer and wild horse.
C. 13 000 to 10 000 BP, Denmark (Aaris Sørensen *et al.*, 1990); a number of species appear to have become (locally) extinct already, such as woolly rhinoceros and musk ox.
D. 9000 to 8000 BP, Denmark (Aaris Sørensen, 1988); the species composition is typical for the rest of the holocene until Recent, when many species went locally extinct due to the spreading of cultivation.

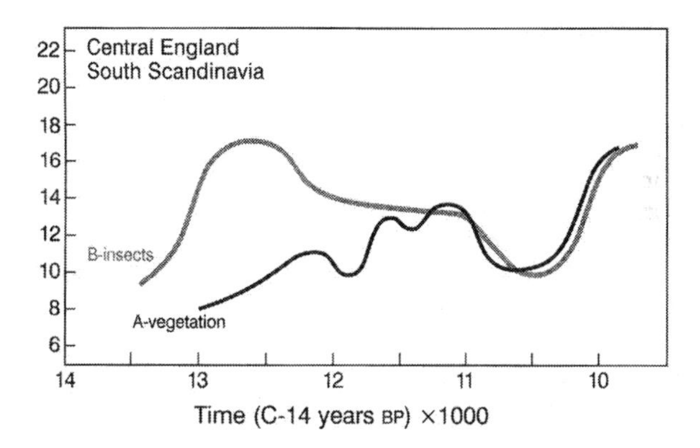

Figure 3.4 Average July temperatures for the time span 13 000–10 000 BP reconstructed from (A) palaeobotanical evidence and (B) palaeoentomological evidence. Both are supposed to be representative for areas with similar climatic regimes (central England and southern Scandinavia). (After Berglund *et al.*, 1984.)

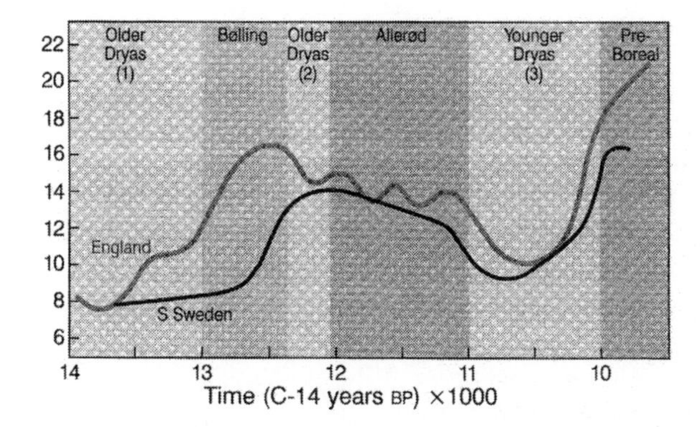

Figure 3.5 Correlation between average July temperatures for the time span 13 500–9500 BP. (After Lemdahl, 1988, referred to in Larsson, 1991.)

microtines during the rapid alterations of the climate during the period 13 000 to about 9000 BP may serve as a warning. From this evidence it is clear that arctic lemming and other vole species that are typical for more temperate conditions alternated in abundance and tracked the changes in climate well (Cordy, 1991). However, it may very well be that large herbivores could not track changes as rapidly as the vole species and, thus, that large mammal assemblages were really disharmonious. By the same token, it is unlikely that the vegetation could respond so quickly to the improved conditions, and it is therefore likely that plant communities were transient and very unstable in time during the transition from the Pleistocene to the Holocene. If the Late Pleistocene faunal and floral assemblages had responded to environmental changes as communities, rather than as individual species, then the large herbivore communities should have been able to track the changes in vegetation communities. Given the expansion of the grassland biome and the decline of the forest biome at the end of the Pleistocene (because of increased continentality of the climate), ecological theory would predict lower mammal extinction rates in the expanding grasslands than in the shrinking forests; this was not the case in North America, where numerous grazing taxa became extinct (Graham and Lundelius, 1984). Alternatively, the evidence points to loss of the coevolutionary adaptations because of drastic changes in the plant communities; hence it appears that the grazing sequence as established for East African savanna ecosystems (De Boer and Prins, 1990) and postulated for the Pleistocene mammoth steppe (Guthrie, 1990) became disrupted. This disruption could have resulted in increased competition between the grazers, causing some species to go extinct, while other species could not assemble the diets they needed for reproduction; this also could have led to extinction (Graham and Lundelius, 1984).

A disruption similar to that of the link between grazers and food plants has been suggested for the link between browsers and trees in the case of the mastodon (*Mammut americanum*), which primarily fed on coniferous trees. During the Pleistocene, large areas were characterized by open *Pinus*-dominated parklands. When pine and spruce forests collapsed into island-like distributions, the mastodon went extinct; apparently, mastodons were not able to adapt to the deciduous forests that succeeded spruce and pine (King and Saunders, 1984). Alternatives to extinction for large herbivores were migration to other communities, as was shown by reindeer, thus causing segregation of species that had once lived together. However, migration also needed adaptation to the new plant communities that emerged. These phenomena may explain the transition from disharmonious assemblages to modern assemblages, and may also explain the high species loss at the beginning of the Holocene.

It is not fully clear when the disharmony of plant communities came to an end, and when 'normal' (i.e. modern) plant communities can be reconstructed or recognized from the fossil record with any degree of precision. While mammal communities were disharmonious during the very last part of the Pleistocene (13 000 to 10 000 BP), they were apparently already harmonious during the Boreal (about 8500 BP) (Table 3.8), and, again, it can be assumed that this also applied to plant communities. The same 'date of transition' can be observed from the North American assemblages (Graham and Lundelius, 1984) and from the African ones (Klein, 1984). Numerous species of the modern North American grassland communities were residents of the Late Pleistocene grasslands (*Stipa comata*, *S. spartea*) but some floral (and faunal) species immigrated into the central grasslands from the eastern forest, western mountains (*Stipa viridula*, *S. agropyron*, *Elymus* spp.) and northern subarctic and by autochthonous evolution (present-day prairie species from the families *Andropogon*, *Bouteloua*, *Buchloe*, *Schizachirium* and *Sorghastrum*). These processes were intrinsically linked with (local) extinction. The culmination of all these events was reached during the transition of the Pleistocene to the Holocene, when the modern grassland communities emerged in North America (Graham and Lundelius, 1984). However, emergence of modernity of these communities does not imply that they remained unchanged, and this will be considered later when discussing chalk grasslands. Taking all the evidence together, it appears reasonable to suggest that from about 10 000 BP plant species and plant assemblages were able to track the climatic change, since from then onward data on the ecology of insect taxa agree with those on plant taxa (Figure 3.4).

The general picture of the vegetation development is that the steppe–tundra expanded when the climate ameliorated at the end of the Pleistocene. Both plant species typical for a modern tundra (such as *Dryas octopetala*, *Saxifraga oppositifolia* and *Salix polaris*) and species typical for steppes (such as *Ephedra*, *Hippophaë* and *Helianthemum*) were growing in most of northwestern Europe (Table 3.7) (see also Welten, 1972) and only in Ireland were there plants that were typical for heaths, such as *Calluna*, *Erica* and *Empetrum* (Firbas, 1949). In the north of Britain the vegetation mainly comprised grasses, *Artemisia* (*norvegica*, *campestris*, *maritima*), *Empetrum*, together with Chenopodiaceae and Caryophyllaceae, *Juniperus* and *Betula* (Tipping, 1991). Tall trees were absent from the landscape of northwestern Europe but *Hippophaë* was present. It is thought that delayed immigration and possibly also insufficiently developed soils were the reason for the lack of deciduous trees (Kolstrup, 1991).

Between the Older Dryas and the Younger Dryas two warmer periods occurred (Bølling and Allerød), interspersed with a very short colder period, Dryas 2 (Figure 3.2), with *Populus*, *Prunus* and *Sorbus* and with grasses mainly on dry sites (Kolstrup, 1991). Most of northwestern Europe became covered by forest dominated by *Pinus* and *Betula* (Table 3.9). Perhaps these forests were not very dense, as the palynological record shows high pollen sums for grasses and sedges, but most of the reported herbs belong to the Umbelliferae, which is an indication that conditions were no longer optimal for large grazers – very few species of this plant family are included in their diet.

During the Younger Dryas (about 10 500 BP), the change towards more reduced tree growth was probably caused by colder conditions. *Betula nana* replaced *B. pubescens* again but this has also been ascribed to drier conditions during this period as compared with the Older Dryas (Kolstrup, 1991). Further south, in southern Germany and northern Italy, pine (*Pinus sylvestris*) and birch forest began to spread over the riverine floodplains during the Older Dryas, to be replaced by oak and mixed oak forest at the end of the Pre-Boreal (9000 BP) (Becker and Kromer, 1991).

In northwestern Europe, the Pre-Boreal starts with the final disappearance of Arctic lemmings which were replaced by rodents typical for grasslands, such as Norway lemming (*Lemnus lemnus*), steppe pika and common hamster. Reindeer and wild horse occurred again during the Younger Dryas and the first part of the Pre-Boreal, but then about 10 000 BP the modern woodland rodent assemblage replaced the grassland assemblage completely (Cordy, 1991). Likewise, reindeer were still present in Schleswig-Holstein about 10 000 BP, apparently in large numbers (Bratlund, 1991), but are gone from the record only 1000 years later in both Denmark (Aaris-Sørensen, 1988) and southern Sweden (Larsson, 1991). The disappearance of reindeer took place at the same date in southern England, coinciding with the disappearance of wild horse (Lewis, 1991). Grassland indicator species disappeared and apparently the dominant vegetation in northwestern Europe became forest.

Table 3.9 Vegetation during the Allerød period (*c.* 11 500 BP) (see Table 3.7 for areas and references)

	Area										
Vegetation	1	2	3	4	5	6	7	8	9	10	11
Grasses and herbs											
Poaceae	+	+	+	+	–	+	–	+	u	+	+
Cyperaceae	+	+	+	+	–	+	–	+	u	+	+
Herbs[1]	–	+	+	+	–	–	–	+	u	+	+
Shrubs and trees											
Hippophaë	–	–	–	+	–	–	–	–	u	+	–
Salix	–	–	–	–	–	–	–	–	u	+	+
Betula	+	+	+	+	+	–	–	+	u	+	+
Pinus	+	+	+	+	+	+	–	+	u	+	+
Picea	–	–	–	–	–	–	–	–	u	+	–
Corylus	–	–	–	–	–	–	–	+	u	–	–
Broad-leaved[2]	–	–	–	–	–	–	+	–	u	–	–

u = unvegetated
[1] Mainly Umbelliferae, such as *Heracleum sphondylium*.
[2] This includes *Carpinus*, *Fagus*, *Quercus*, *Tilia* and *Ulmus*.

3.5 VEGETATION DEVELOPMENT DURING THE HOLOCENE

The present period, the Holocene, began 10 000 BP. Currently it is characterized by the occurrence of many grassland communities, which is in apparent contrast to the previous warm periods during the Pleistocene when forests dominated the landscape. Because the impact of human activities on the landscape became so much more important, it is not unlikely that many of the present grassland communities are the result of these activities. It is important to study the evidence for this during the Holocene.

The first evidence of human occupation in northwestern Europe is from artefacts that can be classified as belonging to the Middle Palaeolithic (about 300 000 to 35 000 BP). Most of these date back to the previous glacial period, the Weichselian, but also to the last interglacial period, the Eemian (130 000 to 110 000 BP). Even during the second-last glacial period, the Saalian, and the preceding interglacial periods (Hoogeveen, about 250 000 BP) humans were already present in, for example, The Netherlands. Even older finds date back to about 400 000 BP from Cagny in northern France and from Boxgrove in southern England and to about 350 000 BP from Bilzingsleben in Germany (Roebroeks, 1990). All ancient sites are from people that lived in northwestern Europe during interglacial periods and the first evidence for survival during a glacial period is from La Cotte de St Brelade, on Jersey (240 000 BP) (Roebroeks, 1990). All evidence so far points to a human culture focused on hunting, scavenging, and gathering during the Palaeolithic. This continued until after the end of the last glaciation and the beginning of the Holocene, and even during the Mesolithic period this was the sole mode of 'production'.

As was shown above, the second part of the Pre-Boreal (10 000 to 9000 BP) coincided with a major change from a fauna and flora typical of grasslands to those of woodland. This period marks the beginning of the Holocene, and forests again dominated the landscape (Table 3.10), with the most important tree species again being *Pinus* and *Betula*. An increasing number of broad-leaved species began to become established (Table 3.10). During the next period, the Boreal (9000 to 8000 BP), *Corylus* (mixed with *Alnus*) and *Picea* increased in importance (Table 3.11), with *Fagus* appearing in Lower Austria. The climax was reached with the Atlanticum (8000 to 5000 BP), when forests dominated by *Quercus robur*, *Q. pubescens*, *Q. petraea*, *Ulmus glabra*, *U. minor*, *U. laevis*, *Tilia platyphyllos*, *T. cordata*, *Fraxinus excelsior*, *Acer pseudoplatanus*, *A. platanoides*, *A. campestre* and *Betula pendula* became widespread in northwestern Europe (Firbas, 1949). Human culture – the Mesolithic – was still characterized by hunting and gathering and the human impact on the landscape was still minimal (Waterbolk, 1985). The Atlanticum, then, is the critical epoch in which to find out whether natural grasslands occurred in the forests of that time or not: the general picture is that of a closed forest in which humans played and had played a minimal role. This hinges critically on the definition of the term 'grassland', which is clearly a physiognomic one (as are forest, bog, etc.) and should be used at the landscape level. This implies that it should be used for units with an area of at least several hectares and with a tree and shrub cover of less than 2% (Loth and Prins, 1986). In the context of this chapter, therefore, the term does not include grassy woodlands, or forests in which small patches of grass occur.

An important corollary to the conclusion that species disequilibrium may have been the major cause of extinction, as discussed above, is that one cannot deduce from the extinction of large mammalian grazing herbivores that

Table 3.10 Vegetation during the Pre-Boreal (*c.* 9500 BP) (see Table 3.7 for areas and references)

	Area										
Vegetation	1	2	3	4	5	6	7	8	9[a]	10	11
Trees											
Betula	+	ND	+	+	+	–	–	+	–	+	+
Pinus	+	ND	+	+	+	+	–	+	–	+	+
Picea	–	ND	–	–	+	–	–	+	–	–	+
Corylus	+	ND	+	–	+	–	+	+	–	+	+
Alnus	–	ND	–	–	–	+	–	+	–	–	+
Quercus	+	ND	–	–	+	–	–	+	–	–	+
Ulmus, *Tilia*, *Carpinus*	–	ND	–	–	+	–	–	+	–	+	–
Populus	–	ND	–	–	–	–	–	–	–	+	+

ND = no data (area 2, Bayerische Voralpen)
[a] In area 9 (Dachstein, central Austria) vegetation spread but was still cold-adapted steppe-like, with *Artemisia*, *Ephedra* and *Juniperus*.

Table 3.11 Vegetation during the Boreal (*c.* 8500 BP) (see Table 3.7 for areas and references)

	Area										
Vegetation	1	2	3	4	5	6	7	8	9	10	11
Trees											
Betula	+	ND	+	+	–	–	ND	ND	+	–	+
Pinus	+	ND	+	+	–	+	ND	ND	–	+	+
Picea	+	ND	+	+	+	+	ND	ND	+	–	+
Corylus	+	ND	+	–	+	+	ND	ND	–	+	+
Alnus	+	ND	–	–	–	+	ND	ND	+	+	+
Quercus	+	ND	–	–	+	+	ND	ND	–	–	–
Ulmus, *Tilia*, *Carpinus*	+	ND	–	–	+	+	ND	ND	+	+	–
Fagus	–	ND	–	–	+	–	ND	ND	–	–	–

ND = no data (area 2, Bayerische Vorlapen; area 7, Southeastern Alps; area 8, Karnische Alps)

grasslands disappeared and that forests became closed (i.e. not offering feeding opportunities to large grazers). However, it may be assumed that, with the near-absence of large grazers during the Holocene after the faunal collapse described above, the forests could become closed since large grazers were no longer present in sufficient variety to maintain grazing swards within these forest. Indeed, during the Boreal only one true grazing large herbivore species was still present in northwestern Europe – namely, the aurochs (*Bos primigenius*). This species was apparently common in the lower Rhineland during the transition of the Late Pleistocene open steppe–tundra to the forests of the Pre-Boreal (Street, 1991). Perhaps it could have become a dominant ungulate of northwestern Europe if agriculture and hunting had not swept it into extinction, just as the African buffalo apparently became dominant in eastern and southern Africa after the demise of *Elephas recki* (Prins, 1996). Additional mixed feeders (i.e. not true grazers) were red deer, wild boar and wisent (*Bison bonasus*). Under present climatic conditions, the latter three species are not able to maintain open swards and it is doubtful whether aurochs would have been able to do so: from a long-term study of African buffalo in East Africa it is clear that buffalo cannot prevent bush encroachment (Prins and Van der Jeugd, 1993). Other grazing species – woolly mammoth, wild horse, giant deer and (to some extent) reindeer – were not able to maintain viable populations in northwestern Europe and thus could not assist the surviving species in maintaining open swards. It is doubtful whether even the aurochs would have been able to maintain itself in the northwestern European forest if open grasslands had not occurred along coasts and large rivers (see below). The American bison (*Bison bison*), even though it survived from the Pleistocene, was not able to maintain open areas in the oak–chestnut (*Quercus–Castanea*) forests of the Appalachians to the east of the Mississippi river, and did not maintain viable populations there, but deer could do so (Guildhay, 1984). The North American data suggest that some of the Late Pleistocene forests were more open and perhaps more patchy than those of today (Graham and Lundelius, 1984), again suggesting that closed forests were the normal climax forests of the Holocene.

Although the occurrence of large herbivores such as roe deer (*Capreolus capreolus*), red deer, aurochs and wild boar and the absence of open grassland species, such as saiga antelope, is normally interpreted as indicating a near-closed forest, some arguments plead against it. The major one is that seedling and establishment ecology, especially of *Quercus* species, indicates that light must have penetrated the forest floor at least occasionally (Vera, 1997). A personal conclusion is that open grasslands will have been restricted to areas above the tree-line, areas above salt domes, salt marshes (Figure 3.6) and some floodplains. Many individual species that are, at present, typical for grasslands will have found their habitat there, while others could survive in small patches or as understorey in the forests (this will be discussed later). The conclusion is that biotic factors appear to have played a minimal role for keeping forests open during the Atlanticum. For this reason, the experiments as conducted with introduced cattle in forest reserves in The Netherlands are of major importance in establishing whether or not large grazers were, at least potentially, able to open forest or to maintain large grassy patches in these forests (WallisDeVries, 1994; Van Wieren, 1995).

It is often assumed that along the large lowland rivers, the floodplains were covered by grassland because the long duration of floods would have prevented tree growth. It is more likely that these plains in the broad-leaved tree zone were forested. These riparian forests (German *Auenwald*, Dutch *ooibos*) (Figures 3.7 and 3.8) can tolerate prolonged flooding even during the growing season (Dister, 1983; Gerken, 1988; Späth, 1988). Three main types are known: *Populus* forest (included in *Alno-Padion*), *Fraxinus–Ulmus* forest (*Alno-Padion*) and *Alnus–Fraxinus* (*Circaeo-Alnion*) forest (names of syntaxa follow Westhoff and Den Held, 1969). *Fraxinus–Ulmus* forests in particular can survive flooding well, for up to 60% of the growing season, and are even resistant to drifting ice (Schmidt, 1969). The *Alno-Padion* communities include the *Salix* communities along rivers. Because of the highly dynamic environment, open places can be found with *Phalaris arundinacea*, *Poa trivialis*, *P. palustris* and many herbs that are known to occur at present in wet meadows used for haymaking (type *Molinio-Arrhenatheretea*) (Ellenberg, 1978):

> *Angelica sylvestris*, *Anthriscus sylvestris*, *Dactylis glomerata*, *Filipendula ulmaria*, *Galium uliginosum*, *Heracleum sphondylium*, *Lathyrus pratensis*, *Lythrum salicaria*, *Lysimachia vulgaris*, *L. nummularia*, *Myosotis palustris*, *Poa trivialis*, *Stachys palustris*, *Taraxacum officinale*.

Genuinely treeless communities along rivers are dominated by reeds (*Phragmitetum* and *Phalaridetum*) or by tall herbs (*Lolio-Potentillion anserinae* = *Agropyro-Rumicion crispi*) and ruderals (*Polygono-Chenopodietum* and *Chondrilletum*) (Figures 3.7 and 3.8). It is surmised that especially the tall herb community was the vegetation type where the remaining large herbivores concentrated during the Holocene when forests were not offering feeding opportunities for wild grazers on a large scale any more (Westhoff and Van Leeuwen, 1966). Especially when grazed, *Lolio-Potentillion anserinae* communities can develop into grasslands that are considered normal at present; relevant plant associations are *Potentillo-Festucetum arundinacea* (with *Festuca arundinacea*) and *Rumici-Alopecuretum geniculati* (with *Alopecurus geniculatus*, *Inula britannica* and *Lysimachia nummularia*), both belonging to the *Arrhenatherion* alliance. With heavy grazing and input from fertilizers these associations develop into *Poo-Lolietum* (with *Lolium perenne*, *Plantago major*, *Poa trivialis*, *P. pratense*, *Taraxacum officinale* and *Trifolium repens*) (Westhoff and Den Held, 1969). This association is nowadays widely distributed over northwestern Europe.

Vegetation types not dominated by trees definitely occurred on salty soils above salt domes in the interior of northwestern Europe (Central Germany,

Figure 3.6 Under present climatic conditions, natural grasslands such as this salt marsh on one of the Friesian islands are uncommon in northwestern Europe.

Poland, Czechoslovakia, Hungary and Rumania) (Ellenberg, 1978). The German and Polish salt-dome vegetation shows affinity with the salt marshes along the North Sea and the Baltic, the other with the area from the Aral to the Caspian Seas (Vicherek, 1973).

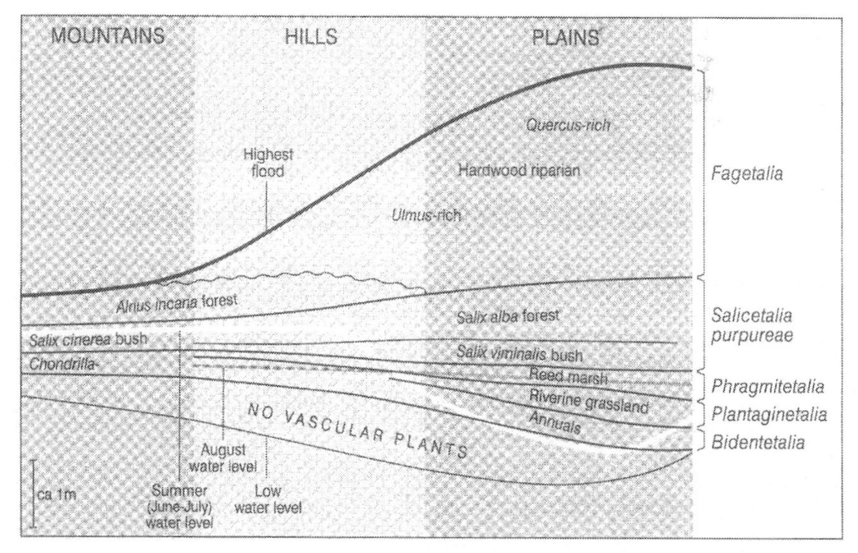

Figure 3.7 The different plant communities along a river course from high (left) to low (right) altitude. (After Ellenberg, 1978.)

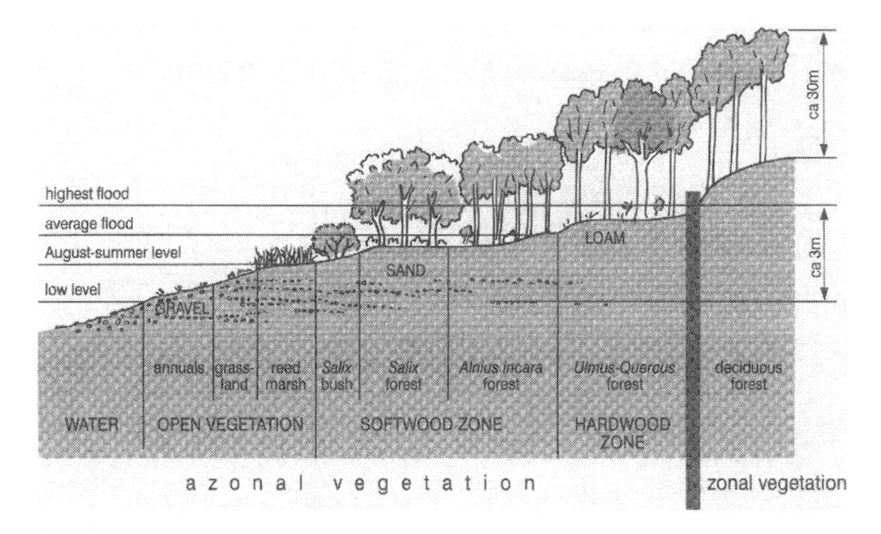

Figure 3.8 A vegetation zonation alongside a river in a montane area. (After Ellenberg, 1978.)

The northwestern European salt marshes have a striking number of species in common with the southeastern steppes; for example:

Armeria maritima, *Artemisia maritima*, *Aster tripolium*, *Atriplex hastata* (= *A. prostrata*), *Blysmus rufus*, *Festuca rubra*, *Juncus maritimus*, *Limonium vulgare*, *Plantago maritima*, *Puccinellia distans*, *Salicornia brachystachya*, *S. stricta*, *Spergularia marina*, *Triglochin maritima*.

Many of these species will have found their niche in the Pleistocene open steppe–tundra, as is known for *Armeria maritima* and *Plantago maritima* (Pennington, 1969), and were restricted with the advance of the forests. Other littoral vegetation types developed on dunes along the coast. The modern plant communities from the *Koelerio-Corynephoretea* (dunes and downs) also have many plant species in common from the more humid steppes of continental Europe (German: *Wiesensteppe*). These are, for example:

Achillea millefolium, *Anthyllis vulneraria*, *Arenaria serpyllifolia*, *Carex praecox*, *Erigeron acer*, *Galium verum*, *Hieracium pillosella*, *Helictotrichon pubescens*, *Helichrysum arenarium*, *Koeleria cristata*, *Potentilla argentata*, *Trifolium campestre*, and *Hippophaë rhamnoides*.

Again, a number of these species are known to have occurred in the Pleistocene steppe–tundra. However, these communities also gave way to forest – that is, to *Fraxino-Ulmetum* or to *Querco-Betuletum*. Whether this happened everywhere is not certain, but in the Sub-Boreal period (about 5000 BP) it appears to have been the normal pattern and only with the advance of human occupation did the *Koelerio-Corynephoretea* grasslands develop again (Zagwijn, 1971; Doing and Doing-Huis in 't Veld, 1971). Some types of these dry grasslands, especially those on soils with a low calcium carbonate content, have a strong affinity with natural grasslands that developed on glacial sand deposits along the large European rivers, such as the Rhine, Ems, Weser, Aller, Elbe, Oder, Netze and Weichsel. Normal vegetation development resulted in *Querco-Betuletum* forests there too, but tree growth was prevented on sands with a very high infiltration capacity (Ellenberg, 1978). Typical for these grasslands are species such as (Ellenberg, 1978):

> *Agrostis canina, Calluna vulgaris, Carex arenaria, Corynephorus canescens, Festuca ovina, Filago minima, Hieracium pilosella, Hypochoeris radicata, H. glabra, Jasione montana, Rumex acetosella, Spergularia morisonii, Teesdalea nudicaulis.*

Large grazing herbivores could not maintain physical condition on these grasslands, as is shown by the WallisDeVries (1994) experiments with cattle. However, where these grasslands were in close proximity to the *Lolio-Potentillion anserinae* communities along the large rivers, it is likely that large grazing herbivores, such as aurochs, could persist if they could exploit both communities within a year (WallisDeVries, 1994).

To summarize, natural grasslands at the beginning of the Neolithic period (i.e. when agriculture started in Europe) will have been restricted to areas where tree growth was prevented by high salinity (above salt domes and on salt marshes), by high water infiltration rates of soils on glacial sand deposits or by long inundations along the lowland rivers. Other types of open vegetation could be maintained where the temperature was too low for tree growth, i.e. above the tree-line in mountains or in the far north. Another niche for grassland species may have been on cliffs (Knapp, 1979/1980), as was also surmised for North America (Marks, 1983).

When the cultural transition of the Mesolithic to the Neolithic took place in northwestern Europe, around 6500 BP (in the Atlanticum), the general picture is one of extensive forests that covered close to 100% of the area. In these forests there was some hunting but its impact will have been rather low since the human density was in the order of about 0.05 humans/km^2 (Waterbolk, 1985). Hunting camps have been found in marshy grounds where people were fishing and fowling. Evidence for human impact on the vegetation is slight but it has been suggested that in order to facilitate hunting, and perhaps even to increase game densities, a form of forest management took place that facilitated the growth of hazel (*Corylus avellana*) or heather (*Calluna vulgaris*) (Simmons *et al.*, 1981). Around camps there is evidence for the increase of *Plantago major* and *Rumex* species and other

Chenopodiaceae (Waterbolk, 1954); and in some forested areas, tree removal may have caused blanket-peat to develop (Simmons, 1981). Manipulation of the vegetation for an increased hunting success has been suggested even for the end of the Pleistocene, in line with the Pleistocene overkill hypothesis (Simmons, 1981). Evidence for the first seems extremely weak (*pace* Bush, 1993), and the postulated Pleistocene overkill of large herbivores does not make sense in the light of hunting strategies that, of necessity, would have conformed with optimal foraging theories.

3.6 HUMAN IMPACT

3.6.1 Neolithic and Bronze Age

In a large part of northwestern Europe, the transition from the Mesolithic to the Neolithic, i.e. with the start of farming and making of pottery, took place during the transition from the Atlantic to the Sub-Boreal period, around 5000 BP (Van Zeist, 1959). Farming is defined as the production of primary foodstuffs and raw materials by means of deliberate cultivation of plants and keeping of animals. Farming started in the Middle East and spread at an average rate of about 2 km per year to northwestern Europe, where it was mainly restricted to loess soils during the Neolithic (Van Zeist, 1980). Humans selected forests on these soils, and started with a slash-and-burn cultivation (for a description from Russia at the end of the last century, see Smith, 1959). This cultivation technique gave quite good results in forests but not in steppes. Soils could be cultivated for some six years before they had to be left fallow. Although hoes have been found, the normal cultivation practice appears to have been use of the ard (scratch-plough), and no evidence for digging sticks has been encountered. Pigs played an important role in the economy and they may have influenced the structure of the surrounding forest. Cattle were becoming important but were not yet used for labour, and there is no evidence of haymaking to maintain them over the winter (Smith, 1959; Ten Cate, 1972; Knörzer, 1975; Ellenberg, 1978; Van Zeist, 1980). In this forest environment there was no place for sheep and goats: they came slightly later and did not become important until the Bronze Age (Clason, 1980). Hunting and fishing remained important in marshy areas, as demonstrated by the very revealing excavations at Bergschenhoek, Swifterbant and Hazendonk (all about 5500 BP) in The Netherlands (Louwe Kooijmans, 1980).

The first farmers cultivated mainly cereals and, because of the small size of the fields and correlated shady conditions, the use of ards, the likely absence of fertilization and the apparent harvesting of ears only, the associated weed community was different from the modern *Secalietea*. The resulting '*Bromo-Lapsanetum praehistoricum*' consisted of species originating from adjacent forests (*Lapsanum communis*, *Poa nemoralis*, *Rumex sanguineus*), species from river sides (*Chenopodium album*, *Polygonum persi-*

caria) and plant species that had migrated from the southeastern steppes (*Galium spurium*, *Polygonum convolvulus* and *Vicia hirsuta*) (Knörzer, 1971; see also Smith, 1959). Also *Plantago major* became important but *P. lanceolata*, indicative of grasslands, remained very rare (Van Zeist, 1959). The cultivated cereals were einkorn (*Triticum monococcum*) and emmer wheat (*T. dicoccum*). From later finds, it is clear that seeds of some 'weeds' were harvested too, such as from *Chenopodium album*, *Polygonum lapathifolium* and *Spergula arvensis* (Van Zeist, 1980). The advance of farming has been associated with the decline of *Ulmus* in the palynological record (Waterbolk, 1954; Van Zeist, 1959; Oldfield, 1967; Godwin, 1975). Because of the absence of evidence for grasslands and haymaking (e.g. Knörzer, 1975), it has been suggested that cattle were fed leaves of elm (*Ulmus*) and ash (*Fraxinus*), and perhaps leaves of lime (*Tilia*), maple (*Acer*) and oak (*Quercus*) as well. The so-called elm decline has been taken as evidence for the Neolithic opening-up of the forest (*landnam*) (for an extensive discussion, see Smith *et al.*, 1981). Evidence for feeding cattle with browse at that time is weak or absent (e.g. Knörzer, 1975) even though broad-leaved tree-fodder feeding has been observed in the last few hundred years (Troels-Smith, 1960 and Iversen, 1973; cited in Peglar and Birks, 1993). However, elm disease has been identified as the very likely cause of the prehistoric elm decline (Perry and Moore, 1987), and the recent work of Peglar (1993) in particular elegantly shows that disease was the most likely cause. Pollarding of trees, perhaps for the production of fodder, may have influenced the rate of transmission of the disease since pollarding instruments could have been contaminated with the fungus (Moe and Rackham, 1992); however, this is linking two assumptions into a chain, if one were to take fodder use as a causal explanation of the elm decline during the Neolithic. This is not to deny the fact that the forest was opened up at the time and that the species composition changed (Peglar and Birks, 1992; Peglar, 1993). Rackham (1980) had already pointed out that for Neolithic man to have an impact on elm, as deducted from the 50% pollen reduction of *Ulmus* during the elm decline, 'requires a human population much larger and more elm-centred than any archaeologist has hitherto proposed' (Rackham, 1980; Peglar and Birks, 1993).

Human density during the Neolithic and the Bronze Age was still very low – for example, in the order of 0.5 humans/km^2 in Drenthe (northern Netherlands) (Waterbolk, 1985): this is about 10 times higher than during the Mesolithic but about 10 times lower than during the Middle Ages. The resultant effect on the vegetation appears to have been mainly the expansion of shrubs. From England, there is evidence that *Crataegus* species, *Prunus spinosa* and *Sambucus nigra* became more important, while new species became established, such as *Acer campestre*, *Buxus sempervirens*, *Euonymus europaeus*, *Malus sylvestris*, *Prunus avium*, *Rhamnus catharticus*, *Salix fragilis*, *Sorbus aria* and *Ulex europaeus* (Godwin, 1975). This 'shrub phase' is also evident from sub-fossil mollusc faunas in southern England (Evans, 1972).

The estimated 0.5 human/km^2 mentioned above is an even lower population density than that of modern Gabon in equatorial Africa. Although the comparison is fraught with difficulties, it is noticeable that in Gabon such a density results mainly in an increased percentage of secondary forest but not in the disappearance of forest (Prins and Reitsma, 1989). Stock-grazing in the secondary forests of northwestern Europe led to good conditions for the establishment of light-demanding tree species such as *Betula* (Turner, 1981; see also Reuter, 1920; Gundermann and Plochmann, 1985) but also *Quercus* (Vera, 1997).

Grass pollen grains are found in increasing numbers with the advance of farming from about 4500 BP. There is evidence from about 5000 BP (beginning of Late Neolithic) that cattle were kept in barns, and *Plantago lanceolata*, indicative of grasslands, becomes increasingly important in the pollen record (Van Zeist, 1959). From about 4000 BP evidence for cattle kraals has been found (Waterbolk, 1954; Brongers and Woltering, 1978), and apparently cattle were used as draught animals with the invention of the plough in eastern Europe (Smith, 1959) but not yet in western Europe. When the Bronze Age began, around 4000 BP, the climate became a bit cooler and especially wetter. Heaths dominated by *Calluna vulgaris* developed, through heavy grazing by livestock, from natural grasslands growing on sandy soils (*Koelerio-Corynephoretea* grasslands) and could spread because shifting agriculture started now on soils other than loess as well, resulting in the demise of forests on higher grounds.

Shifting agriculture – that is, permanent abandonment of farmland on cleared forests after loss of soil fertility – appears to have replaced rotational farming with fallow (shifting cultivation, outfield) that was the norm during the preceding period. Evidence for the absence of fallow has been concluded from the presence of Compositae, Chenopodiaceae, *Spergula* and *Spergularia* in the fossil record (Waterbolk, 1954). Farming practice will have remained more or less the same because of the continuation of the *Bromo-Lapsanetum praehistoricum* (Knörzer, 1971). Baudais-Lundstrom (1978) concluded from excavations in Switzerland that meadows and hayfields were developing around 4000 BP, but her species list contains too many species from forests for such a conclusion. Also, there has been no evidence that forests on heavy soils were felled during the Bronze Age, when soil-tilling was probably too difficult with the available technology.

There is good evidence that chalk grasslands started to develop during the Bronze Age from about 3500 BP onwards. Plant species that were to find a niche in grasslands on chalk uplands (downs; *Mesobromion*) found their original niche in forest fringes, also along cliffs. These forest fringe communities are the *Lithospermo-Quercetum*, *Geranio-Peucedanetum*, and *Seslerio-Fagetum* with species such as (Knapp, 1979/1980):

Anthericum liliago, *Bupleurum falcatum*, *Carlina vulgaris*, *Centaurea scabiosa*, *Cirsium acaulon*, *Euphorbia cyparissias*, *Gentianella ciliata*,

G. germanica, *Leucanthemum vulgare*, *Linum catharticum*, *Origanum vulgare*, *Pimpinella saxifraga*, *Ranunculus bulbosus*, *Salvia pratensis*, *Sanguisorba minor*, *Teucrium chamaedrys*, *T. montanum*, *Thymus praecox*.

Down types in northern France (*Mesobrometum–Seslerio-Polygatelosum*) could date back to the Pleistocene (Stott, 1971), just as the open vegetation from the Upper Teesdale with its many arcto-alpine species (Pennington, 1969) and the *Caricion austroalpinae* from calcareous highlands in northern Italy (Pignatti and Pignatti, 1975). It has been claimed that grasslands on limestone cliffs in England also date back to the Pleistocene ('*Brometalia erecti*': Tansley cited in Shimwell, 1971) but evidence for this does not seem to be available (see also Moore, 1987). Bush (1993) concludes that at least some British chalklands had open grassland habitats continuously from the end of the Pleistocene, but his palaeoecological record shows a gap between 8000 and 4400 BP. This is exactly the time period that is supposed to have known the least human disturbance and Bush's inference that the ensuing vegetation had enough open space for grassland taxa to survive *in situ* does not appear warranted from his data. His conclusion (Bush, 1993) that 'it is clear that chalk grassland is more a landscape type ... than an identifiable community as its component species have changed through time, apparently in response to climatic change' is well supported and important. Evidence for development of the downs of southern England for the Neolithic is weak, and even for the Bronze Age not much proof is available (Smith *et al.*, 1981) because conditions for pollen fossilization were not good (Tinsley and Grigson, 1981). Yet most data point to the development of rather extensive grasslands at that time (Turner, 1981). Some time ago it was thought that in continental Europe a nomadic pastoralist society had already developed by the end of the Neolithic (Standvoetbeker Culture, about 4800 BP) (Waterbolk, 1954; Van Zeist, 1959; Godwin, 1975) but more recently this has been rejected (Brongers and Woltering, 1978). The evidence for such a society appears to be limited to the Celtic Iron Age in Great Britain and Ireland, which is several centuries later (Pennington, 1969; Godwin, 1975); and sagas appear to reflect this life style (see the eighth century Irish *Tain Bo Cuailge*; Kinsella, 1969). Later, during the Iron Age, Julius Caesar would report a similar economy for the Germanic tribes (see below).

Even though the fossil records shows a preponderance of evidence for a pastoral economy during the Bronze Age in southern England (Turner, 1981) this, of course, is not evidence of a pastoral society. Evidence of a purely pastoral society in the British highlands during the Bronze Age is not strong (Tinsley and Grigson, 1981) but extensive clearings developed at that time into heath (moor) and bog which replaced the original oak forest; livestock may have grazed these grassy heaths but cereal cultivation took place as well. A concurrent change from a pig-centred economy to a cattle and sheep economy took place at that time. The end result of the forest clearings, perhaps together with a deteriorating climate, was extensive soil leaching and acidification. This then led to the disappearance of grazing lands, which were replaced by heaths and bogs (Tinsley and Grigson, 1981), but the same happened under forest grazing where forests also became replaced by bogs (Turner, 1981). The same processes were observed on nutrient-poor soils in The Netherlands (Waterbolk, 1954; Van Zeist, 1959).

3.6.2 Iron Age and Middle Ages

During the following Iron Age period, which started about 3000 BP, there was a change in agricultural practice. Arable fields became permanent and were bordered by embankments. Complexes of these fields are called celtic fields, and such complexes can cover more than 100 ha. Firm evidence for soil tilling also dates from this period, because traces of ards are known from the soil in celtic fields (Van Zeist, 1980), and even humus fertilization took place (Brongers and Woltering, 1978). Individual arable fields became larger, and cereal harvesting was now at ground level. Barley (*Hordeum vulgare* and *H. distichum*) and millet (*Panicum miliaceum*) were included in the agricultural practice, possibly as an adaptation to deteriorating soil fertility due to long-term farming but also due to an increased climatic humidity (Van Zeist, 1980). This change in practice led to the disappearance of the *Bromo-Lapsanetum prehistoricum* and species better adapted to the new situation became weeds in the arable fields, such as *Alopecurus myosuroides*, *Anthoxanthum aristatum*, *Apera spica-venti*, and *Poa annua*. This points to the development of modern weed communities associated with cereal production, i.e. the *Secalietea*. It is possible that the first meadow-like vegetation developed at this time (Knörzer, 1971). Indeed, species like *Bromus sterilis*, *Phleum pratense*, *P. nodosum*, *Poa nemoralis*, *P. trivialis* and *P. pratense* found a place in the original cereal weed association, the *Bromo-Lapsanetum prehistoricum* (Knörzer, 1971), but not in the *Secalietea*. During the Iron Age in England, settled agriculture and permanent grazing developed for the first time between 400 BC and AD 100; the open landscape, as we know it, developed at that time (Turner, 1981). Sheep became more important than cattle over most of Britain, which has been taken as evidence for further deterioration of pasture quality (Clark, 1952; Godwin, 1975). These observations from The Netherlands and from England that permanent agriculture developed during the Iron Age have an important corollary: under permanent agriculture, the persistent weed problem increases. This implies that a different cultivation technique had to be used, and, in fact, the ard was replaced by the plough. Good evidence for the use of the plough, i.e. turning of the soil, has been found in, for example, Oudemolen (northern Netherlands) from the Late Iron Age (Waterbolk, 1985). Ploughing can hardly be done without the aid of draught animals, such as cattle (Godwin, 1975), and the result was that there was no place left for grasses in the cereal fields.

More or less simultaneously, the harrow was invented, the function of which is to break clods and to remove grassy weeds. So, stubble fields became less attractive for cattle because of the disappearance of grass from these fields and simultaneously the need for grasslands increased because of the higher demand for labour from draught animals (for which only cattle were used). Hence, it is likely that grasslands became much more important, because now there was an economic need for them. It should be realized that ploughing was often needed in spring, in order to grow summer cereals, which means that peak labour demand was also in spring because that was when the fields had to be prepared. As cattle would have suffered a serious loss of physical condition over the winter (WallisDeVries, 1994), it became of paramount importance to develop a system of storing fodder during winter. This must have led to haymaking, since harvesting grass is a relatively efficient way of collecting food and storing it over winter. Most browse species do provide enough energy to roughage grazers such as cattle in summer but time budgets of labour for the collection of willow leaves and twigs for animal fodder (S.E. Van Wieren, personal communication) show that harvesting grass is at least 10 times more efficient than collecting willow fodder from dense stands of young trees, if efficiency is expressed as dry weight of fodder collected per unit time. Preliminary observations also show that the intake rate of cattle is much higher on hay than on dried leaves of shrubs or trees (S.E. Van Wieren and M.F. WallisDeVries, personal communication). With the invention of the heavy plough in the early Middle Ages the need for an energy source for draught oxen became even more acute (see below). In comparison with horses, cattle are better adapted to work hard on a relative low quality diet and they do not need cereals to do this. Indeed, good quality hay was a prerequisite to cereal production for the human population (Slicher van Bath, 1976). Tüxen (1974) makes mention of mowed *Molinietalia* grasslands dated as 5000 BP, but this is a wrong citation of Janssen (1972): there seems to be no evidence for haymaking from before the Roman occupation in northwestern Europe (e.g. Knörzer, 1975). On calcareous soils in southern England, however, the relative importance of grasslands (as percentage of totally cleared grounds) became less important and the arable field proportion increased concurrent with intensified forest clearance during the Iron Age (Turner, 1981).

Stalling of cattle, which could be taken as evidence for haymaking, is known from the northern part of The Netherlands from the Middle Bronze Age (Emmerhout) but also from the Iron Age (Hijken, Ezinge) (Brongers and Woltering, 1978). Evidence for Iron Age cattle stables has also been found in Jutland (Jankuhn, 1969, cited in Knörzer, 1975). However, the presence of stables may merely indicate that cattle were kept indoors at night and it does not necessarily prove winter feeding. Before the Late Iron Age (about 300 BC) houses were not permanent and they were moved with the shifting agriculture, apparently through the village territory; but during the Late Iron Age and later, houses were permanently located close to the permanent agriculture on the celtic fields (Waterbolk, 1985). In at least some areas of northwestern Europe, human population density had increased to some 10–15 humans/km^2 (calculated from Van Es, 1994, for the areas of the Cananefati and Batavi along the lower Rhine in The Netherlands), which explains the need for novel agricultural techniques. From the Roman period Columella's instructions (in *De Rustico*, about AD 75) are available about how to manage hayland, which includes re-sowing, removal of moss, fertilization with dung and ashes, and exclusion of cattle from the land to preclude sod destruction under wet conditions (Lange, 1976). Even if the Romans had followed this advice in practice, their influence would have been small, as is indicated by the estimate that in the southern Netherlands the total area under cultivation for feeding half the Roman army that occupied Germania Inferior was only 10 000 ha (Van Es, 1972); this is less than 1% of the area. Indeed, Appolinaris Sidonius in his letter to Heronius (AD 476), describes the river banks of a number of rivers in southern Europe as having sandbanks, bordered by reeds, thorn shrubs and *Acer* and *Quercus* (Ter Kuile, 1976) but he does not mention grasslands. Horace (23 BC in his Odes II–5) even mentions heifers 'to play with the calves in marshy willow-brooks' (Shepherd, 1983). Fossil evidence for the occurrence of haylands is very meagre indeed, and plant species typical of the moist, nutrient-rich *Arrhenatheretalia* grasslands have not been identified from the central European excavations (Lange, 1976).

The first evidence for haymaking is known from around AD 200 for the area along the Lower Rhine near Utrecht: haymaking took place in the wet floodplains of the river (Kooistra, 1994). These areas were originally *Phragmitetum* and *Lolio-Potentillion anserinae* communities and not *Alno-Padion* brook forests or other types of forest. Hence hay-making in this first case did not imply felling of trees. Also Knörzer's (1975) study on fossil remains of plant taxa indicative of grasslands in the Rhineland area of Germany reveals that the variety of these taxa is so large that, according to him, it can be concluded that grazing grounds developed during the Roman occupation for sufficiently long duration to set the stage for the formation of modern grassland communities. Plant species in these communities originated mainly from forest floor communities, salt marshes and small open areas on steep slopes, and were even introduced by long-distance dispersal through livestock from steppes and alpine areas. Indeed, all character species of the *Arrhenaterion* grasslands appeared then for the first time in the fossil material and some evidence for dry chalk grassland (*Festuco-Brometea*) has been found, with species such as *Salvia pratensis*, *Scabiosa columbaria*, *Sanguisorba minor*, *Pimpinella saxifraga* and *Carex caryophyllea*. Character species of the nutrient-rich *Arrhenatherion* grasslands and moist *Molinietalia* grasslands were, however, still very rare or even absent (Knörzer, 1975).

During the Iron Age, the salt marshes of northwestern Europe also became occupied. Fossil material from the artificial floodhills (*terpen* or *wierden*) clearly show that the Dutch salt marshes were intensively grazed some 2000 years ago because many species of the *Juncetum gerardii* community, subassociation *leontodontetosum autumnalis*, were found (Van Zeist, 1974). Human occupation of these salt marshes started in only about 500 BC; excavations at Ezinge showed that stables could house over 50 head of cattle (Waterbolk and Boersma, 1976). Caesar's description of the Suebi, a Germanic tribe, as living mainly off milk and off meat obtained through hunting (*Gallic War*, IV-1; Handford, 1951) seems exaggerated when considering the palaeobotanical record (Knörzer, 1975). Likewise Caesar's observation that near the Rhine the Germanic tribes 'are not agriculturalists and live principally on milk, cheese, and meat' (VI-22; Handford, 1951) is refuted (cf. Kooistra, 1994): the observations from the area of the Frisii shows that they grew foodplants, and the experiments of Van Zeist *et al.* (1976) on Dutch salt marshes show that agriculture was possible even without dikes.

Very good evidence for the occurrence of grasslands (*Koelerio-Corynephoretea* and *Festuco-Brometea*) and for moist grasslands possibly used for haymaking (*Molinio-Arrheneatheretea*) dates back to the early Middle Ages, namely from Czech villages from the *Burgwallzeit* (AD 700–900) (Opravil, 1978) and from storage barns from Wrocklaw cathedral (AD 1000) (Kosina, 1978; see also Knörzer, 1975) (Figure 3.9). Written sources also mention haymaking at that time in northern France: 'Sixteen *bonniers* [20 ha] of cropland ... There are 4 bonniers of meadow, from which 30 loads of hay can be taken. There are 3 bonniers of copse' (Pounds, 1974). There is no fossil evidence for real meadows from Germany from before AD 1000 (Willerding, 1978) but evidence becomes increasingly common afterwards (Lange, 1976).

During this period the heavy plough replaced the light plough that was in use during the Roman period. These heavy ploughs needed teams of people and oxen: the coulter made a vertical cut through the soil, while the share cut horizontally, and the mouldboard turned the soil, burying the weeds (Pounds, 1974). In the early Medieval period only oxen were used for ploughing, but with the introduction of the padded collar to the harness, horses could be used additionally in the late Medieval period (Pounds, 1974). 'A chief restraint on animal rearing was the shortage of winter fodder ... [and] there was nothing except hay on which to feed the animals through the winter. Ploughing oxen had a prior claim on the food supply' (Pounds, 1974). Dairy cows were few, except in low-lying lands of Belgium, The Netherlands and northern Germany (Pounds, 1974). Much of the pasture will have been provided by salt marshes until embankments against the sea caused fresh water to replace the salt or brackish water and to alter the vegetation towards ordinary inland grassland dominated by species as *Lolium perenne*.

The fossil evidence thus coincides with historical data, which show that between AD 800 and 1000 and from the year 1400 onward, forests on heavy or wet soils were increasingly cleared (Walter, 1927). This was made possible,

Figure 3.9 Most existing grasslands are 'replacement communities', such as this abundantly flowering, semi-natural hay-meadow in the montane Swiss Alps.

or perhaps necessary, by the increased human density and by better techniques for draining the soils (Pounds, 1974). Since the *Molinietalia* grasslands are a subdivision of the *Molinio-Arrheneatheretea* class, the group of communities that replaced the alder forest and willow stands (namely, the *Calthion palustris*, *Filipendulion* and *Junco subuliflori–Molinion* communities) have developed only since the Middle Ages. The forests from wetter sites were then felled as a result of the development of appropriate technology and a growing demand for land with the rising human population (Pound, 1974). A more detailed account of human impact on plant communities, particularly with respect to grazing by livestock, is given in Chapter 4.

3.7 CONCLUSIONS AND CONSEQUENCES FOR NATURE CONSERVATION

It can thus be concluded that grassland communities in northwestern Europe belong only to a small extent to the natural landscape, in contrast to *Phragmitetea* reed and sedge communities (*Phragmition*, *Oenanthion aquaticae* and *Magnocaricion* communities) which had undoubtedly developed during the previous interglacial periods and perhaps even before the Pleistocene (Table 3.12). The oldest anthropogenic grasslands belong to the *Mesobromion* which started developing in the Bronze Age alongside the *Calluna* heaths. The *Arrhenatherion* grasslands may be seen as a derivative from the natural *Lolio-Potentillion anserinae* communities along the larger rivers, which could spread due to intensified grazing by domestic stock. Moist grasslands used for haymaking, especially the *Molinietalia* grasslands, are of recent origin and date back only to the Middle Ages when forests on poorly drained soils were felled. These anthropogenic grasslands belong to the semi-natural landscape (Chapter 2).

These observations and conclusions have important implications for nature conservation. If nature is defined as the combination of wild native species and the action of natural processes on these wild species, then the aim of nature conservation should only be to maintain the integrity of natural communities and even ecosystems. Nature conservation of grassland communities could then be restricted to the natural grassland types (Table 3.12). Nature management of this latter class should then be aimed at preventing nutrient input but also at the removal of dominance by *Calluna vulgaris*, and grazing as a management technique to suppress grass growth would be inappropriate. To maintain salt marshes one needs flooding by salt water, and grazing or mowing are not necessary as long as mudflats can develop into new salt marshes. Management techniques to maintain the *Mesobromion*, which often include grazing, would be inappropriate, since the natural vegetation there would be broad-leaved forest. By the same token, nature management through haymaking in the *Molinietalia* grasslands would be wrong since the natural vegetation in areas with that type of grassland would be the *Alno-Padion*.

Table 3.12 Overview of the origins of contemporary grassland communities and other open vegetation types from northwestern Europe (syntaxa after Westhoff and Den Held, 1969)

Landscape	Community	Vegetation type
Natural	*Asteretea tripolii*	Salt marshes
	Phragmitetea	Reed and sedge marshes
	Agropyro-Rumicion crispi	Riverine grassland
	Corynephoretalia	Dune grassland from acid soils
	Mesobrometum-Seslerio-Polygaletosum	Mesic calcareous grassland
	Xerobromion	Dry chalk grassland
Anthropogenic (Bronze Age origins)	*Mesobromion*	Mesic calcareous grassland
	Nardo-Calluneta	Heathland from acid soils
	Arrhenaterion elatioris	Fertilized grassland
Anthropogenic (Middle Age origins)	*Molinietalia*	Hay-meadows

Adherents of this view on nature conservation emphasize lack of natural processes and especially the lack of wild large herbivores in the present 'nature reserves'. This leads to the increased desire to reintroduce, for example, beaver and moose, to abandon red deer and wild boar hunting, and to introduce semi-wild grazers to replace the extinct aurochs with Heck cattle and the extinct wild horse with Konik horses. Within the nature conservation movement, adherents of this view have often been denounced as romanticists, implying that what they want is unattainable.

Inter alia, scientists apparently have an urge to discover very ancient remnants of particular vegetation types, remnants of a period long lost but still occurring in the countryside. One example has already been mentioned: the supposed Pleistocene origin of British chalk grasslands, for which the data are definitely too weak to prove their ancient origin. Also some 'ancient' woodlands are supposed to represent a more or less unbroken continuance of forests since the post-Pleistocene reforestation of Britain (Rackham, 1980). Again, the facts appear to be otherwise and a particular ancient woodland which was well studied proved to be 'only' 1000 years old (Day, 1993). Northwestern Europe has surely been used so long and so intensively as to preclude the occurrence of 'original' vegetation in areas that have, or had, agricultural potential. This implies that nature conservationists do not have the natural, wild examples of how northwestern Europe could, and must, look like if and when nature restoration takes place on abandoned agricultural land or even in nature reserves. This presents a major challenge, because now scientists have to find out what natural vegetation types would look like, what vertebrate species assemblages have to be reassembled, and in what densities these animals have to live in areas where nature restoration takes place. Even with the prolonged effort of palaeoecologists the past is still a very imperfect mirror – but this mirror can be used for restoring nature in northwestern Europe.

On the other hand, many conservation organizations in western Europe implicitly define nature as the occurrence of indigenous wild plant species that can maintain themselves without sowing. If at the same time the (plant) communities that have developed over centuries under management techniques from before the Industrial Revolution have to be preserved, then the term 'nature' is used to label the semi-natural landscape that can be described as pre-industrial or low-intensity farm land (Chapter 2). In that case a whole set of management techniques have to be employed to mimic the farming activities of our ancestors. These techniques include: herding sheep, grass and tree suppression, and litter removal on heaths; extensive pastoralism and tree suppression in *Mesobromion* vegetation; and haymaking, extensive grazing, tree suppression and nutrient removal in *Molinietalia* grasslands. If these techniques are to be employed, there should preferably be the same high labour input and low level of mechanization as those of our ancestors, so as to ensure the same impact on the communities. It should be remarked, too, that the costs of this type of nature management may be much higher than the type of management advocated by the 'romanticists'.

Whatever the chosen aim and its dependent set of actions, management activities must always take into account water management to maintain the proper hydrological regime, counter-action against acidification, and prevention of habitat fragmentation and input of nutrients or pollutants. This is a daunting task for which the adherents of the two (often contrasting) views definitely should form a united front against the many forces in European societies that care little for Europe's native fauna and flora or for its landscape.

ACKNOWLEDGEMENT

This chapter is dedicated to Mrs Clara Biewenga who, through her unassuming support to generations of plant ecologists, has been an important factor in the development of science in The Netherlands.

REFERENCES

Aaris-Sørensen, K. (1988) *Danmark's Forhistoriske Dyreverden: fra Istid til Vikingetid*, Gyldendal, Copenhagen.

Aaris-Sørensen, K.J. Petersen and H. Tauber (1990) *Danish Finds of Mammoth (*Mammuthus primigenius *Blumenbach): stratigraphical position, dating and evidence of Late Pleistocene environment*, Danmarks Geologiske Undersøgelse, Copenhagen.

Anderson, E. (1984) Who's who in the Pleistocene: a mammalian bestiary, in *Quaternary Extinctions: a prehistoric revolution*, (eds P.S. Martin and R.G. Klein), pp. 40–89, University of Arizona Press, Tucson.

Atkinson, T.C., Briffa, K.R. and Coope, G.R. (1987) Seasonal temperatures in Britain during the past 22,000 years, reconstructed using beetle remains. *Nature*, **325**, 587–592.

Baudais-Lundström, K. (1978) Plant remains from a Swiss neolithic lake shore site: Brise-Lames, Auvernier. *Berichte der Deutschen Botanischen Gesellschaft*, **91**, 67-84.

Becker, B. and Kromer, B. (1991) Dendrochronology and radiocarbon calibration of the Early Holocene. in *The Late Glacial in North-west Europe: human adaptation and environmental change at the end of the Pleistocene*, (eds N. Barton, A.J. Roberts and D.A. Roe), pp. 22–24, CBA Research Report 77, Council for British Archaeology, London.

Behre, K.E. (1967) The late glacial and early post-glacial history of vegetation and climate in northwestern Germany. *Review of Palaeobotany and Palynology*, **4**, 149–161.

Bell, F.G. (1969) The occurrence of southern, steppe and halophyte elements in Weichselian (last glacial) floras from southern Britain. *New Phytologist*, **68**, 913–922.

Bennett, K.D., Tzedakis, P.C. and Willis, K.J. (1991) Quaternary refugia of north European trees. *Journal of Biogeography*, **18**, 103–115.

Berglund, B.E., Lemdahl, G., Liedberg-Jönsson, B. and Persson, T. (1984) Biotic response to climatic changes during the time span 13,000–10,000 BP – a case study from SW Sweden, in *Climatic Changes on a Yearly to Millennial Basis*, (eds N.A. Mörner and W. Karlén), pp. 25–36, Reidel, London.

Bratlund, B. (1991) A study of hunting lesions containing flint fragments on reindeer bones at Stellmoor, Schleswig-Holstein, Germany, in *The Late Glacial in North-west Europe: human adaptation and environmental change at the end of the Pleistocene*, (eds N. Barton, A.J. Roberts and D.A. Roe), pp. 193–212, CBA Research Report 77, Council for British Archaeology, London.

Brongers, J.A. and Woltering, P.J. (1978) *De Prehistorie van Nederland: Economisch, technologisch*, Fibula-van Dishoeck, Haarlem.

Bush, M.B. (1993) An 11400 year paleoecological history of a British chalk grassland. *Journal of Vegetation Science*, **4**, 47–66.

Clason, A.T. (1980) Jager, visser, veehouder, vogellijmer, in *Voltooid Verleden Tijd?: Een hedendaagse kijk op de prehistorie*, (eds M. Chamaulan and H.T. Waterbolk), pp. 131–146, Intermediair, Amsterdam.

Clark, J.G.D. (1952) *Prehistoric Europe: the economic basis*, Methuen, London.

Coope, G.R. (1987) The response of late Quaternary insect communities to sudden climatic changes, in *Organization of Communities: Past and Present*, (eds J.H.R. Gee and P.S. Giller), pp. 421–438, Blackwell, Oxford.

Coope, G.R. (1994) The response of insect faunas to the glacial–interglacial climatic fluctuations. *Philosophical Transactions of the Royal Society, London, B*, **344**, 19–26.

Cordy, J.M. (1991) Palaeoecology of the Late Glacial and early Postglacial of Belgium and neighbouring areas, in *The Late Glacial in North-west Europe: human adaptation and environmental change at the end of the Pleistocene*, (eds

N. Barton, A.J. Roberts and D.A. Roe), pp. 40–47, CBA Research Report 77, Council for British Archaeology, London.

Cronquist, A. (1968) *The Evolution and Classification of Flowering Plants*, Nelson, London.

Currant, A.P. (1991) A Late Glacial interstadial mammal fauna from Gough's Cave, Somerset, England, in *The Late Glacial in North-west Europe: human adaptation and environmental change at the end of the Pleistocene*, (eds N. Barton, A.J. Roberts and D.A. Roe), pp. 48–50, CBA Research Report 77, Council for British Archaeology, London.

Dansgaard, W., White, J.W.C. and Johnsen, S.J. (1989) The abrupt termination of the Younger Dryas event. *Nature*, **339**, 532–533.

David, A. (1991) Late glacial archaeological residues from Wales: a selection, in *The Late Glacial in North-west Europe: human adaptation and environmental change at the end of the Pleistocene*, (eds N. Barton, A.J. Roberts and D.A. Roe), pp. 141–159, CBA Research Report 77, Council for British Archaeology, London.

Day, S.P. (1993) Woodland origin and 'ancient woodland indicators': a case study from Sidlings Copse, Oxfordshire, UK. *The Holocene*, **3**, 45–53.

De Boer, W.F. and Prins, H.H.T. (1990) Large herbivores that strive mightily but eat and drink as friends. *Oecologia*, **82**, 264–274.

Delcourt, H.R. and Delcourt, P.A. (1991) *Quaternary Ecology: a paleoecological perspective*, Chapman & Hall, London.

Dister, E. (1983) Zur Hochwassertoleranz von Auenwaldbäumen an lehmigen Standorten. *Verhandlungen der Gesellschaft für Ökologie*, **X**, 325–336.

Doing, H. and Doing-Huis in 't Veld, J. (1971) History of landscape and vegetation of coastal dune areas in the Province of North Holland. *Acta Botanica Neerlandica*, **20**, 183–190.

Doyle, J.A. (1978) Origin of angiosperms. *Annual Review of Ecology and Systematics*, **9**, 365–392.

Dudley, N. (1992) *Forests in Trouble: a review of the status of temperate forests worldwide*, WWF, Gland, Switzerland.

Ellenberg, H. (1978) *Vegetation Mitteleuropas mit der Alpen in ökologischer Sicht*, 2nd edn, Ulmer, Stuttgart.

Estes, R.D. (1991) *The Behavior Guide to African Mammals: including hoofed mammals, carnivores, primates*, University of California Press, Berkely.

Evans, J.G. (1972) *Land Snails in Archaeology with Special Reference to the British Isles*, Seminar, London.

Firbas, F. (1949) *Waldgeschichte Mitteleuropas*, Vol. I, Gustav Fisher Verlag, Jena.

Fritz, A. (1972) Das Spätglazial in Karnten. *Berichte der Deutschen Botanischen Gesellschaft*, **85**, 93–100.

Gerken, B. (1988) *Auen: Verborgene Lebensadren der Natur*, Rombach, Freiburg.

Gingerich, P.D. (1984) Pleistocene extinctions in the context of origination-extinction equilibria in Cenozoic mammals, in *Quaternary Extinctions: a prehistoric revolution*, (eds P.S. Martin and R.G. Klein), pp. 211–222, University of Arizona Press, Tucson.

Good, R. (1964) *The Geography of the Flowering Plants*, 3rd edn, Longmans Green, London.

Godwin, H. (1975) *History of the British Flora: a factual basis for phytogeography*, 2nd edn, Cambridge University Press, Cambridge.

Graham, R.W. and Lundelius, E.L. (1984) Coevolutionary disequilibrium and Pleistocene extinctions, in *Quaternary Extinctions: a prehistoric revolution*, (eds P.S. Martin and R.G. Klein), pp. 223–249, University of Arizona Press, Tucson.

Guildhay, J.E. (1984) Pleistocene extinction and environmental change: case study of the Appalachians, in *Quaternary Extinctions: a prehistoric revolution*, (eds P.S. Martin and R.G. Klein), pp. 250–258, University of Arizona Press, Tucson.

Gundermann, E. and Plochmann, R. (1985) Die Waldweide as forstpolitisches Problem im Bergwald. *Forstwissenschaftliches Centralblatt*, **104**, 146–154.

Guthrie, R.D. (1990) *Frozen Fauna of the Mammoth Steppe: the story of blue babe*, University of Chicago Press, Chicago.

Handford, S.A. (transl.) (1951) *Caesar: The Conquest of Gaul*, Penguin, Harmondsworth.

Hess, H.E., Landolt, E. and Hirzel, R. (1967) *Flora der Schweiz und angrenzender Gebiete*, Vol. I, Birkhäuser, Basel.

Housley, R.A. (1991) AMS dates from the Late Glacial and early Postglacial in north-west Europe: a review, in *The Late Glacial in North-west Europe: human adaptation and environmental change at the end of the Pleistocene*, (eds N. Barton, A.J. Roberts and D.A. Roe), pp. 25–39, CBA Research Report No. 77, Council for British Archaeology, London.

Janssen, C.R. (1972) The palaeoecology of plant communities in the Dommel valley, North Brabant, The Netherlands. *Journal of Ecology*, **60**, 411–437.

Janssen, C.R. (1979) The development of palynology in relation to vegetation science, especially in the Netherlands, in *The Study of Vegetation*, (ed. M.J.A. Werger), pp. 229–246, Dr W. Junk, The Hague.

Kahlke, H.D. (1994) *Die Eiszeit*, Urania Verlag, Leipzig.

King, J.E. and Saunders, J.J. (1984) Environmental insularity and the extinction of the American mastodont, in *Quaternary Extinctions: a prehistoric revolution*, (eds P.S. Martin and R.G. Klein), pp. 315–339, University of Arizona Press, Tucson.

Kinsella, T. (1969) *The Tain*, Oxford University Press, Oxford.

Klaus, W. (1972) Spätglazial-Probleme der östlichen Nordalpen Salzburgs – Inneralpines Wiener Becken. *Berichte der Deutschen Botanischen Gesellschaft*, **85**, 83–92.

Klein, R.G. (1974) Ice-Age hunters of the Ukraine. *Scientific American*, **230**, 96–105.

Klein, R.G. (1984) Mammalian extinctions and Stone Age people in Africa, in *Quaternary Extinctions: a prehistoric revolution*, (eds P.S. Martin and R.G. Klein), pp. 553–573, University of Arizona Press, Tucson.

Knapp, H.D. (1979/1980) Geobotanische Studien an Waldgrenzstandorten des Hercynischen Florengebietes. Vol. I, *Flora*, **168**, 276–319. Vol. II, *Flora*, **168**, 468–510. Vol. III, *Flora*, **169**, 177–215.

Knörzer, K.H. (1971) Urgeschichtler Unkräuter im Rheinland: ein Beitrage zur Entsehungsgeschichte der Segetalgesellschaften. *Vegetatio*, **23**, 89–11.

Knörzer, K.H. (1975) Entstehung und Entwicklung der Grülandvegetation im Rheinland. *Decheniana*, **127**, 195–214.

Kolstrup, E. (1991) Palaeoenvironmental developments during the Late Glacial of the Weichselien, in *The Late Glacial in North-west Europe: human adaptation and environmental change at the end of the Pleistocene*, (eds N. Barton, A.J.

Roberts and D.A. Roe), pp. 91–96, CBA Research Report No. 77, Council for British Archaeology, London.

Kooistra, L.I. (1994) Landbouw in een onbedijkt rivierengebied, in *Romeinen, Friezen en Franken in het Hart van Nederland: Van Traiectum tot Dorestad 50 v.C.–900 n.C.*, (eds W.A. Van Es and W.A.M. Hessing), pp. 126–129, Rijksdienst voor het Oudheidkundig Bodemonderzoek, Amersfoort.

Kosina, R. (1978) The cultivated and wild plants from the XIth century granaries on the cathedral-island in Wroclaw. *Berichte der Deutschen Botanischen Gesellschaft*, **91**, 121–128.

Kral, F. (1972) Zur Vegetationsgeschichte der Hohenstufen im Dachsteingebiet. *Berichte der Deutschen Botanischen Gesellschaft*, **85**, 137–152.

Kurtén, B. (1972) *The Ice Age*, Rupert Hart-Davis, London.

Lange, E. (1976) Zur Entwicklung der natürlichen und anthropgenen Vegetation in frügeschichtliger Zeit. *Feddes Repertorium*, **87**, 5–30.

Larsson, L. (1991) The Late Palaeolithic in southern Sweden: investigations in a marginal region, in *The Late Glacial in North-west Europe: human adaptation and environmental change at the end of the Pleistocene*, (eds N. Barton, A.J. Roberts and D.A. Roe), pp. 122–127, CBA Research Report 77, Council for British Archaeology, London.

Lewis, J. (1991) A Late Glacial and early Postglacial site at Three Ways Wharf, Uxbridge, England: interim report, in *The Late Glacial in North-west Europe: human adaptation and environmental change at the end of the Pleistocene*, (eds N. Barton, A.J. Roberts and D.A. Roe), pp. 246–255, CBA Research Report 77, Council for British Archaeology, London.

Lister, A.M. (1991) Late Glacial mammoths in Britain, in *The Late Glacial in North-west Europe: human adaptation and environmental change at the end of the Pleistocene*, (eds N. Barton, A.J. Roberts and D.A. Roe), pp. 51–59, CBA Research Report 77, Council for British Archaeology, London.

Loth, P.E. and Prins, H.H.T. (1986) Spatial patterns of the landscape and vegetation of Lake Manyara National Park. *ITC Journal*, **1986-2**, 115–130.

Louwe Kooijmans, L.P. (1980) De lage landen toen: Prehistorische bewoning van onze kuststreken, in *Voltooid Verleden Tijd?: Een hedendaagse kijk op de prehistorie*, (eds M. Chamaulan and H.T. Waterbolk), pp. 21–46, Intermediair, Amsterdam.

Mai, D.H. (1965) Der Florenwechsel im jüngeren Tertiär Mitteleuropas. *Feddes Repertorium*, **70**, 157–169.

Marks, P.L. (1983) On the origin of the field plants of the northeastern United States. *American Naturalist*, **122**, 210–228.

Moe, D. and Rackham, O. (1992) Pollarding as a possible explanation of the Neolithic elmfall. *Vegetation History and Archaeobotany*, **1**, 63–68.

Mol, D. and van Essen, H. (1992) *De Mammoet: Sporen uit de IJstijd*, BZZTôH, The Hague.

Moore, P.D. (1987) Chalk grasslands in the ice age. *Nature*, **329**, 388–389.

Oldfield, F. (1967) The paleoecology of an early neolithic waterlogged site in north-western England. *Review of Palaeobotany and Palynology*, **4**, 67–70.

Opravil, E. (1978) Synanthrope Pflanzengesellschaften aus der Burgwallzeit (8.–10. Jh.) in der Tschechoslowakei. *Berichte der Deutschen Botanischen Gesellschaft*, **91**, 97–106.

Pastor, J., Dewey, B., Naiman, R.J. *et al.* (1993) Moose browsing and soil fertility in the boreal forests of Isle Royale National Park. *Ecology*, **74**, 467–480.

Peglar, S.M. (1993) The mid-Holocene *Ulmus* decline at Diss Mere, Norfolk, UK: a year-by-year pollen stratigraphy from annual laminations. *The Holocene*, **3**, 1–13.

Peglar, S.M. and Birks, H.J.B. (1993) The mid-Holocene *Ulmus* fall at Diss Mere, South-East England – disease and human impact? *Vegetation History and Archaeobotany*, **2**, 61–68.

Pennington, W. (1969) *The History of the British Vegetation*, English University Press, London.

Perry, I. and Moore, P.D. (1987) Dutch elm disease as an analogue of Neolithic elm decline. *Nature (Lond.)*, **326**, 72–73.

Perschke, P. (1972) Die Vegetationsentwicklung im Waldviertel Niederösterreichs. *Berichte der Deutschen Botanischen Gesellschaft*, **85**, 129–136.

Pignatti, E. and Pignatti, S. (1975) Syntaxonomy of the *Sesleria varia* grasslands of the calcareous Alps. *Vegetatio*, **30**, 5–14.

Pounds, N.J.G. (1974) *An Economic History of Medieval Europe*, Longman, London.

Prins, H.H.T. (1996) *Ecology and Behaviour of the African Buffalo: social inequality and decision making*, Chapman & Hall, London.

Prins, H.H.T. and Van der Jeugd, H.P. (1993) Herbivore population crashes and woodland structure in East Africa. *Journal of Ecology*, **81**, 305–314.

Prins, H.H.T. and Reitsma, J.R. (1989) Mammalian biomass in an African equatorial rain forest. *Journal of Animal Ecology*, **58**, 851–861.

Rackham, O. (1980) *Ancient Woodland: its history, vegetation and uses in England*, Edward Arnold, London.

Ralska-Jasiewiczowa, M. (1972) Remarks on the Late-glacial and Holocene history of vegetation in the eastern part of the Polish Carpathians. *Berichte der Deutschen Botanischen Gesellschaft*, **85**, 101–112.

Reid, C. and Chandler, E.M. (1915) The Pliocene floras of the Dutch–Prussian border. *Mededelingen Rijksopsporingsdienst van Delfstoffen*, **6**.

Reid, C. and Chandler, E.M. (1933) *The London Clay Flora*, British Museum, London.

Reuter, M. (1920) Die Waldweide. *Allgemeine Forst- und Jagdzeitung*, **96**, 40–49.

Roebroeks, W. (1990) *Oermensen in Nederland: De archaeologie van de Oude Steentijd*, Meulenhoff, Amsterdam.

Schmeidl, H. (1972) Zur spät- und postglazialen Vegetationsgeschichte am Nordrand der bayerischen Voralpen. *Berichte der Deutschen Botanischen Gesellschaft*, **85**, 79–82.

Schmidt, G. (1969) *Vegetationgeographie auf ökologisch-soziologischer Grundlage*, Teubner Verlag, Leipzig.

Sercelj, A. (1972) Verschiebung und Inversion der postglazialen Waldphasen am südostlichen Rand der Alpen. *Berichte der Deutschen Botanischen Gesellschaft*, **85**, 123–128.

Shepherd, W.G. (transl.) (1983) *Horace: The complete Odes and Epodes with the Centennial Hymne*, Penguin, Harmondsworth.

Shimwell, D.W. (1971) Festuco-Brometea Br. Bl. and R. TX, 1943 in the British Isles: the phytogeography and phytosociology of limestone grasslands. *Vegetatio*, **23**, 1–28.

Simmons, I.G. (1981) Culture and environment, in *The Environment in British Prehistory*, (eds I.G. Simmons and M.J. Tooley), pp. 282–291, Duckworth, London.

Simmons, I.G., Dimbley, G.W. and Grigson, C. (1981) The Mesolithic, in *The Environment in British Prehistory*, (eds I.G. Simmons and M.J. Tooley), pp. 82–124, Duckworth, London.

Slicher van Bath, B.H. (1976) *Agrarische Geschiedenis van West-Europa, 500–1850*, Spectrum, Utrecht.

Smith, A.G., Grigson, C., Hillman, G. and Tooley, M.J. (1981) The Neolithic, in *The Environment in British Prehistory*, (eds I.G. Simmons and M.J. Tooley), pp. 125–209, Duckworth, London.

Smith, R.E.F. (1959) *The Origin of Farming in Russia*, Mouton, Paris.

Späth, V. (1988) Zur Hochwassertoleranz von Auewaldbäumen. *Natur und Landschaft*, **63**, 312–315.

Stott, P.A. (1971) A Mesobrometum referable to the subassociation *Mesobrometum-seslerio–Polygaletosum* Tx. described for the Somme-valley. *Vegetatio*, **23**, 61–70.

Street, M. (1991) Bedburg-Königshoven: a Pre-Boreal Mesolithic site in the Lower Rhineland, Germany, in *The Late Glacial in North-west Europe: human adaptation and environmental change at the end of the Pleistocene*, (eds N. Barton, A.J. Roberts and D.A. Roe), pp. 256–270, CBA Research Report 77, Council for British Archaeology, London.

Stuart, A.J. (1991) Mammalian extinctions in the Late Pleistocene of northern Eurasia and North America. *Biological Review*, **66**, 453–562.

Stuart, A.J. and Gibbard, P.L. (1986) Pleistocene occurrence of hippopotamus in Britain. With appendix: Pollen analysis from hippopotamus bones from the Cromer Forest Bed series. *Quartärpaläontologie*, **6**, 209–218.

Ten Cate, C.L. (1972) *Wann God Mast gibt: Bilder aus der Geschichte der Schweinenzucht im Walde*, Centre for Agricultural Publishing and Documentation, Wageningen.

Ter Kuile, E.H. (1976) *De Tijd van Rome's laatste Keizers: Veertien brieven van Appollinaris Sidonius uit de jaren 455–475*, De Walburg Pers, Zutphen.

Tinsley, H.M. and Grigson, C. (1981) The Bronze Age, in *The Environment in British Prehistory*, (eds I.G. Simmons and M.J. Tooley), pp. 210–249, Duckworth, London.

Tipping, R. (1991) Climatic change in Scotland during the Devensian Late Glacial: the palynological record, in *The Late Glacial in North-west Europe: human adaptation and environmental change at the end of the Pleistocene*, (eds N. Barton, A.J. Roberts and D.A. Roe), pp. 7–21, CBA Research Report 77, Council for British Archaeology, London.

Turner, C. (1975) Der Einfluss grosser Mammalier auf die interglaziale Vegetation. *Quartärpaläontologie*, **1**, 13–19.

Turner, J. (1981) The Iron Age, in *The Environment in British Prehistory*, (eds I.G. Simmons and M.J. Tooley), pp. 250–281, Duckworth, London.

Tüxen, R. (1974) Synchronologie einzelner Vegetationseinheiten in Europa, in *Vegetation Dynamics*, (ed. R. Knapp), Handbooks of Vegetation Science, Vol. VIII, pp. 267–292, Dr W. Junk, The Hague.

Van den Hoek Ostende (1990) *Tegelen, ons Land 2 Miljoen Jaar geleden*, Teylers Museum, Haarlem.

Van Es, W.A. (1972) *De Romeinen in Nederland*, 2nd edn, Fibula-van Dishoeck, Bussum.

Van Es, W.A. (1994) De Romeinse Vrede, in *Romeinen, Friezen en Franken in het Hart van Nederland: Van Traiectum tot Dorestad 50 v.C.–900 n.C.*, (eds W.A. Van Es and W.A.M. Hessing), pp. 48–63, Rijksdienst voor het Oudheidkundig Bodemonderzoek, Amersfoort.

Van Wieren, S.E. (1995) The potential role of large herbivores in nature conservation and extensive land use in Europe. *Biological Journal of the Linnean Society*, **56** Suppl. A, 11–23.

Van Zeist, W. (1959) Studies on the Post-Boreal vegetational history of south-eastern Drenthe (Netherlands). *Acta Botanica Neerlandica*, 8, 156–185.

Van Zeist, W. (1974) Palaeobotanical studies of settlement sites in the coastal area of The Netherlands. *Palaeohistoria*, **14**, 223–371.

Van Zeist, W. (1980) Prehistorische cultuurplanten, onstaan, verspreiding, verbouw, in *Voltooid Verleden Tijd?: Een hedendaagse kijk op de prehistorie*, (eds M. Chamaulan and H.T. Waterbolk), pp. 147–165, Intermediair, Amsterdam.

Van Zeist, W., Van Hoorn, T.C., Bottema, S. and Woldring, H. (1976) An agricultural experiment in the unprotected salt marsh. *Palaeohistoria*, **18**, 11–153.

Vera, F.W.M. (1997) Metaforen voor de Wildernis: Eik, hazelaar, rund en paard. Doctoral thesis, Wageningen Agricultural University, Wageningen.

Vicherek, J. (1973) *Die Pflanzengesellschaften der Halophyten- und Subhalophytenvegetation der Tschechoslowakei*, Vegetace CSSR, A5, Academia, Praha.

Von Denffer, D., Schumacher, W., Mägdefrau, K. and Ehrendorffer, F. (1976) *Strasburger's Textbook of Botany: New English edition*, Longman, London.

WallisDeVries, M.F. (1994) Foraging in a landscape mosaic: diet selection and performance of free-ranging cattle in heathland and riverine grassland. Doctoral thesis, Wageningen Agricultural University, Wageningen, The Netherlands.

Walter, H. (1927) *Einführung in die allgemeine Pflanzengeographie Deutschlands*, Gustav Fisher Verlag, Jena.

Walter, H. (1968) *Die Vegetation der Erde in öko-physiologische Betrachtung, Vol. II: Die gemäsigten und arktischen Zonen*, Gustav Fisher Verlag, Jena.

Waterbolk, H.T. (1954) *De Praehistorische Mens en zijn Milieu: Een palynologisch onderzoek naar de menselijke invloed op de plantengroei van de diluviale gronden in Nederland*, Van Gorcum, Assen.

Waterbolk, H.T. (1985) Archaeologie, in *Geschiedenis van Drenthe*, (eds J. Heringa, D.P. Blok, M.G. Buist and H.T. Waterbolk), pp. 15–90, Tjeenk Willink, Groningen.

Waterbolk, T.H. and Boersma, J.W. (1976) Bewoning in vøør- en vroeghistorische tijd, in *Historie van Groningen Stad en Land*, (eds W.J. Formsma, M.G. Buist, W.R.H. Koops *et al.*), pp. 13–74, Tjeenk Willink and Bouma's Boekhuis, Groningen.

Webb, S.D. (1977) A history of savanna vertebrates in the new world. Part I. North America. *Annual Review of Ecology and Systematics*, 8, 355–380.

Webb, S.D. (1978) A history of savanna vertebrates in the new world. Part II. South America. *Annual Review of Ecology and Systematics*, 9, 393–426.

Wegmüller, S. (1972) Neuere palynologische Ergebnisse aus den Westalpen. *Berichte der Deutschen Botanischen Gesellschaft*, 85, 75–78.

Welten, M. (1972) Das Spätglazial im nördlichen Voralpengebiet der Schweitz: Veelauf, floristisches, chronologisches. *Berichte der Deutschen Botanischen Gesellschaft*, 85, 69–74.

Westhoff, V. and Den Held, A.J. (1969) *Plantengemeenschappen in Nederland*, W.J. Thieme, Zutphen.

Westhoff, V. and Van Leeuwen, C.G. (1966) Ökologische und systematische Beziehungen zwischen natürlichen und anthropogene Vegetation, in *Anthropogene Vegetation*, (ed. R. Tüxen), pp. 156–172, Dr W. Junk, The Hague.

Willerding, U. (1978) Paläo-ethnobotanische Befunde an mittelalterlichen Pflanzenresten aus Süd-Niedersachsen und dem östlichen Westfalen. *Berichte der Deutschen Botanischen Gesellschaft*, **91**, 129–160.

Zagwijn, W.H. (1960) Aspects of the Pliocene and early Pleistocene vegetation in The Netherlands, 2 vols. Doctoral thesis, University of Leyden, Leijden.

Zagwijn, W.H. (1971) Vegetational history of the coastal dunes in the western Netherlands. *Acta Botanica Neerlandica*, **20**, 174–182.

Zlotin, R.I. and Khodashova, K.S. (1980) *The Role of Animals in Biological Cycling of Forest-Steppe Ecosystems*, (English edn, ed. N.R. French), Dowden, Hutchinson and Ross, Stroudsburg.

4

Effects of human interference on the landscape with special reference to the role of grazing livestock

Richard Pott
University of Hannover, Institute of Geobotany, Nienburgerstrasse 17, D-30167 Hannover, Germany

4.1 INTRODUCTION

Today, European landscapes are typically composed of an anthropogenic mosaic of forests and open spaces, grasslands and fields. Almost everywhere, humans have altered the natural landscape (Chapters 2 and 3). Grazing by livestock is one of the main instruments that forged the traditional pre-industrial landscape. This chapter therefore sets out to elucidate the impact of grazing on the development of semi-natural landscape patterns in prehistoric and historic times. In doing so, the emphasis is placed on the role of domestic herbivores in the landscape as illustrated by the *Hudewälder*, i.e. pasture woodlands or wood-pastures.

In central and northwestern Europe no other cultivation method has had such decisive and long-lasting effects as the grazing of domestic animals in forests (*Waldhude*). Historical developments of the *Waldhude* and the patterns and processes of *Hudewald*-communities are important in the context of grazing management and nature conservation because they reflect the magnitude and diversity of anthropozoogenic impact on woodland ecosystems. These grazing systems are disappearing rapidly, as they are part of low-intensity farming systems, but fine examples of European landscapes shaped by grazing livestock can still be found in, for example, northwest Germany (Pott and Hüppe, 1991), the New Forest in England (Putman, 1986; Tubbs, 1986), the island of Öland in Sweden (Rosén, 1982; Sjögren, 1988) with southern European counterparts in the Spanish *dehesa* of the Extremadura (Ruiz and Ruiz, 1986) and the Camargue in France (Duncan, 1992).

Grazing and Conservation Management. Edited by M.F. WallisDeVries, J.P. Bakker and S.E. Van Wieren. Published in 1998 by Kluwer Academic Publishers, Dordrecht. ISBN 0 412 47520 0.

Grazing livestock effected clearings in the primeval forests and over time contributed to the development of specific semi-natural vegetation types and additional qualitative changes in the remaining wooded areas. Consequences of these alterations are still to be observed in the remnant wooded commons of the modern cultural landscape; their persistence is dependent on the regeneration capacity of many characteristic pasture as well as forest species.

The wood-pasture landscape is often an important management goal for nature conservation because of its high species diversity. One of the intriguing ecological questions that needs to be addressed is how stable it is: can herbivores maintain grassland? Can forest regeneration take place in the presence of herbivores, and if so, will it eventually conquer all the open communities? What, finally, is the human role in the maintenance of this system? The following description of the role of livestock in shaping the cultural landscape of Europe may bring up some answers to these questions.

4.2 HISTORICAL DEVELOPMENT OF THE LANDSCAPE

The present vegetation of northwestern Europe is the result of a long interaction between natural processes and human influences. With the exception of treeless ombrotrophic mires, freshwater rivers and marshes, salt marshes and alpine regions, the Holocene landscape of Europe was once covered by more or less dense primeval forests. The modern landscape has developed from these forests as the area of original deciduous woodland decreased steadily with the rise of human agricultural and settling activities.

The onset of this development dates back some 5000–6000 years, to the first farming activities in the Neolithic era. Humans created a mosaic of semi-natural, half-open landscape structures from the primeval woods. These included pastures, meadows, hedges, scrubland, fields, weed communities, *Juniperus communis* infiltrated heathlands, sand dunes and dry grassland (see also Chapter 3). The transformation of the natural to the semi-natural landscape certainly resulted in an overall loss of species (mainly animal) but also led to a considerable diversification of vegetation patterns as well as an increase in species diversity of the flora at the local and regional scale (Pott, 1996).

Neolithic agriculture and farming with domestic animals demanded – for the first time – that local forest clearance by means of felling and burning took place in order to create, amongst others, cattle yards and fields for the cultivation of cereals. The crops were mainly reserved for human subsistence, whereas the source of nutrition for the cattle was the forest. This dual system of 'wood–field' is not merely characteristic for prehistoric times. Pollen analytical work shows that the wood-pasture began in Mesolithic and early Neolithic times and became widespread in the Middle Ages (twelfth century). Pollen diagrams (Figure 4.1) reflect the first woodland clearances and also the development of grasslands from wood-pasture (Pott, 1995a). Forest grazing during the Neolithic did not differ by much from medieval times. It lasted – in the form of common land and enclosed land – throughout the Middle Ages up to modern times. Only when common land was parcelled out and privatized during the nineteenth century did this system finally come to an end.

The forest meant something completely different to early humans than it does to us today. The utilization of the forest was a general one. Besides providing wood for various purposes, the primary and secondary forest also served as cattle pasture and browse supply through pollarding (Figure 4.2) and shrouding; 'plaggen' (cut sods), litter and manure were also gathered to fertilize the arable fields. In upland regions the woods were even used for rotational cultivation with fire. Amongst all the conventional utilization methods, the wood-pasture is the most ancient and the one which, over the centuries, has had the greatest impact of the landscape. It was the principal form of woodland utilization in nearly all regions and, together with its application to the fattening of pigs by the use of mast, it is mentioned in nearly all written documents as an important and obvious part of woodland management (Pott and Hüppe, 1991; Vera, 1997).

With some exceptions, the methods of traditional woodland utilization led not only to changes and rearrangements of the botanical composition

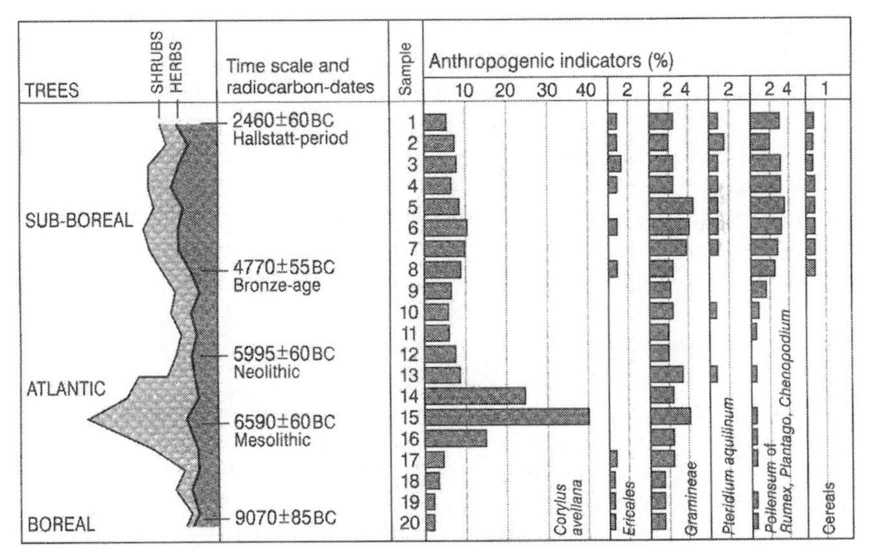

Figure 4.1 Mesolithic and Neolithic forest clearances and the prehistoric exploitation of the land are shown in this pollen diagram from Westfalia, Östinghausen, Germany (from Burrichter and Pott, 1987). Changes in the pollen sum of trees are accompanied by rises in pollen indicative of human activity (e.g. grasses, cereals, *Plantago*, *Rumex* and *Pteridium*).

Figure 4.2 Pollarded mast-bearing oak in the wood-pasture of Borkener Paradies. The main shoot was cut off at a height of about 2 m.

but also to the gradual degradation of the woodland itself. The effects were especially dramatic in the common lands. Prior to the privatization of the commons and the enclosure of land, all that remained of the former woodland, in many regions, was shrubland surrounded by vast tracts of pasture land and heaths.

Other factors governed the private forests of former times, which belonged mainly to the aristocracy or to ecclesiastical institutions. They were either inherited property from the outset or were privatized in the thirteenth and fourteenth centuries or in the sixteenth century. The latter type was parcelled out from the common land on individual estates and various conditions pertaining to trees were imposed, often in the interest of hunting (e.g. in game parks). Depending on the relative rights of possession, all questions concerning rights of usage were decided either by the landowner or by the commoners themselves. Extensive exploitation was thus moderated and the threatening degradation of the forest therefore prevented. Several such wooded areas still exist in Europe, such as the Borkener Paradies, the Bentheimer and Neuenburger Forests in Germany (Pott and Hüppe, 1991), the Bialowieza Forest in Poland (Falinski and Falinska, 1986) and the New Forest (Tubbs, 1986), Epping Forest and forest fragments of southeast Essex in England (Rackham, 1980). Here, many old trees still bear the imprint of conventional wood utilization methods. Their number is constantly decreasing through overmaturity and because land use practices have changed; the time thus draws near when these relicts of natural and cultural history will vanish completely.

4.3 THE ROLE OF HERBIVORY AND THE DEVELOPMENT OF *WALDHUDE* COMMUNITIES

The first clearings and grasslands developed inevitably from the wood-pasture. Cropping, wood cutting and bush clearing by humans and the permanent browsing pressure on the saplings by domestic animals led to a gradual clearing of the wood, even on rich loess and loam soils with *Fagus sylvatica* (beech) or *Quercus petraea* and *Q. robur* (oak) and *Carpinus betulus* (hornbeam) woods. It finally resulted in wood-pasture vegetation in which the open pasture, together with *Corylus avellana* (hazel) bush or heather, expanded or contracted according to the local intensity of grazing.

The effects of grazing in woodlands were especially strong if, in addition to heavy livestock like cattle and horses, smaller animals such as sheep and goats were grazed on the same ground. The grazing of goats had the most devastating consequences for many valued plant species and had therefore already been forbidden by the beginning of modern times. In the woods of central Europe the principal land use, in addition to pasture, was the provision of mast (principally acorns and beech nuts) for the fattening of pigs.

Even today remnant malformations and growth forms of trees bear evidence of their former management. The characteristic broad crowns of solitary growth forms show them to have originally been free-standing trees. Such mast-bearing trees are often more than 500 years old. Their origin can be traced back mostly to the late Middle Ages (thirteenth and fourteenth centuries) and sometimes even as far back as the sixth century. Their development and survival is only explicable if one assumes the former existence of protective shrubs or artificial fences, as was often the case on former common land.

Of crucial importance for the survival of established trees subjected to browsing is their capacity to resist it and to regenerate – that is, the ability to sprout from lateral buds after damage. Yet it is nearly impossible for young trees to establish successfully on an open pasture: they die as a consequence of browsing and cropping while still at the seedling stage. Therefore young trees can only develop when protected by shrubs. The extreme consequence of selective grazing and browsing was that the pasture consisted almost exclusively of barbed, thorny and spiculate or hard, poisonous or strongly aromatic plant species, these being normally disdained by cattle and only eaten in times of food shortage. The thorny bushes then had a constructive effect on vegetation dynamics because, as they protected naturally growing young trees, they were the pioneers of forest regeneration (Figure 4.3).

Figure 4.3 Woodland regeneration complex in a wood-pasture landscape: in the background, the regenerating oakwood; in the foreground, *Prunus spinosa* scrub, invading the pasture by root suckers.

Establishment within a mature shrub complex provides the young trees with optimal protection and allows the development of a normal tree shape. Bushy growth forms are different. The tree germinates only under the protection of initial shrubbery and grows simultaneously with the spiny shrubs (Figure 4.4). Top shoots protruding from the shrub are permanently cropped by animals. The release of apical dominance stimulates the sprouting of lateral buds and leads to the formation of dense bushy ramifications directly from the main stem. The browsed growth forms of oak and beech develop a shrubby and nearly conical habit. They are traditionally referred to as 'cow bushes'. Only when the young tree grows tall or broad enough to be beyond reach of cattle does it carry on growing without hindrance.

Pollarded forms of beech and oak (Figure 4.2) are principally connected with their former utilization as a source of mast (Rackham, 1980; Peterken, 1981). The tree was cut off at a height of 2–3.5 m and thus developed a broad spreading crown yielding a rich and early mast (Pott, 1982).

In many former pasture areas one can find trees with several stems. In beeches the separate trees either form one joint stem or they remain as single trees in a dense, tuft-like group. The explanation for these strange growth forms is that the young trees were planted in bundles of usually seven individuals in one planting hole (Pott and Hüppe, 1991). Flower

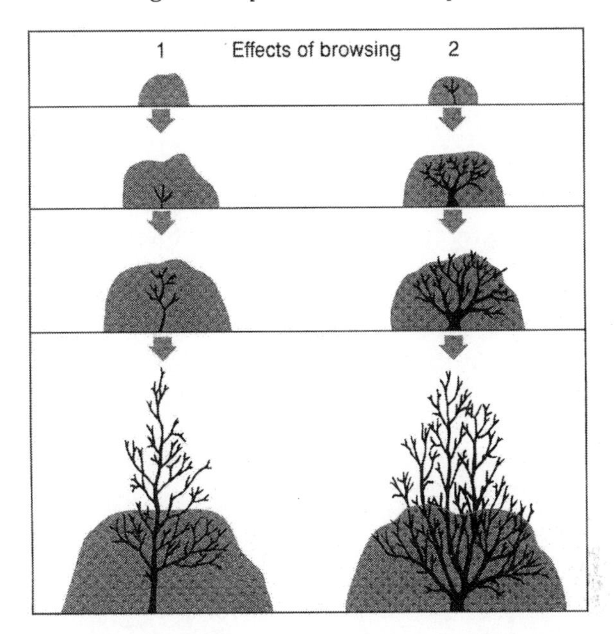

Figure 4.4 Tree forms in wood-pasture landscapes: (1) development of solitary tree completely protected by *Prunus spinosa* scrub against browsing by cattle; (2) development of a bushy growth form: the young shoots are sheared and grow up together with the protective *Prunus spinosa* shrub.

(1980) recorded that the ground around the planted seeds or seedlings was then sown with spiny scrub species – *Crataegus* sp., *Ilex aquifolium*, *Prunus spinosa* or *Rosa* div. spec. – for protection. This system increased the success of establishment and reduced the need for special devices to ward off grazing animals, and the subsequent formation of an overdeveloped crown gave increased mast production.

Mast utilization in woods was advantageous for the maintenance and regeneration of the woodland in two ways: firstly, the direct anthropogenic promotion or planting of oak and beech as mast-bearing trees; and secondly, the loosening of the soil by the grubbing activities of pigs. This created appropriate germination conditions for young trees and assured a permanent regeneration, given controlled cattle grazing.

4.4 VEGETATION PATTERNS AND PROCESSES IN WOOD-PASTURE LANDSCAPES

The vegetation and appearance of wood-pasture areas are shaped by natural variation in habitat conditions in which the respective woodland communities may be regarded as the foundation for the pasture vegetation.

Since, in addition, the impact of grazing shows not only temporal but also spatial variation, there is no uniform pasture type but there certainly are overall similarities in vegetation structure between ecotopes. Grazing- or browsing-tolerant species predominate as zoogenic developments in zonally arranged vegetation units such as drifts, grassland, tall herbaceous fringe communities, scrub and forest remnants (Figure 4.5). The wood's degree of resistance to browsing, conditioned by habitat factors and species composition as well as its regeneration capacity, can be regarded here as the decisive factor which, in interaction with grazing livestock, gives rise to the clusters of degradation and regeneration in the vegetation of wood-pasture areas.

The effects of these degenerative and regenerative processes on community composition is still evident in many forests today (see references in Burrichter *et al.*, 1980; Pott, 1993). Inevitably those species with a stronger regeneration capacity increase at the expense of those having a weaker one.

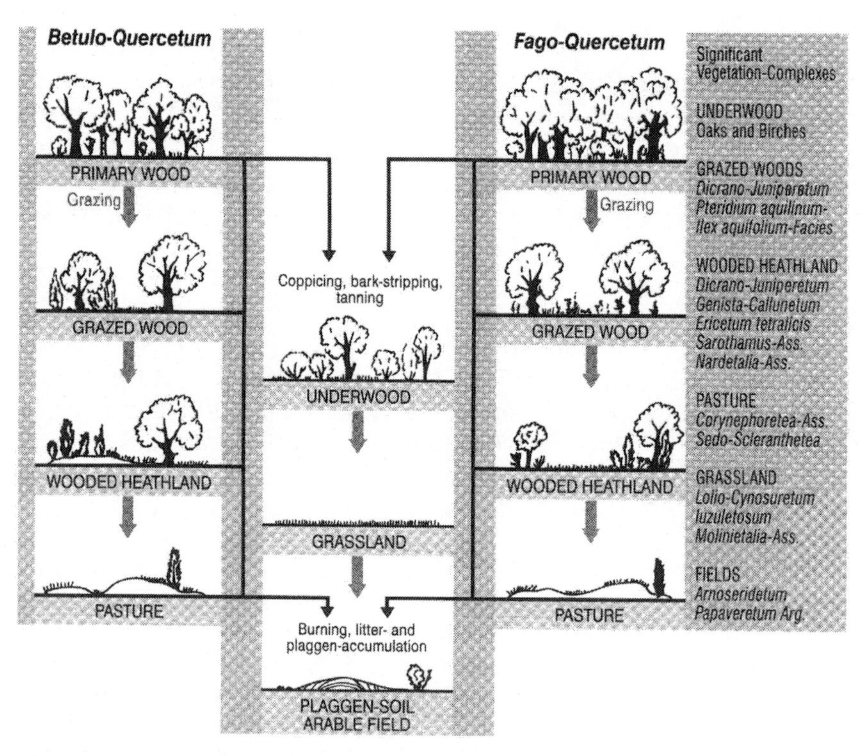

Figure 4.5 Degradation series of lowland woodland communities and their dependent anthropogenic vegetation complexes in northwest European landscapes of the *Quercion roboris* (e.g. beech–oak woods on Pleistocene sandy soils). As effects of grazing, there are zonally arranged vegetation complexes consisting of grassland, tall-herb fringes, scrub and wooded heathland. As anthropozoogenic formations, they consist of positively selected grazing-tolerant plant species.

The selectivity in grazing also has a positive effect on species that are somehow protected against animals. These include, on the one hand, barbed species and, on the other, plants which are either unpalatable, indigestible or even poisonous for the animals. To the first category belong woody species bearing thorns and prickles, such as *Crataegus laevigata*, *Crataegus monogyna*, *Rhamnus catharticus* and *Rosa* spp. *Ilex aquifolium* and *Juniperus communis* are also avoided by cattle because of their spiculate leaves.

Some woodlands have a very low resistance to grazing and extensive utilization, especially the thermophilic mixed *Pinus sylvestris* (pine) and oak woods on calcareous rendzina soils (section 4.4.2b). In the face of overexploitation they are soon replaced by plant communities of mostly dry grassland. Most of the present mesoxerophytic meadows and rough grassland xerothermic plant species owe their existence to extensive grazing and also, depending on the original forest community, their characteristic species combinations.

In northwestern Europe many formerly glaciated regions were changed into heathland. One can recognize the development of an anthropozoogenic vegetation complex with wood-pasture remnants, *Juniperus communis* or *Sarothamnus scoparius* heaths derived from the *Genisto-Callunetum* (Figure 4.5). The permanent overexploitation of the land, partly by the grazing of sheep, degraded the heathlands of early common lands into extensive sand drifts and dune formations (Pott and Hüppe, 1991; Hüppe, 1993) (Figure 4.6).

Human wood-pasture practices over the centuries have thus both diversified and also degraded the vegetation patterns of the European landscape. In doing so this type of land use has constituted a most influential factor in the shaping of the modern landscape. The following section presents an overview of the characteristic vegetation patterns of the wood-pasture landscape and its derivatives in northwestern and central Europe.

4.4.1 Wood-pasture landscapes in the northwestern European lowland

(a) Quercion roboris *landscapes*

Vegetation patterns in the areas of the *Betulo-Quercetum roboris* and the *Periclymeno-Fagetum* (= *Fago-Quercetum petraeae*) are determined to a large extent by soil conditions. Additional variation has been created by grazing influences which can still be clearly distinguished (Figure 4.5).

Various types of grazed beech–oak woods, in which the beech dominates, show a significant cover of *Pteridium aquilinium* in the herb layer. *Pteridium* is avoided by cattle because it is poisonous and untasty and, as a polycormous plant, it can spread rapidly after wood clearance burning and pasture exploitation (Wilmanns *et al.*, 1979; Schwabe-Braun, 1980; Pott, 1982, 1986). Thickets of *Ilex aquifolium* underwood in Atlantic and Sub-Atlantic regions with mild winters are widespread in woodland regions of the

Figure 4.6 Sand dunes with various successional stages of *Corynephoretea* vegetation near Meppen, Germany.

Quercion roboris, *Luzulo-Fagion*, *Carpinion* and *Eu-Fagion*, respectively (Pott, 1982, 1983, 1990, 1995b; Pott and Burrichter, 1983). When used as pasture, beech–oak woods eventually change to heathland, causing a process of anthropozoogenic vegetation differentiation with residual woodland, *Juniperus* and *Sarothamnus* heaths from the *Genisto-Callunetum*. *Sarothamnus* very often indicates the existence of an original *Periclymeno-Fagetum* area. On recent clearings this nitrogen-fixating species forms dense complexes with heather – bushes that are the precursors of tall-herb fringes and scrub communities. Narrow *Teucrium scorodonia* edge communities with acidophytic accompanying species (*Hieracium laevigatum*, *Holcus mollis*, *Agrostis tenuis*) adjoin open cattle tracks.

In the area of grazed *Betulo-Quercetum*, vast *Calluna vulgaris* heaths and regeneration stages with birch, pollarded and coppiced woods, as well as wood remnants, alternate with each other. Residual patches of formerly widespread *Spergulo-Corynephoretum* or *Diantho-Armerietum* associations and their successive stages still belong to the characteristic vegetation scenery of dry inland sand dunes. Lowland sites with poor drainage are characterized by ombrotrophic moorland with mires, *Erica tetralix* heaths and peat bog woodlands. The distribution of these ancient *Quercion roboris* landscapes is restricted to northwestern Europe (Tüxen, 1937, 1967).

The vegetation of Pleistocene sandy soils, together with that of adjacent lowlands and valleys (with potential *Pruno-Fraxinetum*, *Fraxino-Ulmetum* or *Carpinion betuli* forest communities), show a long history of agricultural land use (Preising, 1950; Buchwald, 1951, 1984; Horst, 1964; Burrichter, 1973; Gimingham, 1972; Gimingham and De Smidt, 1983). Since the early Neolithicum the richer sandy soils of the beech–oak woodland were used for settlements and agricultural fields; the poor areas of the oak–birch woodland were mainly left for traditional pasture and common land utilization. Overexploitation through the grazing of breeds of heathland sheep, like the Heidschnucke of northwestern Germany, and especially the permanent removal of plaggen, litter usage and burn cultivation, led to the deterioration of heathlands and the formation of sand dunes. The open sand is colonized and fixed by *Corynephorion canescens* communities. On somewhat richer sandy material a mosaic of different successional stages often develops with pioneer *Corynephorion* communities, *Thero-Airion* associations and the *Diantho-Armerietum* community (Burrichter *et al.*, 1980; Jeckel, 1984; Pott and Hüppe, 1991). These dry sandy soils also provide local refuges for submediterranean and subcontinental geobotanical elements such as *Dianthus deltoides*, *Galium verum*, *Euphorbia cyparissias*, *Veronica spicata*, *Armeria elongata*, *Vicia lathyroides* and *Ranunculus bulbosus*. These form an obvious floristic east–west gradient within the northwestern German vegetation. On dry and bleached, poor sandy soils the vegetation of the *Spergulo-Corynephoretum* type is replaced in the course of succession by *Nardus stricta* communities of the *Nardo-Juncetum squarrosi* with alternating heath fragments. Its characteristic species, *Nardus stricta* and *Juncus squarrosus*, are unpalatable to grazing animals.

Characteristic features of fixed humous sands in pasture land are the *Juniperus* populations, which can be classified as *Dicrano-Juniperetum* (Barkman, 1985) (Figure 4.7). With the gradual clearing of the woods the heliophilic and soil-indifferent *Juniperus communis* could spread time and again on the burnt or stripped plaggen soils. Due to its aromatic substances and prickly needles, *Juniperus* is avoided by cattle; its fruit is mainly eaten and dispersed by thrushes and pigeons. It can form dense bushes, in the protection of which oak and birch can establish. The *Juniperus* shrubs show an interesting variation in their individual habitus (Pilger, 1931; Tüxen, 1974) which can be categorized into three main growth types, with many transitions, after Barkman *et al.* (1977) and Barkman (1985). Due to the relatively rapid nitrogen mineralization of the *Juniperus* needles and the accumulation of animal dung, nitrophilic species such as *Urtica dioica*, *Sambucus nigra*, *Humulus lupulus* and *Galium aparine* often accompany the *Juniperus* bushes. However, the germination and growth of *Juniperus* is hindered on shifting sand dunes; it establishes mainly after the cessation of intensive grazing. This explains its sporadic former occurrence on heathland and its recent spreading on abandoned pasture land, also on loamy soils and chalk (Hüppe, 1995).

Figure 4.7 *Juniperus communis* heathland in the Bockolter Berge near Münster (Westfalia, Germany) as a characteristic vegetation complex after cessation of grazing on sandy soils. It originated from acidophytic mixed oakwoods.

(b) Carpinion betuli *landscapes*

The nutrient availability in the lowland pastures of river valleys is much higher than in the *Quercion* area. In northwestern Europe these areas were originally dominated by *Fraxino-Ulmetum* woodlands. In the richest parts *Carpinus betulus* and *Fraxinus excelsior* (ash) are present, so that sometimes these woodlands closely resemble the adjacent oak–hornbeam woodland (*Stellario-Carpinetum*), especially under moderate grazing. These woods owe their preservation to a superior regenerative capacity and to the favourable productive potential of these sites.

In the grazed *Carpinion* landscapes, *Prunus spinosa* plays a pioneer role in forest establishment due to its vegetative spreading into open pastureland from the bush zone (Figure 4.8). In the traditional pasture of the Borkener Paradies in the Emsland near Meppen, Germany, Burrichter *et al.* (1980) and Pott and Hüppe (1991) showed that the presence of young trees is restricted to those well defended bush complexes (*Corno-Prunetum*) with *Prunus spinosa*, *Crataegus laevigata*, *Rhamnus catharicus* and *Rosa canina*. Grazed *Carpinion* woodlands show a typical structure of irregularly distributed tree clusters and former mast-bearing oaks, interspersed by clearings, tall-herb fringe communities, nitrophilic shrub zones and pasture grasses.

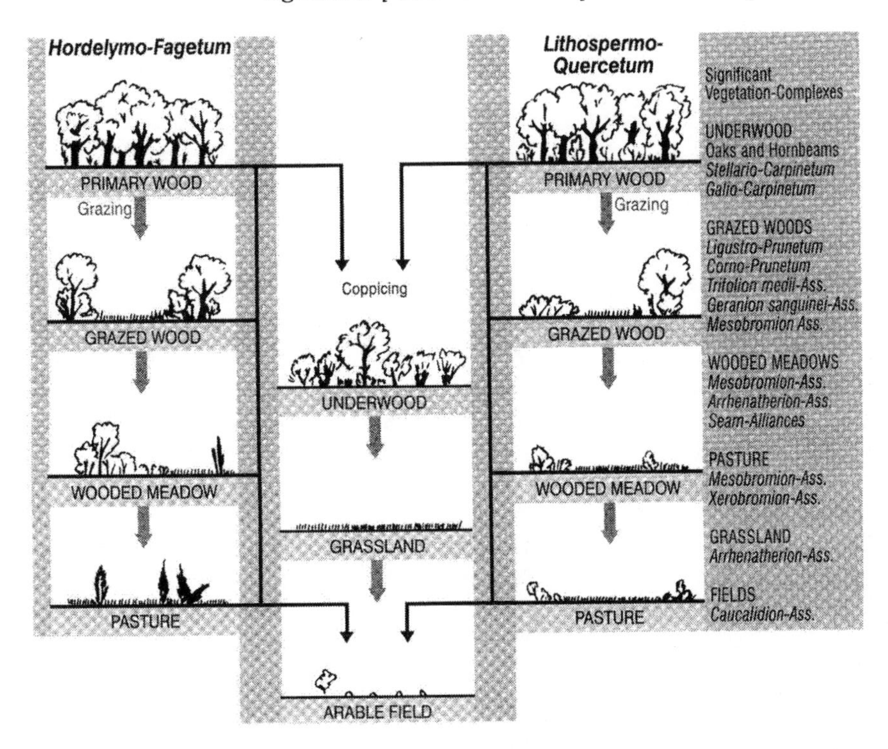

Figure 4.8 Degradation series of *Fagion sylvaticae* woodlands. Several vegetation complexes are derived from the beechwoods (*Hordelymo-Fagetum*): grassland, tall fringe communities, scrub and woodland.

4.4.2 Wood-pasture landscapes of northwestern and central European hills

The potential beech regions, especially those on lime and loess soils, were already the preferred settling areas for prehistoric men. The first traces of Neolithic agricultural settlements were recorded from the loess areas of central Europe by pollen analysis in the form of cereal pollen and dated at about 6500 BP (Pott, 1993). These settlements belong to the linear Pottery culture, which spread over the primeval woodlands, just after the immigration of the beech (Pott, 1992, 1993).

With respect to the distribution of beech, soil moisture is more important than soil type, as soil types where beech occurs range from slightly loamy sands to loess and calcareous loamy soils, corresponding to parabrown earths, brown soils and rendzinas. The southerly exposed lime soils of potential *Hordelymo-Fagetum* and *Carici-Fagetum* habitats also offer a suitable microclimate for submediterranean and continental species. The geobotanical gradient within the continental *Festucetalia vallesiacae* steppes and the more submediterranean *Xerobromion* and *Mesobromion* communities shows an extremely high species diversity of light-demanding, drought-tolerant and often grazing-resistant plants. Xerothermic vegetation units show an island-like distribution in the region of (mixed) beechwoods, which can be explained as a consequence of secondary wood clearing and land-use (Figure 4.9).

There is a geographical discontinuity in the distribution of this mixture of contrasting geobotanical elements. A relatively large proportion of thermophilic species has its furthest occurrence on extensively used pasture land on limestone in northwestern Germany (Pott, 1985; Garve, 1987; Raabe, 1987). Among the continental species, *Helianthemum nummularium* and *Pulsatilla vulgaris* reach as far as northwestern Germany, The Netherlands and Great Britain; the predominantly submediterranean *Xerobromion* species, however, such as *Hippocrepis comosa*, *Koeleria pyramidata* and *Orchis purpurea*, reach their northwestern limit further south.

(a) Fagion sylvaticae *landscapes*

The main characteristic vegetation complexes of wood-pasture within the potential beech-dominated *Hordelymo-Fagetum* and *Carici-Fagetum* woodlands are the *Gentiano-Koelerietum* pasture, the *Trifolio-Agrimonietum* herb zones and especially *Ligustro-Prunetum* bushes (Figures 4.5 and 4.8); in the absence of grazing a secondary progressive succession will lead back to the regeneration of woodland.

Figure 4.9 Grazed *Juniperus communis* heathland with mesoxerophytic grassland as replacement vegetation instead of calcareous beechwoods.

Pasture vegetation on these mostly calcareous soils consists mainly of mesoxerophytic grasslands and their contact communities (Pott, 1993, 1995b). Under the influence of extensive land use these expanded not only on the dry sun-exposed habitats but also on moister habitats, which normally have forest-bearing potential. Wherever connected areas were used as pastures, especially for sheep, this has led to the formation of vegetation mosaics of *Mesobromion* pasture in combination with *Juniperus communis* communities, tall-herb fringes and scrub communities, and wood islands. The same wood-pasture landscape develops as in the *Carpinion* landscape, similar in vegetation structure but different in species composition.

Besides grazing, summer drought also plays an important role in the formation of *Mesobromion* communities by lowering microbial activity in the soil and thereby reducing mineralization and plant productivity (Gigon, 1968; Hakes, 1987). As a mown meadow this grassland type is cut once or twice per year, which favours relatively tall species with outstanding regenerative power. A comparison between *Mesobromion* vegetation and fertilized meadows, which are mown twice or three times a year (e.g. Mahn, 1957), reveals a remarkably greater underground root mass on the poorer grassland as an adaptation to nutrient limitation and drought.

The *Onobrychido-Brometum* (often simply called *Mesobrometum*: Oberdorfer, 1978; Pott, 1995b) is the classic colourful community of mown meadows on unfertilized calcareous soils in the colline or montane regions. The extensively mown *Mesobromion* meadows of central Europe differ in their species composition from those of northwestern Europe by the dominance of species that readily regenerate after being mown but are sensitive to trampling (e.g. *Onobrychis viciifolia*, *Anthyllis vulneraria*, *Dianthus carthusianorum* and numerous orchids). When the *Bromus* communities are overexploited, the sward opens up and the topsoil is washed away; as a consequence, more xerophytic *Mesobrometum globularietosum* or *Xerobromion* communities develop. Such secondary anthropogenic communities are restricted to extremely small areas and can be clearly distinguished within the surrounding *Mesobrometum* vegetation. It is very doubtful whether they persist if the human influence is stopped.

The dominant species of the central European *Mesobrometum* is *Bromus erectus*, which also suffers from grazing and is obviously favoured by mowing. In northwestern Europe, however, *Bromus erectus* is absent from *Mesobrometum* sites, as these were occupied by grazed pasture communities (Bornkamm, 1960; Dierschke, 1985), belonging to the *Gentiano-Koelerietum*. In contrast to the mown grasslands, poisonous and unpalatable plants and rosette-forming species, as well as thorny species, are abundant in traditionally grazed sheep pastures. Besides *Juniperus communis*, which is found on almost every extensive pasture, thistles (*Carlina aucaulis*, *Cirsium acaule*), gentians (*Gentiana ciliata*, *G. germanica*) and *Euphorbia cyparissias* are the characteristic species of grazed *Mesobrometum* communities. As a pasture weed, *Brachypodium pinnatum* is the antagonist of *Bromus erectus*, particularly as it is able to spread vegetatively by means of far-reaching rhizomes, whereas *Bromus erectus* only propagates generatively and is preferentially grazed. Only in recent times, after the abandonment of the pastures, did *Bromus erectus* successfully extend its range. As a result, the species of formerly grazed or mown areas increasingly mix with each other. Together with *Brachypodium pinnatum*, *Bromus erectus* forms long-lasting populations, though they are quite often competitive. Such ungrazed areas do not become bushy for a certain time due to a massive litter production (see also Bornkamm, 1961, 1974; Kienzle, 1979, 1984; Schiefer, 1981; Stöcklein and Gisi, 1985). These fallow meadows are poor in species compared with the former *Mesobromion* communities (Wilmanns and Kratochwil, 1983; Kienzle, 1984). Thorough grazing or mowing may push such fallow states back again to a point where bush establishment can reoccur and the landscape again takes on its typical pasture appearance.

The bush community of the *Ligustro-Prunetum* often forms a characteristic mosaic with the open pasture land due to the invasive properties of *Prunus spinosa*, which can quickly develop into pioneer bushes with equally thorny *Rosa* and *Crataegus* species. Under a light grazing pressure the development of tall herbs from *Trifolion medii* (*Trifolio-Geranietea*) fringe communities will lead to an increase in the diversity. Such biotopes are refuges for numerous endangered species (especially pollinating insects) because of their ecotone structure and species composition (Kratochwil, 1983; Wilmanns and Kratochwil, 1983). However, such communities, especially the *Trifolio-Agrimonietum*, can also get rapidly overgrown by encroaching bush. The *Corylo-Rosetum* scrub from central Europe is comparable to the *Ligustro-Prunetum*, but it is an exclusively anthropogenic community (Witschel, 1980). Clearly, these vegetation types are intimately connected with each other in space and time and correspond to different levels and modes of exploitation (Zoller, 1954; Bornkamm, 1958, 1960; Zoller *et al.*, 1984).

The characteristic species composition can only be maintained when the pasture remains unfertilized. Over the past decades, many traditional grassland areas have been fertilized to enhance their productivity. Numerous nitrophilic plant species have invaded those areas, gradually replacing the typical grassland species combination. Furthermore, the territory of dry grassland formations is rapidly dwindling because these economically unprofitable areas are reafforested, frequently with *Pinus nigra*. Leaving these areas fallow eventually also leads to forest, following the succession of herbaceous plant associations and bushes. The preservation of dry and unfertilized grassland formations therefore appears to depend not only on the maintenance of suitable habitat conditions but also on the application of traditional, low-intensity methods of land use (Bignal and McCracken, 1996).

(b) Quercion pubescentis *landscapes*

In central Europe, in the Donau and upper Rhine region as well as the middle German Thüringer Becken, there are small areas of thermophilic woodlands on dry, sun-exposed rocky sites that have been considered as relict habitats of a warmer period (Gradmann, 1898; Ellenberg, 1954; Pott, 1993). This woodland is dominated by the oaks *Quercus pubescens* and *Q. petraea*, and their hybrids. Other components of this xerothermic vegetation complex are extrazonal *Xerobromion* communities and tall-herb fringe communities and scrub.

On cliffs and rocky slopes the trees show bent, low-growing 'fighting forms'. This woodland type is so open that it is infiltrated with bushes and perennial plants of southern continental geobotanical elements, e.g. *Helianthemum nummularium*, *Pulsatilla vulgaris*, *Hippocrepis comosa* and *Orchis purpurea*. During the post-glacial warm periods, these species immigrated into central Europe. In the Atlanticum their distribution was restricted to scattered open refuges, such as dry sandy areas or rocky cliffs, by the spreading of oak forests (Pott, 1995a). The vegetation mosaic of these relict habitats includes natural bush communities (e.g. *Cotoneastro-Amelanchieretum*, *Prunetum mahaleb*; Müller, 1986) and open vegetation (e.g. *Geranio-Dictamnetum*).

The *Lithospermo-Quercetum* thermophilic woods often show all sorts of anthropozoogenic influences. They turn out to be very sensitive to grazing and open up because the young trees are eaten by livestock. The wood is replaced by bush vegetation of *Berberidion vulgaris* communities (*Ligustro-Prunetum* and *Prunetum mahaleb*) with submediterraneanly influenced geographical vicariants. Closely connected with them are *Geranion sanguinei* fringe communities such as the *Geranio-Dictamnetum*, *Geranio-Peucedanetum cervariae* and *Geranio-Anemonetum sylvestris*. On the secondary habitats, as in the *Fagion* landscape, one can differentiate grazing dependent *Mesobromion* communities of the *Gentiano-Koelerietum* type or the *Mesobrometum erecti* type, respectively.

This anthropozoogenic vegetation complex with its grassland, fringe communities, scrub and thermophilic wood remnants, has brought about an incorrect expansion of the steppe–heath concept (Witschel, 1980; Pott, 1995a). At this point it needs to be emphasized that the immigration process of xerothermic plant species was not governed purely by edaphic and climatic conditions, as in the primeval landscape. When humans settled down and started raising cattle and farming, many regions of central Europe were gradually cleared of forest. Prehistoric man did not settle down arbitrarily, but actually selected favourable areas. Gradmann (1901, 1906) found in southern Germany that those landscapes with frequent xerothermic floristic elements coincided with high numbers of Neolithic findings. He called these areas steppe heaths and hypothesized that they dated from a warm post-glacial period, rendering the growth of forests locally impossible. So, an open landscape, steppe or woodland steppe, must have developed in the otherwise densely forested regions. Gradmann (1906) assumed that primeval forests were unfavourable for primitive farming, because prehistoric man lacked the means to clear the forest. In consequence, humans would have settled in the unforested areas – before the forest, favoured by increasing humidity of climate, could conquer the open spaces. In this respect, the steppe–heath theory could not be confirmed (e.g. Ellenberg, 1939; Tüxen, 1939; Pott, 1996).

Firstly, the coincidence of xerothermic vegetation and abundant Neolithic findings is restricted to southern Germany. Secondly, the Neolithic period shows no drought period of the dimensions that Gradmann postulated. Thirdly, the concept of the forests preventing primitive settling had to be abandoned altogether. Evidently, Neolithic humans were able to clear the forest, at least to thin it out; and grazing livestock furthered the process of clearing.

The different forest associations show a varying resistance against anthropozoogenic influences, determined by the productivity of their habitat. Depending on the character of the landscape and the level of civilization of the settlers, considerable differences existed in the way of *landnam*, i.e. the cultivation of the land. As a rule, the habitats of those plant associations best suited for farming were occupied first. In particular, the mixed oak forest was a varied and productive source of food. When excessively used as pasture and a source of mast, the forest thins out until a mosaic of bush and shrub vegetation, grassland, woodland and nitrophilic herbaceous plant associations has developed. Thus, at a very early stage human influence probably has considerably enlarged the territories of xerothermic grassland vegetation, especially by destroying thermophilic forests growing in habitats prone to erosion and sand drift.

After the Neolithic Age, extensive rocky heathland and poor grassland formations gradually developed and were kept open by grazing, wood cutting or burning. Anthropozoogenic grassland even developed on more fertile soils with a more balanced water supply. Here the *Mesobrometum* associations developed: in contrast to the poor dry grassland association *sensu stricto* (the *Xerobrometum*) they contain fewer thermophilic species of mediterranean origin and more mesophyllous species (e.g. *Dactylis glomerata*, *Lotus corniculatus*, *Plantago media*, *Daucus carota*).

Some *Xerobromion* species are able to spread rapidly over cleared areas, such as fallow land and burned patches. In most cases, grazing cattle play an important role in such secondary successions. Without cattle biting off the young shoots, shrubs would soon establish, especially on deep, formerly tilled soils, and bush and forest vegetation would close before true grasslands could develop. On the other hand, as a consequence of deforestation, erosion can destroy the soil layer on steep slopes to a degree that hinders the growth of trees even when human influences ceases.

Single trees found in regions bearing rocky steppe vegetation (e.g. the Valais, Switzerland) indicate that the natural vegetation of such habitats is a woodland formation – possibly a sparse oak or pine forest. In the warm southwestern German river valleys, thermophilic oak forests have been widely replaced by viniculture. In the late Middle Ages, wine was grown even in the arid regions of central and eastern Germany (Mecklenburg, Pomerania and Silesia). But it took great efforts, because the conditions were unfavourable: each vine had to be protected separately against winter frosts. All these outpost cultures became abandoned, being economically unprofitable compared with other wine regions. The abandoned, fallow vineyards were soon covered with secondary dry grassland, which was commonly used as sheep pasture. In eastern central Europe, these slopes are covered with steppe-like meadows containing *Anthericum*, *Adonis* and *Salvia* species or with steppe heavily dominated by *Stipa capillata*, *Stipa pinnata* and *Festuca valesiaca*, insofar as they are kept open by grazing or burning. About 100 years ago, however, cattle were gradually kept more and more indoors and fed with hay; also, the tending of sheep has become much less important over the last 50 years. As a consequence of these structural changes in agriculture, the pastures that could be used for haymaking were excluded from grazing.

4.4.3 Wood-pasture relicts in the uplands

In the uplands traditional land use has created a vegetation pattern which is regionally distinguishable. Grazing impact has been one of the main vegetation-shaping factors in these areas (Pott, 1985, 1988a,b). A characteristic example is provided by the Siegerland Hauberg region in the Eifel, Black Forest and Vosges. In these ares of potential montane beech forests, the specific land use practice traditionally included a cycle of coppicing, charcoal burning (for ore smelting), bark stripping, integrated rotational slash-and-burn culture (*Reutberge*) (Figure 4.10), agriculture and cattle grazing. Palynological investigations and radiocarbon dating indicate the appearance of the first anthropozoogenic influences at about 2000 BC. The palynological evidence points to ore smelting exploitation and farming for the Hallstatt and Latene period in the Iron Age, i.e. since 700 and 450 BC. The influence of this drastic human impact on the beech forest can be recognized by the decline of *Fagus sylvatica* pollen and the increasing pollen proportion of anthropogenic indicators and those trees which have a high regenerative ability after coppicing (*Quercus*, *Betula*) (Pott, 1985, 1986).

The typical vegetation of the *Hauberg* landscape can be classified by phytosociological methods into the following dominant vegetation units:

- Coppiced woodland with oak and birch underwood instead of beech woods. Coppicing rotations happened every 18–22 years for bark-stripping and charcoal production.

Figure 4.10 Ancient cultural landscape (*Reutberge*) in the area of potential *Luzulo-Fagetum* or *Luzulo-Abietetum* woodland in the Black Forest.

- Cereal fields and tilled land with buckwheat (*Fagopyrum esculentum*) between oak and birch trees.
- *Sarothamnus scoparius* communities as a result of burning practice and successional heliophilic communities dominated by *Epilobium angustifolium*, *Senecio fuchsii* and *Digitalis purpurea* after slashing.

- Grazed dwarf-shrub heaths with *Vaccinio-Callunetum*, *Juniperus communis* shrubs and grass-dominated areas.

Besides the quantitative reduction of woodland area by deforestation, more qualitative alterations of the residuary forests have occurred since prehistoric land use. The original vegetation of the forests disappeared, depending on the intensity and duration of anthropozoogenic influences and the woodland's regenerative capacity. The physiognomy and the phytosociological structure of anthropogenic plant communities can only be understood as a result of the specific human interactions and cultivation methods. An essential consequence of human impact was the clearance of former deciduous primeval forests. Numerous secondary biotopes, such as hedges, meadows and pastures, fallow land, fields and cultivated areas, could develop as new secondary biotopes.

The importance of the selective effect of grazing is particularly clear in the vegetation mosaic of pastureland in the *Luzulo-Fagion* range on silicate soils. In particular, the Franconian period of *landnam* (from 900 BC) led to an area reduction of beechwoods and favoured the growth of coppices and dwarf-shrub heaths with *Juniperus* in the higher parts of the uplands (Pott, 1986, 1988a,b).

(a) Pastures and heathlands in Luzulo-Fagion *landscapes*

In potential *Luzulo-Fagion* areas wood–field cultivation and wood utilization during a regular cyclic *Hauberg* system (for an exact description see Pott, 1985, 1986) created a semi-natural landscape with coppices, *Epilobion angustifolii* communities, forest precursor units and grazed heaths with *Sarothamnus scoparius* and *Juniperus communis*.

The floristic aspect of the *Nardus stricta* pastures from poor soils is fairly uniform; they can be classified as *Polygalo-Nardetum* due to the almost permanent occurrence of *Nardus stricta*, *Polygala vulgaris*, *Arnica montana* and *Genista germanica*. As a heathland community the *Vaccinio-Callunetum* is especially well developed on humous soil in the shelter of rocks and stones which protect the plants from grazing. This so-called mountain heath, with its differential species *Vaccinium uliginosum* and *Genista pilosa*, is characterized by the dominance of *Vaccinium myrtillus*, *Vaccinium vitis-idaea* and *Calluna vulgaris*, in contrast to the purely *Calluna vulgaris*-dominated lowland heathland (*Genisto-Callunetum*) (Figure 4.11).

Especially remarkable relicts of slash-and-burn cultivation are the thickets of *Sarothamnus scoparius* in rocky places. Starting from natural indigenous *Sarothamnus* heaths on rocky slopes (Lohmeyer, 1986), the predominantly subatlantic *Sarothamnus scoparius* expanded immensely when the anthropozoogenic plant communities of the montane *Reutberge* and *Hauberg* areas developed. Phytosociologically this vegetation can be classified as *Rubo plicati-Sarothamnetum* (Pott, 1985, 1995b; Weber, 1987). Occasionally, *Sarothamnus* plains are infiltrated with light-demanding *Rubus plicatus* and other forest pioneer species such as *Betula pendula*, *Sorbus aucuparia* and *Frangula alnus* (= *Rhamnus frangula*).

Figure 4.11 Open grazing land in Gräftenberg (Sauerland, Germany) in the area of potential *Luzulo-Fagetum* beechwoods, with *Vaccinio-Callunetum* dwarf-shrub heather, *Juniperus communis* bushes and remnants of grazed woodland. (From Pott, 1995b.)

When livestock density is low enough, beech can re-establish. Under the influence of browsing the beech become shrubby and develop a bushy, sometimes conical, habitus. At higher altitudes they can be bent down horizontally due to the pressure of the snow. In the former areas of wood–field cultivation, such mast-bearing beeches were not only protected but also deliberately replanted.

(b) Wood-pasture vegetation in the mountains of central Europe

In the montane and subalpine regions of central Europe, grazing has followed a transhumance pattern since the Middle Ages, with livestock gradually moving upwards with the advancing growing season to the (sub)alpine summer pasture and returning to the stables in autumn.

With respect to the impact of browsing on montane woodland, deciduous and coniferous trees have to be considered separately. Conifers, especially fir (*Abies alba*), have little or no regenerative capacity (Ellenberg, 1986; Holtmeier, 1986, 1989). If the young trees are bitten off near the ground they will not regenerate, whereas deciduous trees may still develop into 'cow bushes'. As deciduous trees get rare anyway at higher altitudes, the forests can clear up easily or even turn into grassland if the duration of grazing is prolonged.

The extensive pasture communities, which are several centuries old, show numerous formations which are primarily differentiated by edaphic factors – in particular, the base saturation – and secondarily by land use influences. The *Festuco-Genistetum sagittalis* is the characteristic extensive pasture of the more acidic mountains up to heights of about 1200 m. It bears resemblance to the heathlands of lower elevations. Due to relatively intensive but irregular grazing of the pasture with heavy livestock, not only the unpalatable *Nardus stricta*, *Carlina acaulis*, *Arnica montana* and *Juniperus communis* increased over the centuries as 'grazing weeds', but also the characteristic *Genista sagittalis* with its hard, partially lignified shoots (Schwabe and Kratochwil, 1987). Luxurious *Genista sagittalis* garlands may be seen along the cattle tracks, which run parallel to the slopes.

If the pastures are grazed very lightly or abandoned, *Calluna vulgaris* and *Vaccinium* species can become dominant as mountain dwarf shrub-heaths of the *Vaccinio-Callunetum* type, especially around gneiss and granite blocks or stone piles. Here, beech can regenerate very successfully. On the pasture borders *Teucrium scorodonia* zones occur, often in connection with extensive *Sarothamnus scoparius* bushes.

4.5 CONCLUSION

The vegetation patterns of the traditional semi-natural landscape of central and northwestern Europe have been described here as the product of a long history of human land use, with livestock grazing as a major influential factor. Starting with a more or less closed forest in the post-glacial Atlanticum, humans and their livestock gradually destroyed the forest, thus opening up the landscape and shaping new semi-natural plant communities: woodland pasture, scrub, herbaceous fringe communities, open pastures, meadows, heathlands and – as an extreme consequence of overexploitation – sand drifts. In the wood-pasture landscape the dynamic balance between grazing impact and forest regeneration is illustrated most clearly.

The conservation merits of the communities of the grazed wood-pasture landscape are evident in its high species richness and diverse vegetation structures. The degradation of the forests has undoubtedly caused a loss of species, especially animals; it has also resulted in a remarkable increase of vegetation diversity at a small level of spatial scale. Moreover, the clearance of the forest has caused the expansion of the distribution of continental and submediterranean species.

While it is clear that grazing has played an important role in creating these vegetation patterns, it is doubtful whether it is the sole factor by which they can be maintained. The presented evidence suggests that human beings were the necessary facilitating agent that allowed the characteristic high grazing pressure of the wood-pasture landscape. This agrees with the conclusion reached by Prins in Chapter 3. Yet it has also been argued that the full native species assemblage of wild herbivores could exert a sufficient impact to maintain a substantial degree of openness in the landscape (Vera, 1997). I do not share this view and conclude that the study of vegetation history in combination with phytosociological methods shows that the combination of anthropogenic and zoogenic influences was essential to create the wood-pasture landscape. If this is indeed true, its preservation demands the maintenance of remaining low-intensity farming systems and their revival where they have disappeared.

REFERENCES

Barkman, J.J. (1985) Geographical variation in associations of juniper scrub in the central European plain. *Vegetatio*, **59**, 67–71.

Barkman, J.J., Masselink, A.K. and De Vries, B.W.L. (1977) Über das Mikroklima in Wacholderfluren, in *Vegetation und Klima*, (ed. R. Tüxen), pp. 35–80, Den Haag, The Netherlands.

Bignal, E.M. and McCracken, D.I. (1996) Low-intensity farming systems in the conservation of the countryside. *Journal of Applied Ecology*, **33**, 413–424.

Bornkamm, R. (1958) Standortbedingungen und Wasserhaushalt von Trespen-Halbtrockenrasen (*Mesobromion*) im oberen Leinegebiet. *Flora*, **146**, 23–67.

Bornkamm, R. (1960) Die Trespen-Halbtrockenrasen im oberen Leinegebiet. *Mitteilungen der Floristisch-soziologischen Arbeitsgemeinschaft*, **8**, 181–208.

Bornkamm, R. (1961) Zur Konkurrenzkraft von *Bromus erectus*. Ein sechsjähriger Dauerversuch. *Botanisches Jahrbuch*, **80**, 466–479.

Bornkamm, R. (1974) Zur Konkurrenzkraft von *Bromus erectus II*. *Botanisches Jahrbuch*, **94**, 391–412.

Buchwald, K. (1951) Wald- und Forstgesellschaften der Revierförsterei Diensthoop, Forstamt Syke bei Bremen. *Angewandte Pflanzensoziologie*, **1**, 1–72.

Buchwald, K. (1984) Zum Schutze der Gesellschaftsinventare vorindustriell geprägter Kulturlandschaften in Industriestaaten. Fallstudie Naturschutzgebiet Lüneburger Heide. *Phytocoenologia*, **12**, 395–432.

Burrichter, E. (1973) Die Potentielle Natürliche Vegetation in der Westfälischen Bucht. *Siedlung und Landschaft in Westfalen*, **8**, 1–58.

Burrichter, E. and Pott, R. (1987) Zur spät- und nacheiszeitlichen Entwicklungsgeschichte von Aüen-Ablagerungen in Ahse-Tal bei Soest (Hellwegbörde). *Münstersche Geographische Arbeiten*, **27**, 129–135.

Burrichter, E., Pott, R., Raus, T. and Wittig, R. (1980) Die Hudelandschaft 'Borkener Paradies' im Emstal bei Meppen. *Abhandlungen Landesmuseum für Naturkunde*, **42**(4), 1–69.

Dierschke, H. (1985) Experimentelle Untersuchung zur Bestandsdynamik von Kalkmagerrasen (*Mesobromion*) in Südniedersachsen. Vegetationsentwicklung auf Dauerflächen, 1972–1984, in *Sukzession auf Grünlandbrachen* (ed. K.F. Schreiber). *Münstersche Geographische Arbeiten*, **20**, 9–24.

Duncan, P. (1992) *Horses and Grasses: The Nutritional Ecology of Equids and their Impact on the Camargue*, Ecological Studies 87, Springer Verlag. New York.

Ellenberg, H. (1939) Über Zusammensetzung, Standort und Stoffproduktion bodenfeuchter Eichen- und Buchen-Mischwaldgesellschaften Nordwestdeutschlands. *Mitteilungen der Floristisch-soziologischen Arbeitsgemeinschaft*, **5**, 3–135.

Ellenberg, H. (1954) Steppenheide und Waldweide. *Erdkunde*, **8**, 188–194.

Ellenberg, H. (1986) *Vegetation Mitteleuropas mit den Alpen in ökologischer Sicht*, 4th edn, E. Ulmer, Stuttgart.

Falinski, J.B. and Falinska, K. (1986) *Vegetation Dynamics in Temperate Lowland Primeval Forests: Ecological Studies in Bialowieza Forest* Geobotany 8, W. Junk, Dordrecht.

Flower, N. (1980) The management history and structure of unenclosed woods in the New Forest, Hampshire. *Journal of Biogeography*, 7, 311–328.

Garve, E. (1987) *Atlas der gefährdeten Gefääpflanzenarten in Niedersachsen und Bremen*, Parts I and II, Niedersächsische Landesanstalt für Ökologie, Hannover.

Gigon, A. (1968) Stickstoff- und Wasserversorgung von Trespen-Halbtrockenrasen (*Mesobromion)* im Jura bei Basel. *Berichte der Geobotanischen Institut ETH*. **38**, 26–85.

Gimingham, C.H. (1972) *Ecology of Heathlands*, Chapman & Hall, London.

Gimingham, C.H. and De Smidt, R.J.T. (1983) Heaths as natural and seminatural vegetation, in *Man's Impact on Vegetation*, (eds W. Holzner, M.J.A. Werger and I. Ikusima), pp. 185–199, W. Junk, The Hague.

Gradmann, R. (1898) *Das Pflanzenleben der Schwäbischen Alb*, Vol. 1, Stuttgart.

Gradmann, R. (1901) Das mitteleuropäische Landschaftsbild nach seiner geschichtlichen Entwicklung. *Geographische Zeitschrift*, 7, 361–377, 435–447.

Gradmann, R. (1906) Beziehungen zwischen Pflanzengeographie und Siedlungsgeschichte. *Geographische Zeitschrift*, **12**, 305–325.

Hakes, W. (1987) Einfluss von Wiederbewaldungsvorgängen in Kalkmagerrasen auf die floristische Artenvielfalt und Möglichkeiten der Steuerung durch Pflegemassnahmen. *Dissertationes Botanicae* **109**, 151S.

Holtmeier, K.F. (1986) Die Waldgrenze unter dem Einfluss von Klima und Mensch. *Abhandlungen Westfalisches Museum für Naturkunde*, **48**(2/3), 395–412.

Holtmeier, K.F. (1989) Ökologie und Geographie der oberen Waldgrenze. *Berichte der Reinhold Tüxen-Gesellschaft*, **1**, 15–45. Hannover.

Horst, K. (1964) Klima- und Bodenfaktoren in Zwergstrauch- und Waldgesellschaften des Naturschutzparks Lüneburger Heide. *Naturschutz und Landschaftspflege Niedersachsen*, **2**, 1–60.

Hüppe, J. (1993) Entwicklung der Tieflands-Heidelandschaften Mitteleuropas in geobotanisch-vegetationsgeschichtlicher Sicht. *Berichte der Reinhold Tüxen-Gesellschaft*, **5**, 49–75.

Hüppe, J. (1995) Zum Einfluss von Wildkaninchen auf Wacholder (*Juniperus communis*) in Nordwestdeutschland. *Zeitschrift für Ökologie und Naturschutz*, **4**, 1–8.

Jeckel, G. (1984) Syntaxonomische Gliederung, Verbreitung und Lebensbedingungen nordwestdeutscher Sandtrockenrasen (*Sedo-Scleranthetea*). *Phytocoenologia*, **12**, 9–153.

Kienzle, U. (1979) Sukzessionen in brachliegenden Magerwiesen des Jura und des Napfgebietes. Dissertation University of Basel, Sarnen, Switzerland.

Kienzle, U. (1984) *Origano-Brachypodietum* und *Colchico-Brachypodietum*, zwei Brachwiesen-Gesellschaften im Schweizer Jura. *Phytocoenologia*, **12**, 455–478.

Kratochwil, A. (1983) Zur Phänologie von Pflanzen und blütenbesuchenden Insekten eines versaumten Halbtrockenrasens im Kaiserstuhl. *Beiheft Veröffentlichungen Naturschutz und Landschaftspflege Baden-Württemberg*, **34**, 57–108.

Lohmeyer, W. (1986) Der Besenginster (*Sarothamnus scoparius*) als bodenständiges Strauchgehölz in einigen natürlichen Pflanzengesellschaften der Eifel. *Abhandlungen Westfalisches Museum für Naturkunde*, **48**, 157–174.

Mahn, E.-G. (1957) Über die Vegetations- und Standortverhältnisse einiger Porphyrkuppen bei Halle. *Wissenschaftliches Zeitschrift der Martin Luther Universität Halle-Wittenberg*, **6**, 177–207.

Müller, Th. (1986) *Prunus mahaleb*-Gebüsche. *Abhandlungen Westfalisches Museum für Naturkunde*, **48**, 143–155.

Oberdorfer, E. (1978) *Süddeutsche Pflanzengesellschaften*, Part II, 2nd edn, Gustav Fischer Verlag, Stuttgart.

Peterken, G.F. (1981) *Woodland Conservation and Management*, Chapman & Hall, London.

Pilger, R. (1931) Die Gattung *Juniperus L. Mitteilungen der Deutschen Dendrologischen Gesellschaft*, **43**, 255–269.

Pott, R. (1982) Das Naturschutzgebiet Hiddeser Bent – Donoper Teich in vegetationsgeschichtlicher und pflanzensoziologischer Sicht. *Abhandlungen Westfalisches Museum für Naturkunde*, **44**, 1–108.

Pott, R. (1983) Geschichte der Hude- und Schneitelwirtschaft Nordwestdeutschlands und deren Auswirkungen auf die Vegetation. *Oldenburger Jahrbuch*, **83**, 357–376.

Pott, R. (1985) Vegetationsgeschichtliche und pflanzensoziologische Untersuchungen zur Niederwaldwirtschaft in Westfalen. *Abhandlungen Westfalisches Museum für Naturkunde*, **47**, 1–75.

Pott, R. (1986) Der pollenanalytische Nachweis extensiver Waldbewirt-schaftungen in den Haubergen des Siegerlandes, in *Anthropogenic Indicators in Pollen Diagrams*, (ed. K.-E Behre), pp. 125–134, Balkema, Rotterdam.

Pott, R. (1988a) Extensive anthropogene Vegetationsveränderungen und deren pollenanalytischer Nachweis. *Flora*, **180**, 153–160.

Pott, R. (1988b) Impact of human influences by extensive woodland management and former land-use in North-West Europe, in *Human Influence on Forest Ecosystem Development in Europe*, (ed. F. Salbitano), pp. 263–278, ESF-FERN-CNR, Pitagora Editrice, Bologna, Italy.

Pott, R. (1990) Die nacheiszeitliche Ausbreitung und heutige pflanzensoziologische Stellung von *Ilex aquifolium* L. *Tuexenia*, **10**, 497–512.

Pott, R. (1992) Man–ecosystem interactions in the beginning of human civilisation. History and influence of human impact on the vegetation in Early Neolithic landscape of NW Germany. *Annali di Botanica*, **50**, 97–118.

Pott, R. (1993) *Farbatlas Waldlandschaften. Ausgewählte Waldtypen und Waldlandschaften unter dem Einfluss des Menschen*, E. Ulmer, Stuttgart.

Pott, R. (1995a) The origin of grassland plant species and grassland communities in Central Europe. *Fytosociologia*, **29**, 7–32.

Pott, R. (1995b) *Die Pflanzengesellschaften Deutschlands*, 2nd edn, E. Ulmer, Stuttgart.

Pott, R. (1996) *Biotoptypen Deutschlands*, E. Ulmer, Stuttgart.

Pott, R. and Burrichter, E. (1983) Der Bentheimer Wald – Geschichte, Physiognomie und Vegetation eines ehemaligen Hude- und Schneitelwaldes. *Forstwissenschaftliches Centralblatt*, **102**, 350–361.

Pott, R. and Hüppe, J. (1991) *Die Hudelandschaften Nordwestdeutschlands*, Westfalisches Museum für Naturkunde, Münster.

Preising, E. (1950) Nordwestdeutsche Borstgras-Gesellschaften. *Mitteilungen der Floristisch-soziologischen Arbeitsgemeinschaft*, 2, 33–41.

Putman, R.J. (1986) *Grazing in Temperate Ecosystems: Large Herbivores and the Ecology of the New Forest*, Croom Helm, London.

Raabe, E.W. (1987) *Atlas der Flora Schleswig-Holsteins und Hamburgs*, Wacholtz, Neumünster.

Rackham, O. (1980) *Ancient Woodland. Its History, Vegetation and Uses in England*, Edward Arnold, London.

Rosén, E. (1982) Vegetation development and sheep grazing in limestone grasslands of south Öland, Sweden. *Acta Phytogeographica Suecica*, 72, 1–104.

Ruiz, M. and Ruiz, J.P. (1986) Ecological history of transhumance in Spain. *Biological Conservation*, 37, 73–86.

Schiefer, J. (1981) Bracheversuche in Baden-Württemberg. *Veröffentlichungen Naturschutz und Landschaftspflege Baden-Württemberg*, 22, 1–328.

Schwabe-Braun, A. (1980) Weidfeld-Vegetation im Schwarzwald. Geschichte der Nutzung – Gesellschaften und ihre Komplexe – Bewertungen für den Naturschutz. *Urbs et Regio*, 18, 1–212.

Schwabe, A. and Kratochwil, A. (1987) Weidbuchen im Schwarzwald und ihre Entstehung durch Verbiss des Weideviehs. *Beiheft Veröffentlichungen Naturschutz und Landschaftspflege Baden-Württemberg*, 49, 1–120.

Sjögren, E. (ed.) (1988) Plant cover on the limestone Alvar of Öland: ecology – sociology – taxonomy. *Acta Phytogeographica Suecica*, 76, 1–160.

Stöcklin, J. and Gisi, Ü. (1985) Bildung und Abbau der Streu in bewirtschafteten und brachliegenden Mähwiesen, in *Sukzession auf Grünlandbrachen*, (ed. K.F. Schreiber), *Münstersche Geographische Arbeiten*, 20, 101–109.

Tubbs, C.R. (1986) *The New Forest. A Natural History*, Collins, London.

Tüxen, R. (1937) Die Pflanzengesellschaften Nordwestdeutschlands. *Mitteilungen der Floristisch-soziologischen Arbeitsgemeinschaft Niedersachsen*, 3, 1–170.

Tüxen, R. (1939) Die Pflanzendecke Nordwestdeutschlands in ihren Beziehungen zu Klima, Gesteinen, Boden und Mensch. *Deutsche Geographische Blätter*, 42, 1–8.

Tüxen, R. (1967) Die Lüneburger Heide. Werden und Vergehen einer Landschaft. *Rotenburger Schriften*, 26, 1–52.

Tüxen, R. (1974) Die Haselünner Kuhweide. Die Pflanzengesellschaften einer mittelalterlichen Gemeindeweide. *Mitteilungen der Floristisch-soziologischen Arbeitsgemeinschaft*, 17, 69–102.

Vera, F.W.M. (1996) Metaforen voor de Wildernis: Eik, hazelaar, rund en paard. Doctoral thesis, Wageningen Agricultural University, Wageningen.

Weber, H.E. (1987) Zur Kenntnis einiger bislang wenig dokumentierter Gebüschgeseilschaften. *Osnabrücker Naturwissenschaftliche Mitteilungen*, 13, 143–157. Osnabrück.

Wilmanns, O., Schwabe-Braun, A. and Emter, M. (1979) Struktur und Dynamik der Pflanzengesellschaften im Reutwaldgebiet des Mittleren Schwarzwaldes. *Documents phytosociologiques N.S.*, 4, 984–1024.

Wilmanns, O. and Kratochwil, A. (1983) Naturschutz-bezogene Grundlagen – Untersuchungen im Kaiserstuhl. *Beiheft Veröffentlichungen Naturschutz und Landschaftspflege Baden-Württemberg*, **34**, 39–56.

Witschel, M. (1980) Xerothermvegetation und dealpine Vegetationskomplexe in Südbaden. *Beiheft Veröffentlichungen Naturschutz und Landschaftspflege Baden-Württemberg*, **17**, 1–212.

Zoller, H. (1954) Die Arten der *Bromus erectus*-Wiesen des Schweizer Juras. *Veröffentlichungen der Geobotanischen Institut Rübel*, **28**, 1–284.

Zoller, H., Bischof, N., Ehrhardt, A. and Kienzle, U. (1984) Biocoenosen von renzertragsflächen und Brachland in der Berggeschichte der Schweiz. Hinweise zur Sukzession, zum Naturschutz und zur Pflege. *Phytocoenologia*, **12**, 373–394.

Part Two

Impact of Grazing on Community Structure

5

The impact of grazing on plant communities

Jan P. Bakker
Laboratory of Plant Ecology, University of Groningen, PO Box 14, 9750 AA Haren, The Netherlands

5.1 EFFECTS OF EXCLUDING GRAZING ANIMALS

The simplest way to discuss the effects of grazing on plant communities is to compare grazed and ungrazed situations. The results of a number of studies in which large herbivores were excluded from previously grazed landscapes/plant communities are summarized in Table 5.1. This list is certainly not complete but it allows some generalizations. Grazed areas harbour pioneer species, including annuals and biennials that have to establish from seedlings, low-stature species and rosette plants. A similar conclusion was reached by Scherfose (1993) in a literature review on the impact of grazing on plant species in salt marshes in the Wadden Sea area. The ratio of species suffering from grazing and species promoted by grazing turned out to be about 1 : 1 at the lower and middle salt marsh. At the higher salt marsh more species were promoted by than suffering from grazing. After the exclusion of grazing, the aforementioned groups of species are often replaced by tall grasses and herbs accompanied by litter accumulation, and by shrubs and trees. The general pattern shows a higher above-ground standing crop in exclosures than in continuously grazed plots, as reported in a review by Milchunas and Lauenroth (1993).

These changes result in a decrease of the number of species under eutrophic and mesotrophic soil conditions. The composition of the communities and their species richness show hardly any change under oligotrophic soil conditions in calcareous grasslands or in *Nardus stricta* stands. This confirms theoretical considerations by various authors. Al-Mufti *et al.* (1977) and Grime (1973, 1979) collected numerous samples over a wide

Grazing and Conservation Management. Edited by M.F. WallisDeVries, J.P. Bakker and S.E. Van Wieren. Published in 1998 by Kluwer Academic Publishers, Dordrecht. ISBN 0 412 47520 0.

Table 5.1 Effects of exclusion of large herbivores in different landscapes/plant communities

Landscape/ plant community	Changes	Diversity Com.	Diversity Spp.	*N* years	Country	Assessment
Salt marshes	Tall grass *Elymus athericus* becomes dominant	–	–	22	NL	Bakker, 1989; Van Wijnen *et al.*, 1997
	Elymus athericus spreads into various communities	?	?	15	D	Andersen *et al.*, 1990; Bakker *et al.*, 1997
	Phragmites becomes dominant at low salinity	?	?	?	D	Jeschke, 1987
	Tall grasses and herbs become dominant at higher salinity	?	?	?	D	Jeschke, 1987
	Tall grasses become dominant	?	?	?	D	Schmeisky, 1977
	Phragmites becomes dominant	?	?	?	SF	Siira, 1970
	Elymus repens/Phragmites become co-dominant	?	–	8	NL	Bakker *et al.*, 1997
	Tall grass becomes dominant	?	–	4	DE	Berg *et al.*, 1997
Dry grasslands	Tall herbs become dominant	?	?	24	UK	Hill *et al.*, 1992
	Tall herbs become dominant	?	–	4	UK	Smith and Rushton, 1994
	Increase in litter, tall herbs and grasses	?	–	9	D	Schreiber and Schiefer, 1985
	Tall herbs and grasses become dominant	?	–	9	D	Schreiber and Schiefer, 1985
	Tall grasses stay dominant	?	–	20	D	Schreiber, 1997
	Tall herbs and shrubs become dominant	?	–	12	NL	Bakker, 1989
Calcareous grasslands	Little change (oligotrophic)	?	?	12	UK	Elkington, 1981
	No changes (oligotrophic)	?	–/=	20	S	Bakker *et al.*, 1996a
	Litter accumulation and bush encroachment	?	–	?	S	Rejmánek and Rosén, 1988
	Tall grasses become dominant	?	–	6	UK	Gibson and Brown, 1991
	Bush encroachment	?	–	10	NL	Willems, 1983
	Litter accumulation and decrease in short species	?	=/–	4	D	Schiefer, 1981
	Tall grasses become dominant	?	–	9	D	Schreiber and Schiefer, 1995
	Woodland formation	?	–	20	D	Schreiber, 1997
	Bush encroachment	?	?	40	F	Dutoit and Allard, 1995
	Bush encroachment	?	?	30?	D	Poschlod *et al.*, 1991
Heathlands	Woodland formation	?	?	30	UK	Hester *et al.*, 1991
	Tree invasion	?	–	10	NL	Bakker, 1989
	Increase in ericoids and tall grasses	?	?	25	UK	Rawes, 1981
Nardus stricta stands	No changes	?	?	?	UK	Rawes, 1981
	No changes	?	=	4	D	Schiefer, 1981
	Grasses and herbs increase	?	=	4	D	Schiefer, 1981
Molinia caerulea stands	Tall grasses become dominant	?	–	6	UK	Grant *et al.*, 1996a
Wet grasslands	Tall herbs and trees invade	–	–	25	S	Persson, 1984
	Tall herbs become dominant	?	–	6	NL	Bakker, 1989
	Tall herbs become dominant	?	–	9	D	Schreiber and Schiefer, 1985

Com., community diversity
Spp., species diversity
***N* years**, time after grazing ceased

range of herbaceous vegetation types near Sheffield (England) in one particular period. They observed that species diversity can be plotted as bell-shaped curves along gradients of maximum standing crop, including litter. There is evidence from a large number of observations that the number of plant species is a unimodal function of habitat productivity or other measures of nutrient supply rates. In fact no cases are known in which plant species diversity is a simple increasing function of productivity or nutrient supply (Tilman and Pacala, 1993). On the one hand, low maximum standing crop and low species diversity might be correlated with high 'environmental stress' and/or high 'disturbance' (Grime, 1979). Stress is defined as the sum of external constraints limiting the rate of dry matter production of all or part of the vegetation, e.g. very dry, dark, saline or nutrient-poor conditions. Disturbance is defined by Grime as the sum of mechanisms which limit plant biomass by causing its partial or total destruction (e.g. burning, grazing, cutting) and thus can be considered as an intensity of management. On the other hand, a high standing crop is likely to mean a high productivity of some species that might limit species diversity by competitive exclusion. At high levels of productivity, plant growth is primarily limited by light. Hence species diversity is bound to decline under these conditions (Tilman and Pacala, 1993).

Grime's (1973) original model interpreted the axis of maximum standing crop (including litter) as environmental stress and as disturbance (which equals intensity of management). Huston (1979) considered their effects simultaneously. He replaced Grime's stress gradient by a 'rate of displacement' and interpreted it as a gradient of growth rate. This interpretation is consistent with observations on the application of fertilizers. Huston replaced Grime's disturbance gradient by a 'frequency of reduction'. Huston's model, based on computer simulation, showed that moderate levels of both disturbance and environmental stress are prerequisites for a high species diversity (see also Peet *et al.*, 1983). The occurrence of the highest number of species in dunes at intermediate grazing pressure by rabbits supports the outcome of such a model (Zeevalking and Fresco, 1977).

To summarize, one can expect high species diversity in mesotrophic environments where the above-ground biomass is removed frequently. Species diversity can also be high in more oligotrophic environments, where the above-ground biomass is removed occasionally.

Although frequently recorded, bush encroachment and woodland formation do not always take place after the cessation of grazing. Various successional patterns were seen after a 20-year period of abandonment in 15 study sites in Germany. Many sites indeed showed a unidirectional succession to woodland, but others not at all or only to a minor extent (Schreiber, 1997). The history of the fields may play a role. Secondary succession after abandonment of chalk grassland in France indicated encroachment by *Cornus sanguinea* and *Crataegus monogyna* after about 40 years. It turned out that the agricultural exploitation history since the early nineteenth century had been related to the speed of bush encroachment. Fields that had been used as arable fields at least once showed an encroachment of 45% cover on average (with some fields having more than 80% cover), whereas fields which always had been grazed showed a smaller encroachment – 30% on average (Dutoit and Alard, 1995). Long-term studies in both grazed and ungrazed sites may reveal unexpected results. An open heathland in Australia, fenced out since 1945, showed a decline in shrub cover in the 1980s, whereas the shrubs continued to increase in the grazed plot (Wahren *et al.*, 1994). The most plausible reason for this seeming discrepancy may be found in the history of the site, and differences in the life history of shrub species and their palatability to cattle. The vegetation on both plots was burned in 1939. The shrubs *Phebalium squamulosum* (emerged from seeds) and *Prostanthera cuneata* (vegetative resprouting) became established in the decade following the fire. Shrub cover increased on both plots until 1979. The continued shrub cover on the grazed plot resulted partly from the expansion of the long-living *Prostanthera*, which is unpalatable to cattle and can form closed heaths 1–2 m tall. The decrease in shrub cover on the ungrazed plot was largely due to the senescence of the majority of *Phebalium* shrubs, which were replaced by a dense sward of *Poa hiemata* that prevented seedling establishment (Wahren *et al.*, 1994).

Considerable information is omitted if only species diversity is measured. Parts of the ancient grazing area in the salt marsh at the island of Schiermonnikoog, The Netherlands, have been excluded from cattle grazing since 1973. Although the number of species decreased after excluding livestock, the decline has been slow (Bakker, 1989). This may be due to the relatively short period of observation, as it takes a long time before all individual tillers of perennial species disappear. The relative equitability diminished quicker than the number of species (Bakker, 1985, 1989), due to the dominance of a single species. Bobbink and Willems (1987) also reported that the equitability decreased prior to the number of species in an exclosure experiment on chalk grassland in The Netherlands.

Another important issue is the scale at which species diversity is measured. In a chronosequence representing a series of grazed, 20 years ungrazed, 55 years ungrazed and 80 years ungrazed 'alvar' limestone grasslands on the Baltic island of Öland in southern Sweden, the total number of species on 40 m^2 remained at about 60. However, the mean number of species on 4 m^2 declined from 39 to 25, and it declined from 23 to 10 on $0.4 m^2$. This implies that many species occur sparsely with a low frequency of occurrence over a large area (Bakker *et al.*, 1996a). Excluding cattle from a moist to wet grassland in southern Sweden which had been grazed for centuries showed a decline in average species number from 35 to $18/m^2$. The total number of species inside the exclosure of 810 m^2 revealed a much lower decrease from 113 to 89 species. The much greater relative reduction at the smaller scale than for the entire exclosure indicates a development from a fine-grained mixture of species towards a more patchy vegetation (Persson, 1984).

Not only may the species diversity decrease after cessation of grazing. A succession scheme was derived from long-term studies of vegetation changes after abandonment of grazing and after resumption of grazing (at a stocking rate of 1.6 heifers/ha) on an abandoned salt marsh on the barrier island of Schiermonnikoog. After resumption of grazing, the return to the initial succession stage generally took 5–10 years. After cessation of the grazing regime, it required 10–20 years to reach a new equilibrium (Bakker, 1989; Bakker *et al.*, 1997). During this period *Elymus athericus* outcompeted most other species on the higher and middle marsh with a subsequent decrease of the number of plant communities (Figures 5.1 and 5.2).

Scherfose (1993) stressed that it should not be concluded that grazing on higher salt marshes is always the best management practice. He suggested that the proportion of endangered plant communities and species should also be taken into account. In salt marshes, 28 'red list' species occur, of which 18 are promoted by grazing and five suffer from grazing; no data are available for the five other species. Eleven out of 27 plant communities of salt and brackish marshes are endangered; three of these are promoted by grazing, whereas eight suffer from grazing. According to Scherfose (1993), plant community diversity should be taken as a criterion for nature conservation purposes instead of species diversity.

In conclusion, it is important to realize that both the number of species in each community and the number of plant communities of open vegetation may decrease when grazing ceases. The first years after exclusion may be mis-

Figure 5.1 Ungrazed abandoned salt marshes in the Wadden Sea area gradually become dominated by stands of *Elymus athericus* and *Juncus maritimus* (dark).

Figure 5.2 After resumption of grazing by cattle, the dominance of *Elymus athericus* may be reversed and replaced by a short turf of *Festuca rubra*.

leading, as these show an increase in flowering plants. This turns out to be a short-term effect (Bakker *et al.*, 1997). The resumption of grazing after abandonment or starting extensive grazing after the cessation of fertilizer application may show the opposite effects; this depends on local abiotic conditions, stocking rate and local grazing intensities. Within a fenced area with low stocking density, very low grazing intensities may occur locally with subsequent development of tall-herb communities with low species richness. Only a few studies have compared changes in species richness at different spatial scales (e.g. Persson, 1984; Gibson and Brown, 1991a) and these show that changes at a small scale are bigger than at a large scale. This agrees with the carousel model as proposed by Van Der Maarel and Sykes (1993). At a very small scale they found that the species number on 1 m^2 remained the same during a number of years, but that individual species disappeared and emerged at the scale of 1 cm^2, and 'travelled' through the bigger plot.

5.2 EFFECTS OF INTRODUCING GRAZING ANIMALS

5.2.1 Effects on plant communities and species richness

So far we have seen the effects of grazing by excluding large herbivores. This section deals with the effects of grazing after abandonment or after intensive agricultural exploitation, as often seen in conservation management. The effects of introducing large herbivores into various areas are summarized in Table 5.2. Moist and wet sites show an increase in tall herb communities, and a subsequent decrease in the number of plant communities and species. This may be attributed to a relatively high productivity in combination with a low stocking rate. The herbivores then graze selectively and avoid the wetter sites. In a wet abandoned area dominated by *Phragmites australis*, the dominant species decreased and species richness increased. These effects were less clear at the very wet sites (Rozé, 1993). At higher stocking rates, wet sites were indeed also grazed. Thus, cattle and horses exploited the whole area of 160 ha in an alluvial floodplain along the river Thames, England, at a stocking density of 1.6 animal/ha (Putman *et al.*, 1991).

The initial high productivity of abandoned agricultural areas is reflected by plant species indicating eutrophic conditions. Over time a shift towards species of mesotrophic sites is observed in some studies. This was the main successional trend in the sheep-grazed grassland in the Westerholt, The Netherlands, during the first 15 years (Bakker, 1989). Species indicating oligotrophic conditions, i.e. heathland communities that existed before the reclamation to grassland, hardly (re-)appeared. Seeds of heathland species did occur in the persistent soil seedbank, as could be seen after removal of the topsoil (Bakker, 1989). Twenty-five years after the start of extensive grazing, the proportion of heathland species in the vegetation had increased to about 5%. It is thus questionable whether grazing can reduce soil fertility to a significant degree.

Erica tetralix heathlands are gradually overgrown by *Molinia caerulea* as a result of ceasing the cutting of sods and sheep grazing, in combination with an increased input of nitrogen from atmospheric deposition. Sheep grazing is not expected to be effective in shifting the balance to *Erica* again. Nitrogen loss from the ecosystem by grazing may not exceed 2 kg N/ha per year, whereas input from atmospheric deposition may amount to 20–40 kg N/ha per year. Grazing may even enhance nitrogen availability by replacing *Molinia* litter (high C : N ratio) with faeces (low C : N ratio). Where nutrient availability determines the outcome in the competition between the two species, only sod removal will result in such low nitrogen availability that grazing can shift the balance between the two species in favour of *Erica tetralix* (Berendse, 1985). Bobbink *et al.* (1992) compiled data on critical levels of nitrogen deposition for different plant communities. Changes in the communities occur above these levels, which are 15–20 kg N/ha per year for various heathland communities, 14–25 kg N/ha per year for calcareous grasslands, 20–30 kg N/ha per year for grasslands on neutral to acid soils and 20–35 kg N/ha per year for mesotrophic fens. The natural background level is 5 kg N/ha per year or lower (Erisman, 1990). It is therefore unlikely that grazing reduces soil fertility to such an extent that plant communities on oligotrophic soil can be maintained or emerge from mesotrophic soil conditions.

Table 5.2 Effects of introduction of large herbivores in different landscapes/plant communities that had not been grazed before (embankments), had not been grazed for some time (salt marshes, dunes, heathlands, *Molinia caerulea* stands), or were taken out of intensive agricultural exploitation (dry grasslands, calcareous grassland, wet grasslands)

Landscape/ plant community	Changes	Diversity Com.	Spp.	*N* years	Country	Assessment
Embankments	Bushes controlled	?	+	11	NL	Slim and Oosterveld, 1985
	Tall grasses not controlled	?	?	5	NL	Drost and Muis, 1988
	Phragmites controlled	?	?	5	NL	Drost *et al.*, 1990
	Phragmites controlled	?	?	2	NL	Van Deursen and Drost, 1990
	Tall grass controlled in tidal flats, not in sedimentation fields	?	?	2	NL	Van Deursen *et al.*, 1993
Salt marshes	Dominance of tall herbs and grass controlled	+	+	13	NL	Bakker, 1989
	Trampled soil, dominance of tall herbs controlled	+	–	13	NL	Bakker, 1989
Dunes	Tall grasses controlled	+	?	5	NL	Ehrenburg *et al.*, 1995; Mourik *et al.*, 1995; Van Til, 1996
	Tall grasses controlled	?	?	5	NL	De Bonte and Boosten, 1996
	Bushes controlled	?	+	7	NL	Van Dijk, 1992
	Bushes controlled in wet	?	+	5	NL	Van Djik, 1992
	sites, change eu- to mesotrophic, increase *Ammophila arenaria* in dry sites	?	+	7	NL	Van Dijk, 1992
Heathlands	*Calluna vulgaris* and grasses decreased	?	+	10	NL	Van Der Bilt and Nijland, 1993
	Grasses decreased, *Calluna* increased	?	?	5	NL	Bokdam and Gleichman, 1989
	Grasses decreased	?	?	?	DK	Bülow-Olsen, 1980a
Molinia caerulea stands	Grasses decreased	?	+	6	UK	Grant *et al.*, 1996a
Dry grasslands	Tall herbs increased in moist sites	–	=	10	NL	Bakker, 1989
	Tall grasses controlled, changes eu- to mesotrophic	+	+	10	NL	Bakker, 1989
	Litter accumulation	–	–	10	NL	Bakker, 1989
	Local increase of tall herbs changes eu- to mesotrophic	?	–/=	8	NL	Bakker and Grootjans, 1991
	Litter accumulation	+	=	8	NL	Bakker and Grootjans, 1991
Calcareous grasslands	No changes in trophic conditions	?	+	6	UK	Bullock *et al.*, 1994
	Decrease eutrophic species	?	+	7	NL	Willems, 1983
	Patterning in vegetation	+	?	3	NL	Hillegers, 1984
Wet grasslands	Tall herbs increased	?	?	4	D	Kaiser, 1995
	Tall herbs increased	?	–	8	NL	Bakker and Grootjans, 1991
	Phragmites decreased	?	+	4	F	Rozé, 1993

Com., community diversity.
Spp., species diversity.
***N* years**, time after grazing started.

With the introduction of grazing, the number of plant communities and plant species increased in many study sites but not in all (Table 5.2). Assessments of diversity or species richness must be interpreted with caution. Grazing for conservation and restoration management never takes place in small and homogeneous areas; therefore one has to cope with large and heterogeneous areas. In such areas, diversity can only be compared within a single large area with locally different grazing intensities and often no possibilities for replicates.

Starting extensive year-round grazing (3 sheep/ha on 11 ha) in the Westerholt, an area previously treated with artificial fertilizers after reclamation from heathland, revealed local differences in grazing intensity after some years (Bakker, 1989; Bekker and Bakker, 1989). The moist sites dominated by *Juncus effusus* were hardly grazed, places dominated by *Agrostis capillaris* were moderately grazed, and places dominated by *Holcus lanatus* were relatively heavily grazed. The areas of plant communities found in 1972 (at the start) and in 1977, 1982 and 1987 are indicated in Table 5.3. The picture of plant communities dominated by *Holcus lanatus* with some *Agrostis capillaris* in 1972 completely reversed in the 15-year period of extensive grazing. The community of *Poa pratensis* increased after five years, but then disappeared. The community dominated by *Juncus effusus* gradually spread, and a community with *Juncus effusus* and *J. acutiflorus* appeared. A new plant community characterized by *Leontodon autumnalis* and other rosette plants appeared in 1977 and formed mosaics with the communities dominated by *Holcus lanatus* and by *Agrostis capillaris*. Most striking was the transition from grassland with a few plant communities covering relatively large uniform areas to a grassland with an intricate pattern of smaller patches. Species richness showed a positive relationship with grazing intensity in the grassland area. Preferential grazing transformed an initially uniform *Holcus lanatus*-dominated sward into a pattern of three stands after 15 years: a lightly grazed *Agrostis capillaris* stand with 7 species/4 m^2, a moderately grazed *Holcus lanatus* stand with 17 species/4 m^2 and a heavily grazed *Leontodon autumnalis* stand with 23 species/4 m^2 (amongst them six rosette species). The grazing pattern emerged within 10 years and proved to be constant (section 5.3). The number of plant species in 400 cm^2 samples was significantly negatively correlated with the total aerial biomass in the three stands together, but not within each individual stand (Ter Heerdt *et al.*, 1991). This means that the correlation can be attributed to the developed pattern of plant communities (see also Moore and Keddy, 1989).

The effects of grazing in an abiotically different landscape may differ correspondingly, as illustrated by a study in The Netherlands in the brook valley of the Heest (*c.* 25 ha) in the middle course of the reserve of the Drentsche A river. The study area is located at the junction of two rivulets. Part of the area features wet meadows on peat soil, which were acquired by the State in 1971 and have since been grazed or cut without fertilizer appli-

Table 5.3 Area (ha) of plant communities in the grazed grassland of Westerholt, The Netherlands, at 5-year intervals (1972–1987); in parentheses is area included in mosaic communities (after Bekker and Bakker, 1989)

	Year			
Vegetation type	1972	1977	1982	1987
1. *Poa pratensis*	0.01	0.63	0	0
2. *Holcus lanatus*	5.47	2.29	0.13 (0.35)	0.01 (0.06)
3. *Leontodon autumnalis/Hypochaeris radicata*	0	0.12	0 (0.98)	0 (1.06)
4. *Agrostis capillaris/Holcus lanatus*	0	1.49	0.32	0.14
5. *Agrostis capillaris*	0.61	1.63	3.71 (4.46)	3.40 (4.42)
6. *Juncus effusus/J. acutiflorus/Agrostis capillaris/A. stolonifera*	0	0.05	0.31	0.20
7. *Juncus effusus/Carex nigra*	0.84	0.72	0.75	1.00
Mosaic of types 3 and 5	0	0	1.51	2.02
Mosaic of types 3 and 2	0	0	0.44	0.10
Total	6.93	6.93	6.87	6.87

cations. Another part of the area is a sandy plateau, which was used as fertilized farmland until 1976. On the plateau, rainwater infiltrates; whereas seepage occurs in the rivulet valleys. The brook valleys border upon the sandy plateau and are therefore affected by a strong mineral-rich groundwater stream, which can reach the ground level without impediment. In 1976 the whole area, including the wet meadows, started to be grazed in the summer with sheep and tall herbs were cut locally. In 1979 the grazing was taken over by cattle (approximately 1/ha) and no more cutting took place (Bakker and Grootjans, 1991).

The vegetation on the wetter peatland was, in fact, hardly grazed any more. Cattle only entered the wetter area late in the growing season. The *Lolio-Cynosuretum* and large parts of well-developed *Calthion palustris* communities with *Carex acutiformis* transformed into a species-poor *Carex acutiformis*–tall-herb community. Many short hayfield species disappeared, whereas tall herbs have greatly increased in abundance. Relics of the former *Calthion palustris* are only found at the junction of the two brooks, where the mineral-rich groundwater rises high up in the soil profile. At this point even orchids hold their ground amidst the tall *Filipendula ulmaria* (Bakker and Grootjans, 1991). On the dry parts of the plateau, the *Poö-Lolietum* transformed into a *Lolio-Cynosuretum*, which points to impoverishment of the soil. This was not a spatially homogeneous development. The cattle rested on the driest parts but did not graze there, resulting in a great density of dung patches and the establishment of tall herbs. These species displaced the low-growing species. Intensive grazing occurred at the western edge of the plateau. In 1989, 90% of the vegetation there was shorter than 5 cm and was intensively grazed. The impoverishment of the soil manifested itself in a reduction in species of eutrophic soils and an increase in *Agrostis capillaris*, which is an indicator of mesotrophic soils. An indication of the high grazing intensity was the increase in rosette plants (Van den Bos and Bakker, 1990). On the relatively wet eastern edge of the plateau moderate grazing took place. In 1989, 50% of the vegetation was shorter than 5 cm.

Grazing in a previously abandoned salt marsh resulted in an increase of the number of species. At the higher salt marsh, grazing led to a higher species richness than haymaking (Bakker, 1989; Bakker *et al.*, 1997), although it took five years before the species number in the grazed sites exceeded that in the mown sites. Cattle grazing prevented competitive exclusion of small-statured species (Olff *et al.*, 1997). On the low salt marsh with high salinity soil, however, cattle grazing resulted in a low species number (Figure 5.3) (Bakker, 1989; Olff and Ritchie, 1998). Plants on high salinity soils are small in stature and water-stressed. These stress factors may relax competitive interactions, as resource supply rates exceed demands, so that biomass reduction of dominant species by grazing does not benefit subordinate species. Salinity stress also improves forage quality by preventing 'dilution' of nitrogen in plant biomass. The selection for plants in saline parts of the salt marsh may even cause excessive damage to certain species, which reduces diversity (Olff and Ritchie, 1998). Soil compaction of wet soils at the lower salt marsh and trampling by large herbivores can also inhibit plant diversity (Bakker, 1989).

It can be concluded that abiotic factors, such as atmospheric deposition, groundwater level and soil type, often determine the nature of the impact of

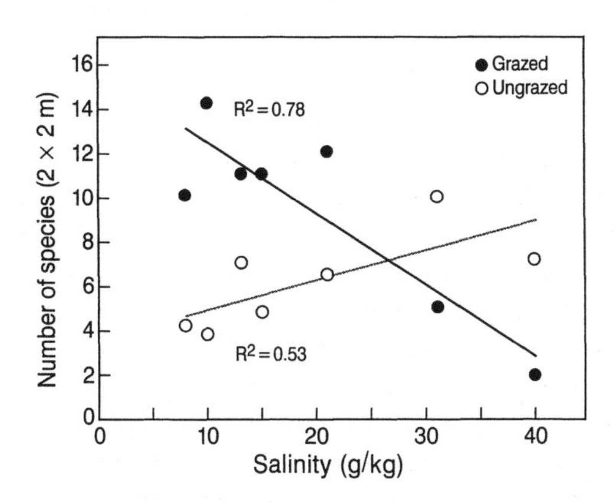

Figure 5.3 Shifting effects of herbivores on plant diversity across an environmental gradient at a salt marsh in The Netherlands. (After Olff and Ritchie, 1998.)

grazing on plant communities. Seemingly contradictory effects of grazing on the community and species diversity may be attributed to the fact that many study sites are heterogeneous and include different plant communities, which are not uniformly attractive to the herbivores. This is the subject of section 5.3.

5.2.2 Different herbivore species and stocking rates

Different herbivore species stocked at different densities may also cause widely dissimilar impacts on the vegetation, but methodological obstacles have hampered the accumulation of insight in this domain. Some experiments comparing the effects of different herbivore species are known. A comparative study of diet selection by cattle and sheep grazing *Nardus stricta* grassland in the growing season (Grant *et al.*, 1985; Hodgson *et al.*, 1991) showed that cattle ingested more *Nardus* than did sheep. Another comparative study in paddocks (0.15–2.3 ha) in Scotland (Grant *et al.*, 1996b) made clear that the utilization of *Nardus* increased as the height maintained of between-tussock grasses *Agrostis* spp., *Deschampsia flexuosa* and *Festuca ovina* was reduced. Cattle and goats utilized more *Nardus* than did sheep. Under cattle grazing the cover of *Nardus* decreased from 55 to 30% over five years, whereas it increased to about 80% under sheep grazing. Broad-leaved grasses tended to increase under cattle grazing. Utilization by cattle or goats may lead to a reduced cover of *Nardus* and an increased cover of *Agrostis* and *Festuca* species (Grant *et al.*, 1996b).

Studies on grazing of heather moorland in northeast Scotland suggested that the stocking rates above which *Calluna vulgaris* declines was estimated at 0.2 cattle and 2.7 sheep/ha with a normal amount of light grazing by other herbivores, like red deer, rabbit, hare and red grouse (Welch, 1984a,b). These figures only hold for a certain level of plant production, however, as indicated by the observation that *Calluna* cover increased under a year-round stocking rate of 1.2 sheep/ha (Hewson, 1977), but was checked at a density of 0.7 sheep/ha on slow-growing blanket bog (Welch and Rawes, 1966). Records of the botanical composition over periods of 4–11 years in many moorland sites in northeast Scotland indicated that heavy grazing (> 240 ml dung/m^2 per year) favoured grasses and herbs, whereas light grazing (< 60 ml dung/m^2 per year) favoured, amongst others, *Erica cinerea*, *E. tetralix* and some lichens (Welch, 1984c). Twenty-year trends in botanical composition revealed that heavy grazing resulted in a decline of *Calluna*, other ericoids, lichens and *Deschampsia flexuosa*, whereas significantly increasing species included *Agrostis capillaris*, *Anthoxanthum odoratum*, *Festuca ovina*, *Galium saxatile*, *Luzula multiflora*, *Nardus stricta* and the bryophyte *Rhytidiadelphus squarrosus*. At a site close to agricultural reseeding, *Cynosurus cristatus*, *Dactylis glomerata* and *Lolium perenne* invaded. The unpalatable graminoids *Juncus squarrosus*,

Molinia caerulea and *Nardus stricta* did not increase. Light grazing resulted in an increase of *Calluna* and several bryophyte species, whereas graminoids and herbs declined significantly, especially when the *Calluna* sward was continuous (Welch and Scott, 1995). At sites with patchy occurrence of *Calluna,* species growing outside the heather patches were not (yet?) affected by the tall and dense heather. Nevertheless, *Agrostis capillaris* and *Festuca rubra* declined significantly (Welch and Scott, 1995).

The effects of four years of grazing by red deer (*Cervus elaphus*) were studied in paddocks in northeast Scotland. Old *Calluna vulgaris* was less able to withstand grazing than young heather, and heather cover was reduced where the stocking rate was above two hind-equivalents/ha. Reduction in heather cover was accompanied by a rapid increase of *Deschampsia flexuosa* and a slow increase of *Agrostis* spp. and *Vaccinium myrtillus* (Grant *et al.*, 1981). These effects seem similar to the aforementioned effects of heavy sheep grazing. Such similarities could well be attributed to heavy grazing giving little opportunity for differences in diet selection between herbivore species, so that differential effects on the vegetation will be limited (Grant *et al.*, 1981).

The effects of different stocking rates of sheep were studied in an experiment on an artificial salt marsh in Germany. The intensively grazed paddock (10 sheep/ha) was covered by a short monotonous sward of *Puccinellia maritima* after five years. *Aster tripolium* and *Atriplex portulacoides* were rare. In the paddocks stocked at 1.5 and 3 sheep/ha, *Aster* and *Atriplex* had increased – especially further away from the dike (Figure 5.4). Close to the dike these species were still lacking as a result of local heavy grazing (Kiehl *et al.*, 1996). This shows the problem of comparing the effects of different stocking rates: variation in grazing intensity inside a paddock may be larger than between paddocks. More attention should therefore be given to vegetation patterns resulting from patterns in grazing intensity.

5.3 VEGETATION PATTERNS AT VARIOUS SPATIAL SCALES

In large areas grazed for nature conservation purposes, vegetational patterns include patterns of plant communities with different species composition and patterns in vegetation structure. These differences can be due to abiotic conditions and/or differences in grazing intensity resulting in short and taller stands. We will discuss both types of patterns and try to relate them to patterns in grazing intensity.

To distinguish between the specific use of the word grazing (which refers only to the defoliation process) and its more general use (which includes the associated effects of treading and of dung and urine return), the term 'terrain use' or 'occupancy' is preferred for grazing in general, and 'foraging' for grazing denoting consumption (see also Hodgson, 1979). Grazing intensity then implies terrain use or occupancy. It is often

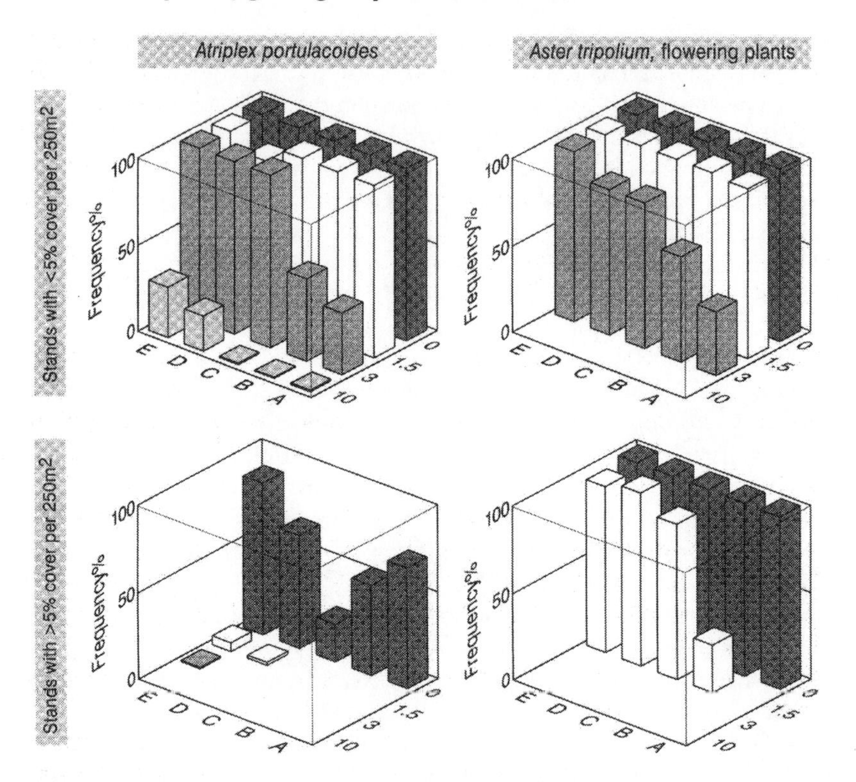

Figure 5.4 Frequency of occurrence of *Atriplex portulacoides* and *Aster tripolium* at various distances from the dike, ranging from A to E, in the artificial mainland salt marsh of Sönke-Nissen-Koog (Schleswig-Holstein, Germany) after five years of sheep-grazing with various stocking rates ranging from 0 to 10 sheep/ha. (After Kiehl *et al.*, 1996.)

recorded by direct observations or by the amount of dung voided. Some rough indications are used to assess foraging intensity, like the area of short turf, the height of the canopy and the amount of litter. A more sophisticated measurement to estimate foraging intensity is the utilization factor (Esselink *et al.*, 1991), which is the herbage utilization ('t Mannetje, 1978) per unit area per unit time expressed as a proportion of the net herbage accumulation (Hodgson, 1979) per unit area per unit time, when herbage utilization is the sum of: the herbage consumed by the animals; subsidiary losses of herbage due to spoiling by trampling and fouling; and decomposition. Herbage accumulation is the primary production minus decomposition and consumption by invertebrates and small vertebrates measured in temporarily (4–6 weeks) excluded areas. Hence the term 'utilization factor' is analogous to Hodgson's 'efficiency of grazing'

(Hodgson, 1979), being the herbage consumed expressed as a proportion of the net herbage accumulation (see also De Leeuw and Bakker, 1986). Estimates of the herbage consumed as a percentage of the herbage utilized range from 30 to 50% for cattle (Esselink *et al.*, 1991). Parsons *et al.* (1984) reported an almost 80% overestimation when using the exclusion method in a perennial ryegrass sward as opposed to a measured intake. Herbage utilization, therefore, is not the same as consumption.

5.3.1 Vegetation patterns in relation to abiotic conditions and uneven terrain use

Differences in grazing intensity, whether in terrain use or in herbage utilization, are particularly interesting for nature conservation if they result in patterns in the structure of the sward. This is only possible if patterns of uneven grazing intensity are constant for several years. Only plant communities dominated by annual plants can follow swiftly changing patterns, provided these plant species have a persistent seedbank or good dispersal capacities. Vegetational patterns can of course only be found where herbivores do not utilize the entire the annual production. Herewith extensive grazing is defined. This definition implies that extensive grazing can only be defined in terms of the ratio of foraging, including trampling, and the local plant production.

Patterns in grazing intensity are the result of spatial differences in abiotic conditions and plant characteristics causing selection by the herbivores. Selectivity in terrain use is a well known phenomenon (Chapter 9 and section 5.7). In the Scottish uplands, Charles *et al.* (1977) found that deer preferred feeding on grasslands rather than heaths and bogs. Wielgolaski (1975) in Norway and Job and Taylor (1978) in Wales reached similar conclusions for upland sheep grazing. Hunter (1962) demonstrated selectivity for the more productive grasslands in upland communities in Scotland with various sheep-grazing intensities. Thalen *et al.* (1987) observed a pattern of terrain use four years after ponies started year-round grazing in a mosaic (100 ha) of *Pinus* plantation, heathland and abandoned arable fields in The Netherlands. The pattern appeared relatively stable through the years, since a similar pattern was found 10 years after the start of the grazing regime. The grazing patterns of cattle and sheep in the previously discussed areas (section 5.2) of Heest (Bakker and Grootjans, 1991) and Westerholt (Bakker, 1989) also turned out to be constant for a series of years. In a former arable field grazed by cattle, the pattern of terrain use seemed to reach relative stability within four years (Van den Bos and Bakker, 1990).

These patterns in grazing intensity often seem to reflect the preference for or avoidance of the dominant species of different plant communities. Grazers typically prefer open grasslands, if these are not too wet. Many additional habitat factors may cause uneven terrain use (Chapter 9). It can be concluded that the observed patterns of grazing intensity at a larger spatial scale are controlled by the patterns of plant communities, which in turn are determined by abiotic conditions.

5.3.2 Macro-patterns induced by herbivores

At an intermediate spatial scale, vegetational patterns may be induced by differences in local grazing intensity, even on more or less homogeneous substrate. Van den Bos and Bakker (1990) found a clear patterning in the grazed vegetation of a former arable field. Heavily grazed communities of *Hypochaeris radicata*, *Lolium perenne* and *Trifolium repens* developed, as well as moderately grazed *Agrostis stolonifera* grassland with *Cirsium arvense* and *Rumex acetosella* and a very lightly grazed *Juncus effusus* community.

An example from areas with a longer grazing history concern artificial salt marshes (Dijkema, 1983). These are characterized by a relatively flat topography and a dense artificial drainage system. In the Leybucht salt marsh, Germany, vegetation height increased from the dike towards the intertidal flats, which reflected a decrease of cattle grazing intensity in this direction; the gradient in vegetation height was indeed not found in ungrazed sites (Andresen *et al.*, 1990). A similar pattern of terrain use was found in other areas in Germany (Dierssen *et al.*, 1996) and in The Netherlands (Bakker *et al.*, 1997). The species gradients from the dike towards the tidal flats found in the aforementioned sheep-grazing experiment of Kiehl *et al.* (1996) (Figure 5.4) were apparently caused by the position of freshwater points for the grazing animals, as these are situated close to the dike or even further inland. The presence of a stable or favoured resting site can also be a point from which trampling and dunging gradients originate. Rosén (1973) described a gradient in plant communities in a 400 m transect on the 'alvar' limestone grassland of Öland, Sweden. A few ruderals occurred close to the stable, mosses were found from 70 m onwards, whereas lichens occurred more than 200 m from the stable. Total species richness reached an optimum at a distance of 100 m from the stable.

5.3.3 Micro-patterns induced by herbivores

Apart from macro-patterns at the plant community level, grazing selectivity may result in a small-scale mosaic or micro-pattern (Figure 5.5). The creation and maintenance of micro-patterns in grasslands by grazing can be illustrated with the example of the Westerholt (Bakker *et al.*, 1984). In the initially ungrazed uniform *Holcus lanatus* community, a mosaic of taller tufts interspersed with shorter vegetation emerged under grazing. The diameter of the tufts ranged from 0.5m to a maximum of 3 m (Figure 5.6). Areas with a canopy height < 10 cm and little litter accumulation were recorded as heavily grazed, patches with a taller canopy and more litter accumulation as lightly grazed (Table 5.4). Biomass samples from lightly grazed and heavily grazed patches confirmed the higher quality of the latter (Bakker, 1989). From repeated mapping it was shown that centres are apparent both in the lightly grazed taller tufts and in the heavily grazed short turf. The boundaries between the centres with tall tufts and short turf apparently shifted within an edge zone. The diameter of lightly grazed patches increased during spring and early summer because of a higher herbage accumulation than utilization, and decreased again during autumn and winter when herbage utilization exceeds accumulation, resulting in a system of 'pulsating patches'. The resemblance between patterns was stronger at a short interval of one year than over two years. Nevertheless, the resemblance did not decrease further after two years and remained significant even after four years (Bakker, 1989). Hence it may be concluded that the micro-pattern did not change randomly from year to year, but was fixed under the grazing regime. Cutting and exclosure experiments suggested that the micro-patterns apparently developed due to 'random grazing' (Bakker, 1989), which should only be expected in a uniformly palatable vegetation, as was the case in this community.

An experiment in paddocks of 12.6 ha in native mixed-grass prairie in Kansas, United States, compared seasonal cattle grazing during 150 days at 0.7 cattle/ha with intensive early stocking during the first 75 days at 2.1 cattle/ha, for two consecutive seasons. Intensive early stocking resulted in heavily grazed areas and undergrazed patches. The rather uniform vegetative cover at the beginning of the second season hardly affected the devel-

Figure 5.5 Micro-patterns in *Agrostis/Festuca* grassland under grazing at low stocking rate.

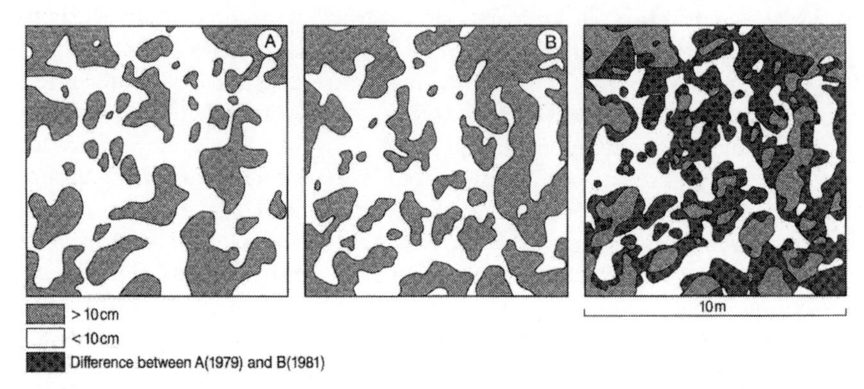

Figure 5.6 Micro-pattern (10 × 10 m²) in grassland vegetation of Westerholt, The Netherlands, in (A) 1979 and (B) 1981, and the difference between the two years. (After Bakker *et al.*, 1984.)

Table 5.4 Live and dead aerial biomass (g/m², ±S.E.) in heavily grazed areas and lightly grazed patches of Westerholt, The Netherlands, measured in sample plots of 78.5 cm² (N = 5) in September 1981 (after Bakker *et al.*, 1984) (the number of species was 11 in both types)

Species	Heavily grazed	Lightly grazed
Agrostis capillaris	15.4±3.2	306.6±36.7
Holcus lanatus	15.7±5.6	71.7±14.6
Anthoxanthum odoratum	13.6±4.8	5.6±3.6
Poa pratensis	5.6±1.9	7.0±1.8
Cynosurus cristatus	0.1±0.1	4.4±3.8
Lolium perenne	0	1.1±1.1
Festuca rubra	0	0.9±0.8
Total monocots	50.4±7.2	379.3±44.6
Trifolium repens	19.2±4.0	11.1±2.6
Ranuculus acris	3.7±1.1	15.9±8.2
Leontodon autumnalis	27.5±8.0	36.4±18.9
Taraxacum spec.	0.5±0.3	0
Hypochaeris radicata	5.5±4.4	0
Trifolium pratense	0.3±0.3	0
Stellaria media	0	0.5±0.5
Total dicots	56.7±9.8	63.9±24.6
Green aerial biomass	107.1±10.9	461.2±31.8
Dead plant material	93.2±6.6	393.1±66.1
Bryophytes	10.6±4.7	0
% Monocots	47.0±5.4	86.1±6.7

opment of patches in that season. Low-stocking seasonal grazing, however, resulted in re-establishment of undergrazed patches (13 cm height and 2450 kg/ha, in contrast to 4.6 cm height and 700 kg/ha in heavily grazed patches) in the second year. Of the undergrazed points in the first year, 86% remained undergrazed in the second year, and hence locations of patches in two consecutive years were significantly associated (Ring *et al.*, 1985).

Hunter (1962) and Nicholson *et al.* (1970) found micro-patterns in upland vegetation in the United Kingdom due to the avoidance of species. They described the development of *Nardus stricta* tussocks in a *Nardus stricta/Festuca ovina/Molinia caerulea/Anthoxanthum odoratum* sward under grazing with a stocking density of 2–5 sheep/ha. Single plants or clumps of *Anthoxanthum odoratum* or young *Molinia caerulea* shoots provided the loci that were initially moderately grazed. Because of subsequent grazing around the edges of these patches, the initial centres expanded outwards, producing a patchwork of short-tufted areas. Closely grazed areas eventually joined and grazing pressure over the entire area became relatively uniform, except for isolated rejected areas – usually *Nardus* tussocks. In a formerly intensively exploited fen meadow in Germany, stocking density was reduced to one animal/ha. A pattern then developed within one grazing season by the avoidance of *Cirsium arvense*, *Phalaris arundinacea* and *Deschampsia cespitosa* (Fischer, 1995).

The relationship between large- and small-scale vegetation patterns may be illustrated by an example from wet heathland, dominated by *Erica tetralix* and *Molinia caerulea* and grazed by sheep in the Dwingelose Heide nature reserve (1300 ha), The Netherlands (Figure 5.7). Several hundred sheep leave the stable in the morning and return in the afternoon, tended by a herdsman. The first site described was at a distance of 150 m from the stable and was heavily grazed and trampled by sheep; *Molinia caerulea* covered 85%, *Erica tetralix* 1%. Mean canopy height was 4 cm, litter was lacking, 90% of incoming light reached the soil, whereas hardly any spatial pattern of light interception existed. The second site was at 1000 m from the stable and intermediately grazed by sheep; *Erica tetralix* covered 55% and *Molinia caerulea* 35%. Mean canopy height was 15 cm, litter thickness was 2 cm, 15% of incoming light reached the soil, and a small-scale spatial pattern of light interception was found. The third site was at 2000 m from the stable and hardly grazed at all by sheep; *Molinia caerulea* covered 50%, *Sphagnum* spp. 30% and *Erica tetralix* 1%. Mean canopy height was 25 cm with large variations, litter thickness was 7 cm, 15% of incoming light reached the soil, and a large-scale pattern of light interception was found. We conclude that grazing may influence vegetation patterns at a range of spatial scales.

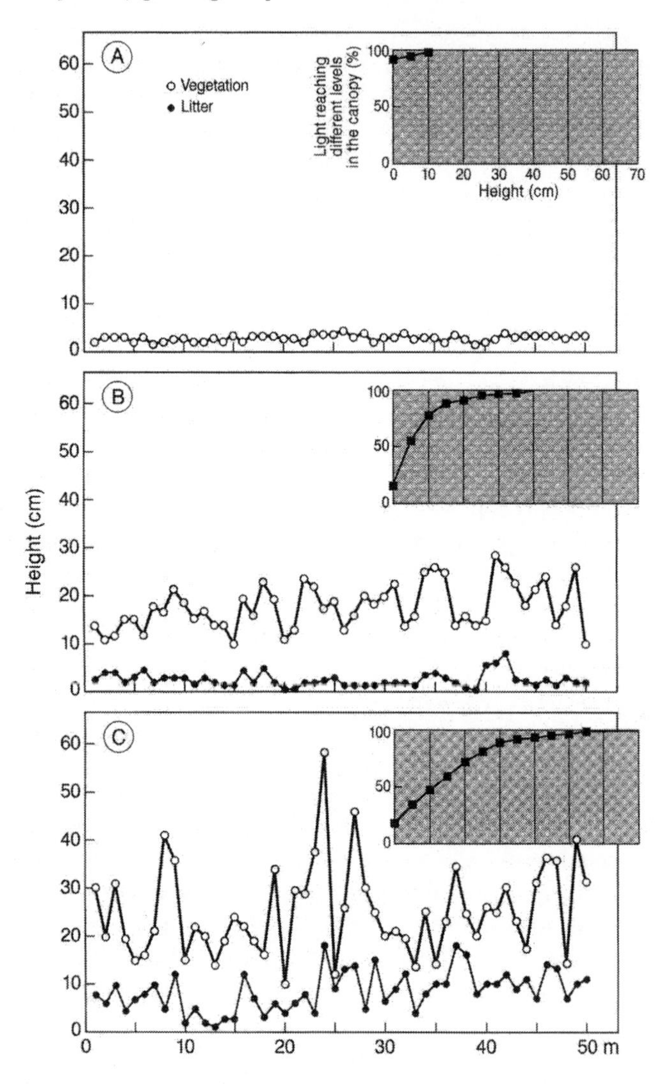

Figure 5.7 Height of the canopy and litter layer along transects of 50 m, and percentage of light reaching different levels in the canopy, in a sheep-grazed gradient in a heathland in the Netherlands at different distances from the stable: (A) 150 m; (B) 1000 m; (C) 2000 m.

5.3.4 Mechanisms behind vegetation patterns

To explain the creation and the maintenance of micro-patterns the question to be answered is: what makes grazing animals return to previously depleted patches? Changes in forage quality and canopy structure, particularly the height of the vegetation, are likely answers. Herbivores may select forage on the basis of attainable intake rate, which is related to the structure of the sward, namely, tiller density and canopy height (Black and Kenney, 1984; Kenney and Black, 1984; Penning, 1986; Drent and Prins, 1987) or cover instead of biomass of favoured food species (Van der Wal *et al.*, 1998). The trade-off between quality and quantity can be explained by the senescence of plant material and the presence of stems in tall swards. The proportion of cell wall increases during senescence (Osbourn, 1980) and stems have a lower digestibility than leaves (Deinum and Van Soest, 1969; Hacker and Minson, 1981). To reach the same intake level, grazing animals have to spend more time in short swards than in tall swards but they are rewarded with a higher quality that may yield a greater daily energy intake because of a higher food processing rate in the gut (Wilmshurst *et al.*, 1995). WallisDeVries and Daleboudt (1994) found that cattle in *Agrostis/Festuca* and in *Lolium* grassland may maximize their daily energy intake by selecting for the short patches that offer a high digestibility. They concluded that, since the degree of selection was moderate, cattle were probably forced to include tall patches with lower digestibility, due to foraging-time limitation. In an experiment with wapiti (*Cervus elaphus*) in Canada, Wilmshurst *et al.* (1995) indeed found a selection for patches of intermediate plant biomass, as predicted by their model. Grazing thus enhances forage quality in comparison with ungrazed areas (Ydenberg and Prins, 1981; Cargill and Jefferies, 1984) in such a way that the animals may return and harvest regrowth in a cyclic pattern (see also Drent and Van der Wal, 1998). Cyclic grazing over 4–5-day periods has been observed in several wild herbivores, ranging from geese (Prins *et al.*, 1980) to wildebeest (*Connochaetus taurinus*) (McNaughton and Banyikwa, 1995) and African buffalo (*Syncerus caffer*) (Prins, 1996). The preference of herbivores for patches with different vegetation structure in relation to forage quality will be discussed further in Chapter 9.

5.4 VEGETATION PATTERNS IN TIME

All vegetation patterns discussed so far seem stable in time. Interesting additional information on vegetation patterns came from a study in a uniform salt marsh vegetation dominated by *Festuca rubra* in an artificial marsh along the German Wadden Sea (Berg *et al.*, 1997). Five years after the start of a seasonal sheep-grazing experiment, micro-patterns were quantified for the first time (mapped at 10 × 2 m). No pattern was found in the heavily, traditionally grazed (10 sheep/ha) and the ungrazed sites, because of their uniform short and tall canopy, respectively, but a micro-pattern did develop at intermediate stocking rates (1.5, 3 and 4.5 sheep/ha). The spatial diversity index was significantly highest at the 3 sheep/ha stocking rate (Berg *et al.*, 1997). Consecutive mapping during three years revealed that the observed micro-patterns were not stable, in contrast to the previously mentioned situations. The moderately grazed transects showed a hummocky topography with the highest spatial diversity at 1.5 sheep/ha, in contrast to the heavily grazed transect. Most of the sediment during inundation will be trapped in the taller stands (section 5.5). Marsh elevations were on average up to 3 cm lower in the short than in the taller stands, suggesting that somewhat lower-lying patches were grazed and trampled more than elevated patches. The explanation for the unstable micro-patterns in the vegetation might come from the existing macro-patterns of heavy grazing close to the dike and lower grazing intensity further towards the tidal flats (Dierssen *et al.*, 1996). Moist periods will result in higher plant production in salt marsh systems as a result of desalination (De Leeuw *et al.*, 1990). This will allow the sheep to concentrate their grazing near the dike, resulting in a higher proportion of tall stands at a greater distance from the dike. One year with a wet spring coincided with high incidence of tall stands in the transects. In dynamic coastal environments, abiotic processes like rainfall and winter sedimentation apparently control summer grazing in determining vegetational patterns (Berg *et al.*, 1997).

In the long run, vegetation succession on the salt marsh leads to dense stands of *Elymus athericus,* which is unpalatable to the small native herbivores (geese, hares and rabbits). The geese abandoned older salt marsh areas and turned to newly formed areas. The succession can only be reversed by cattle; and geese and hare may then return (Olff *et al.*, 1997). In inland areas, large herbivores do not seem able to stop the succession by bush encroachment in the long term, unless they are facilitated by megaherbivores, such as the elephant (Chapter 3). This has not been truly tested with a native herbivore assemblage.

By combining the outcome of a number of studies, Davidson (1993) tried to relate the effects of herbivory in old-field succession in temperate North America to resources for plant growth, palatability and herbivore preferences, and plant defences in response to herbivory (Figure 5.8). Initially nitrogen is limiting, but the increase of organic matter in the soil enhances nitrogen availability, which becomes limiting again by occlusion in tree biomass. As the canopy closes during succession, light becomes increasingly limiting. Although annuals are poorly defended, they can escape from herbivores by their short life cycle. Herbaceous perennials often produce nitrogen-based toxins and feeding deterrents, which make them unpalatable to herbivores. Hence herbivores prefer graminoids, palatable shrubs and pioneer trees, and thus retard succession. Removal of shrubs and pioneer trees by herbivores gives way to lower graminoids that store a considerable fraction of their resources below-ground. The species with rapid regrowth on fertile soils allocate much carbon to regrowth. Invading late-succession trees allocate carbon to carbon-based defences. Moreover, they store nutrients, thus rendering the environment more nutrient-limited and unfavourable for fast-growing species of the mid-successional series. Grazing of the understorey will accelerate succession to the climax stage. In summary: the effects of herbivory on succession may be a predictable outcome of plant resource distribution, and its influence on the evolutionary plant defence characteristics and palatability to herbivores (Davidson, 1993). Specific responses of individual plant communities may differ from this general picture. Edaphic factors may affect vegetation structure, quality and the course of succession. Also, below-ground herbivores may have an impact of which little is known so far (Davidson, 1993).

A successional gradient often represents a productivity gradient. Huisman *et al.* (1998) modelled the effects of competition for light between two plant species. The low-growing species will win in the long term at low levels of nutrient availability; the tall species will win at high levels of nutrient availability. These results remain unchanged when a herbivore that prefers the short species is introduced. Modelling and field experiments with two species of herbivores and with palatable and unpalatable plant species would render valuable information for the management of plant

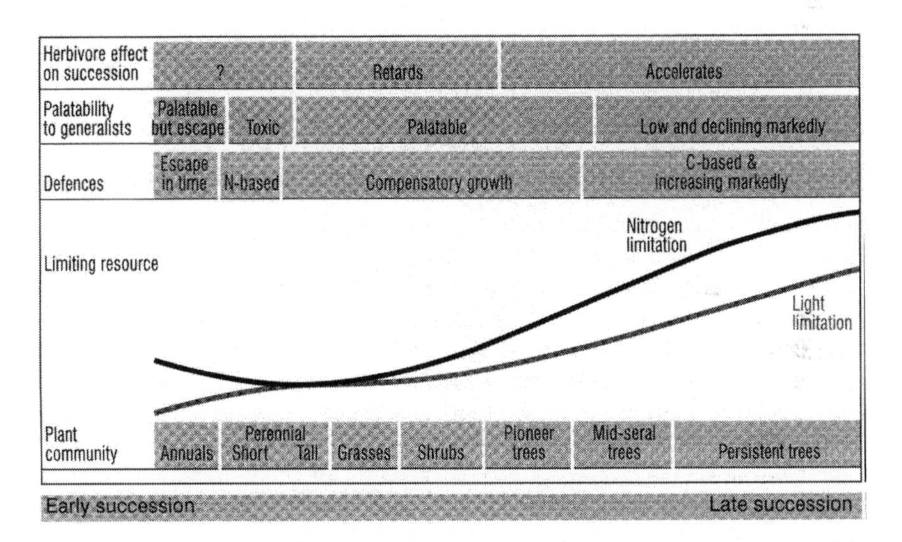

Figure 5.8 Herbivory effect on vegetation succession in old fields. Relatively favourable nitrogen levels early in many old-field secondary successions are correlated with the presence of rapidly colonizing herbs, which may be comparatively undefended (especially annuals) or defended with N-based secondary compounds (some annuals and herbaceous perennials, often those of smaller stature). Herbivory affects principally grasses, some of the taller herbs and deciduous shrubs and trees. Because these palatable plants comprise the intermediate successional seres, herbivory on these plants retards succession from earlier seres, but expedites succession to later seres that include species with a greater investment in plant defences. (After Davidson, 1993.)

diversity. With respect to the changes in time involving plant resources and preference by herbivores, it is interesting to consider the species richness of many grazed grasslands. Mitchley (1988) studied the turf of an ancient species-rich and 'fine-grained' chalk grassland. The more abundant species were more abundant because of their taller stature, yet the lower-growing, sparse species apparently were not shaded out. It is likely that grazing removes proportionally more from the leaf canopy of the taller species. Under a given grazing regime, a steady-state system may exist between tall species (promoted between grazing events) and short species (favoured at grazing events). The balance will depend on the morphology and phenology of each pair of species, the timing and degree of defoliation, regrowth characteristics and palatability. At low grazing intensity the balance will benefit the species with tall growth forms; at high grazing intensity the low-growing species will win.

5.5 EFFECTS ON ABIOTIC CONDITIONS

Micro-patterns with large differences in abiotic conditions and plant communities between heavily and lightly grazed patches were mentioned by Pigott (1956) in Upper Teesdale (UK) and Regnéll (1980) in southern Sweden. They described peaty tussocks on a calcareous and loamy subsoil. The tussocks were considered to be the residual parts of an initially continuous turf, split up by trampling and subsequent water erosion. Large abiotic differences were found between tussocks and depressions. This finding, in combination with the age of the tussocks, explained their divergent plant composition. Once tussocks are formed they perpetuate and govern the trampling pattern of the cattle. Out of 117 steps of freely grazing cattle, 106 were placed in depressions, 5 at an intermediate position and only 6 on tussocks (Regnéll, 1980). Salt marsh at transitions to dune may show accumulation of organic matter due to freshwater seepage from the dune hinterland. Such wet organic soils are often puddled by trampling in a way that creates a similar tussocky vegetation with deep cattle hoofmarks at Skallingen salt marsh, Denmark (Jensen *et al.*, 1990).

Differences in grazing intensity on sandy soil in the Westerholt led to variation in soil resistance, as measured with a penetrometer, with soil compaction in heavily grazed areas (11.5 ± 2.3 kg/cm^2) in comparison with lightly grazed patches (5.5 ± 0.7kg/cm^2) (Bakker, 1989). The different components of the micro-pattern in the Westerholt harboured different dominant plant species (Table 5.4). The considerably larger amount of green aerial biomass and dead plant material in the lightly grazed patches indicates litter accumulation. It probably limited the occurrence of mosses. The C : N ratio of fresh dead plant material in September was significantly lower in the heavily grazed patches than in the taller stands, and the decay rate of dead material was correspondingly higher in the heavily grazed patches (Ter

Heerdt *et al.*, 1991). Differences in species dominance in short and tall patches were also reported by Silvertown *et al.* (1994). Different C : N ratios and subsequent rates of decomposition of dominant plant species, and differences in soil aeration as a result from soil compaction, may eventually induce differences in soil conditions (Miles, 1987).

Recent studies in temperate salt marshes, enriched in nutrients by marine sedimentation, indicate higher N-mineralization rates in areas excluded from small herbivores (Van Wijnen *et al.*, 1998) and sheep (Vivier, 1997) during a few years than in grazed areas. Apparently, the cessation of grazing resulted in an accumulation of easily decomposable litter that enhances nutrient availability. Nevertheless, the decomposition of litter is slower than that of faeces (Perkins *et al.*, 1978). Hence, grazing generally enhances nutrient cycling, sometimes up to 10-fold, of N (Floate, 1970), P (Harrison, 1978), Ca, Mg, Mn and P (Bülow-Olsen, 1980b). Grazing – in particular, extensive grazing – results in uneven deposition of dung. Enhancement of nutrient cycling therefore only occurs locally by concentration of nutrients, and does not necessarily contradict the finding that extensive grazing prevents litter accumulation and thus may lower nutrient availability at a larger scale.

A dramatic example of a keystone herbivore affecting abiotic conditions is the lesser snowgoose (*Anser caerulescens caerulescens*). Experiments suggested that intermediate levels of goose grazing pressure, including the return of nutrients by faeces, resulted in greatest enhancement of plant production in the extremely nutrient-poor subarctic salt marshes (Hik and Jefferies, 1990). This suggests enhanced cycling of nutrients released from the faeces of the geese. The effects of intensive foraging result in the progressive destruction of subarctic salt marsh vegetation through a positive feedback that generates unfavourable soil conditions for plants. A 10-fold increase of breeding pairs over 25 years at La Pérouse Bay, Canada, as a result of favourable conditions in the winter-staging areas, caused an intensification of foraging on salt marsh vegetation (foraging by geese includes both grazing and grubbing). Both the aerial biomass in intact swards and the area of intact swards decreased. Grubbing for roots and rhizomes of graminoid species caused the destruction of salt marsh swards (Jefferies, 1988). In sites with reduced biomass, soil salinity increased due to evaporation and thereby reduced plant growth (Srivastava and Jefferies, 1996). Inland from the salt marsh, *Salix* stands turned into bare soil as grubbing around the bushes increased soil temperature and soil salinity. This process started a positive feedback, which eventually caused the death of the *Salix* bushes (Iacobelli and Jefferies, 1991).

A similar positive feedback may be found between plant density and water infiltration in semi-arid grasslands. A herbivore-induced decrease of plant biomass can result in soil degradation and reduced plant growth. Positive feedback between reduced plant density and deteriorating soil may then contribute to irreversible vegetation destruction (Van de Koppel *et al.*, 1997; see also Chapter 10).

Several authors have reported higher soil salinity in grazed than in abandoned salt marshes (Kauppi, 1967; Siira, 1970; Schmeisky, 1977; Hansen, 1982). Due to evapotranspiration in grazed areas with little plant cover, capillary rise may result in increasing soil salinity. Soil compaction and the consequent reduction in water infiltration rate and retention capacity also contribute to a comparatively slow desalination in areas affected by grazing (Slager *et al.*, 1993). This phenomenon was attested by the occurrence of species from the lower salt marsh in grazed salt marshes, and the maintenance of halophytic plant species in a grazed sward on desalinating embankments (Joenje, 1985; Slager *et al.*, 1993), whereas a tall herb community of glycophytic species had developed in the abandoned area nearly 50 years after embankment (Westhoff and Sykora, 1979). The proportion of characteristic species of the lower salt marsh (e.g. *Aster tripolium*, *Glaux maritima*, *Limonium vulgare*, *Plantago maritima*, *Spergularia maritima*, *Salicornia europaea*, *Suaeda maritima*) was higher in the grazed than in abandoned sites in the salt marsh, and higher even than in mown sites (Bakker, 1989). No differences in soil salinity were found here. A sowing experiment revealed that the higher interception of light in the taller canopy was responsible for the absence of lower salt marsh species in the mown and abandoned sites (Bakker and De Vries, 1992). Variation in soil salinity is therefore not the only explanation for the more frequent occurrence of halophytic species in embanked salt marshes under grazing.

Gray and Scott (1977) showed that above the main zone of distribution *Puccinellia maritima* was restricted to local hollows (hoofmarks?) or areas opened up by turf-cutting. Driftline material can also generate gaps by covering established vegetation, which then dies off. Species that normally thrive lower down the salt marsh can temporarily colonize these gaps higher on the salt marsh (Bertness and Ellison, 1987). Upper salt marsh species cannot spread down towards the lower salt marsh because of their low salt tolerance. Lower salt marsh species, however, can invade the upper salt marsh since they grow very well in less saline environments, provided that there is no light interception by a taller canopy (Bakker, 1989). A similar picture emerges from an experimental field study in an Alaskan salt marsh, where many species could potentially survive. Species occurring in zones along a physical gradient are limited by physiological tolerance towards one end of the gradient, and by competitive ability as influenced by grazers towards the other end of the gradient (Snow and Vince, 1984). Bertness and Ellison (1987) reached a similar conclusion on the basis of transplantation experiments.

Litter accumulation not only contributes to the interception of light reaching the soil but may also be a mechanical barrier for the establishment of plant species. Five years after the start of cattle grazing on part of the Schiermonnikoog salt marsh, most of the litter layer of about 5 cm had disappeared. Detailed measurements indicated that it takes several years before grazing significantly reduces the amount of litter. This seems to be in agreement with the slow increase of the number of species in formerly abandoned salt marshes due to grazing, as compared with the swift increase in grazed sites that were formerly mown with removal of standing crop and litter after cutting (Bakker, 1989).

A study on the Dollard marsh in the Ems estuary, at the border between The Netherlands and Germany, examined the effects of management of both abiotic and biotic conditions (Esselink *et al.*, 1998). After designation as a nature reserve, grazing with cattle continued, but at reduced stocking rates (0.5–1 cattle/ha), and the maintenance of the drainage system was abandoned. Vertical accretion rates were generally negatively correlated with the initial marsh elevation, the distance from the tidal flats, the distance from main creeks and, in many cases, the distance from minor creeks. Because of a gradient in grazing intensity, vegetation density decreased from the outer marsh towards the dike, and was probably one of the causes in the aforementioned accretion patterns (section 5.4). After the abandonment of drainage, the number of levees along former ditches increased, as did the elevation differences at many already existing levees. The levee development was more pronounced towards the dike. This could be explained by the greater differences in vegetation structure between the levees and the marsh interiors towards the dike, and indirectly by the gradient in grazing intensity (Esselink *et al.*, 1998). Higher rates of sedimentation at levees than in the marsh interiors were also reported by Leonard *et al.* (1995). The role of the vegetation in catching sediment was confirmed by the finding that water flow speeds in less densely vegetated interiors may exceed speeds on the more densely vegetated levee. This implies that vegetation structure is a more important control over free surface flow speed than proximity to the creek. The possible effect of grazing on vegetation structure and, hence, on sedimentation rate was also demonstrated in the mainland marshes of the Leybucht, Germany. After four years of measurements a rate of sedimentation of 2.1 cm/year was found in ungrazed sites at 40 cm above MHT, and 1.6–1.7 cm/year in grazed sites (Erchinger *et al.*, 1996).

It can be concluded that grazing has little direct impact on abiotic conditions by soil compaction and nutrient cycling. The main effects are indirect, by affecting the structure of the vegetation canopy and thereby influencing light availability and opportunities for germination and seedling establishment, sediment rate in periodically inundated areas, and litter accumulation. Catastrophic events induced by grazing, such as in subarctic and semi-arid regions, are not reported from temperate regions.

5.6 DISPERSAL OF DIASPORES

5.6.1 Endozoochorous and epizoochorous transport

Seed dispersal by grazing domestic animals is mentioned by Klapp (1965) and Ellenberg (1978) (Figure 5.9). They suggested that the dispersal of seeds of clover and grasses with agricultural value was facilitated by alternatively grazing agriculturally improved land and rough grazing areas. On the other hand, dispersal of diaspores is discussed in agricultural references in the context of contamination by weeds. The study of endozoochorous transport of seeds by livestock has a long tradition. Cattle were fed seed-containing material and viability of seeds in the dung was tested after excretion (Kempski, 1906; Gardener *et al.*, 1993). Hansen (1911) reported huge amounts of seeds from *Matricaria chamomilla* (198 000, 27% germination capacity) and *Plantago* species (85 000, 58% germination capacity) dispersed by dung of a single cow. Cattle dung in Switzerland and Germany contained seeds from, amongst others, *Agrostis capillaris*, *Poa annua*, *P. pratensis*, *Lolium perenne*, *Festuca rubra*, *Trifolium pratense*, *T. repens*, *Prunella vulgaris*, *Plantago lanceolata*, *P. major* and *Luzula campestris* (Müller, 1955; Boeker, 1959). Viable seeds of *Plantago lanceolata*, *Rumex acetosa* and *R. acetosella* were found in sheep droppings by Müller (1955). In Canada, Dore and Raymond (1942) recorded viable seeds of 49 species in cattle manure. A formerly intensively exploited fen meadow became extensively grazed at a stocking rate of one animal/ha (Scholtz, 1955). The existing *Elymus repens* stand was invaded by *Cirsium arvense* from the second year onwards; *Urtica dioica* decreased as a result of trampling and was replaced by *Poa pratensis* and *P. trivialis*. After inclusion of the adjacent field, *Rumex crispus* spread strongly into the grazed area. This spreading was facilitated by the grazing cattle.

Grazing with a nature conservation aim often includes relatively uniform grasslands that should be transformed into communities with more conservation interest, e.g. heathland. Welch (1985) recorded many seeds of *Agrostis capillaris*, *Cerastium fontanum*, *Poa pratensis*, *Rumex acetosella*, *Sagina procumbens*, but also of *Calluna vulgaris* dispersed by dung of cattle, sheep and non-domestic herbivores in heather moorland in Scotland. He studied the dung of six herbivore species and found 55 species from moorland communities and 18 species that were absent or scarce in moorland. He also found that non-heathland species invaded and spread into the moorland. Bakker (1989) showed that viable seeds of species from a grassland area may have been transported to the heathland via dung pellets. Seeds of those species were also dispersed within a grassland area. Only the seeds of *Calluna vulgaris* were transported from the heathland area into the grassland area. In a salt marsh area, viable seeds of *Juncus gerardi*, *Spergularia salina*, *Agrostis stolonifera*, *Poa pratensis* and *Festuca rubra* were found in cattle dung (Bakker, 1989). Over 20 calcareous grassland species were dispersed by the dung of sheep in Germany (Fischer *et al.*, 1995). Malo and Suarez (1995) reported the establishment of plant species dispersed in cattle dung in dehesa systems in Central Spain. Brown and Archer (1987) mentioned that cattle dung deposition increased species diversity and spatial heterogeneity of the herbaceous vegetation, and con-

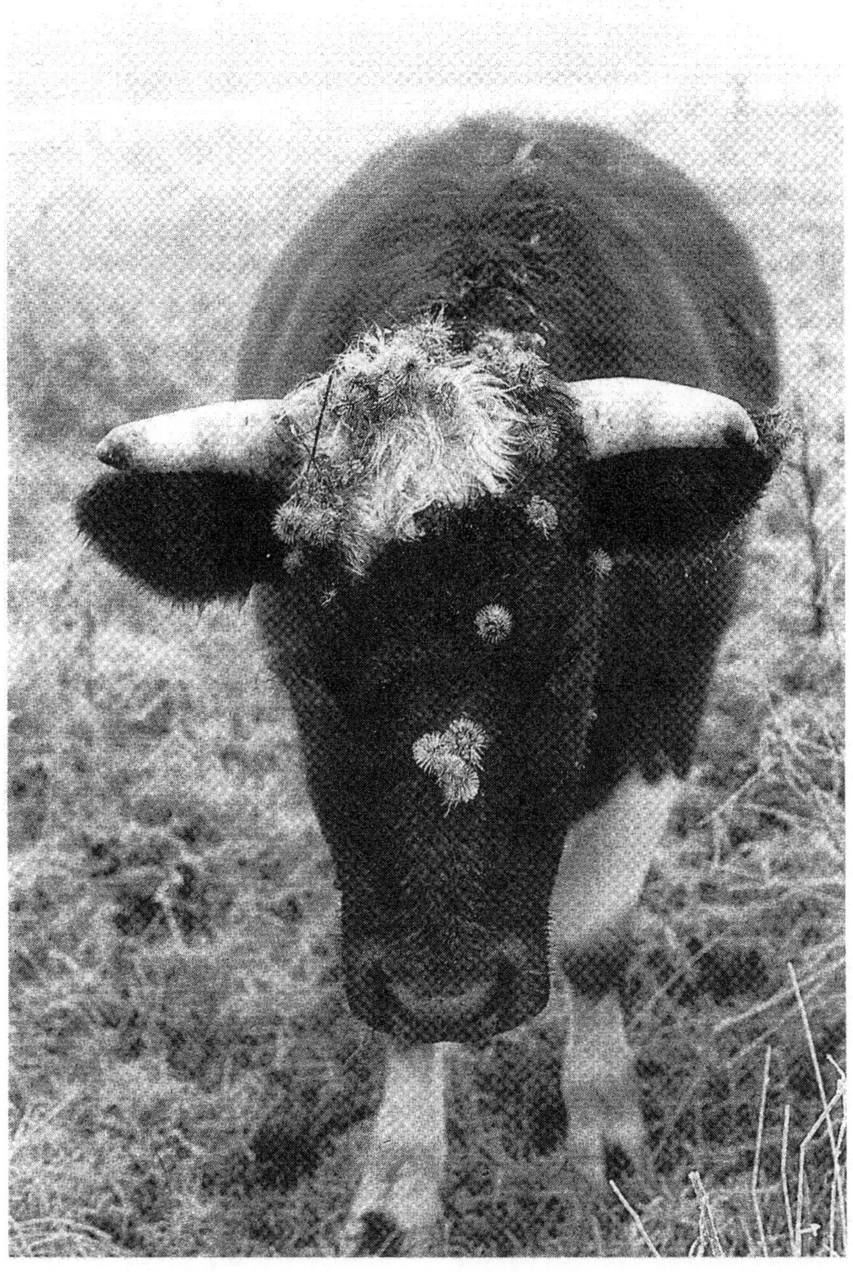

Figure 5.9 An example of eipzoochorous dispersal of *Arctium lappa* seed heads by cattle.

tributed to the development of a fine-grained mosaic vegetation in subtropical North American savanna woodland. Bülow-Olsen (1980a) surmised that *Pimpinella saxifraga*, *Campanula rotundifolia*, *Galium verum*, *Festuca ovina* and *Agrostis capillaris* had established in a former heathland overgrown with *Deschampsia flexuosa* in Denmark, after the cattle in the area had grazed a species-rich grassland containing these species.

With respect to epizoochorous transport, Hillegers (1985) found seeds of the genera *Agrimonia*, *Carduus*, *Cirsium*, *Galium*, *Lappa* and *Cynoglossum* attached to the long fleece of Mergelland sheep in The Netherlands. Fischer *et al.* (1995) studied the fur of a tame sheep in a flock at regular intervals from the end of June until the end of September. The flock was herded and travelled over calcareous grasslands, fertilized grasslands and fallow arable fields in Germany. In this study 8511 diaspores belonging to 85 plant species were recorded. The majority (70%) were grasses – in particular, *Dactylis glomerata* and *Bromus erectus*. Many dicots were also recorded, although at a lower frequency. The majority of occurring species with hooked, bristly and coarse surface structure of diaspores were found in the fleece, as might be expected, and their numbers were also comparatively high. Interestingly, 50% of the occurring species with a smooth diaspore surface were also found in the fleece but their number represented not more than 1% of all recorded diaspores (Fischer *et al.*, 1996). Tall species had a great chance of being picked up by the sheep. Nearly 90% of the species with a height over 60 cm were found in the fleeces. In contrast, the number of diaspores of species lower than 40 cm was only 2.5%. Experiments with a sheep dummy demonstrated that species shorter than 20 cm (e.g. *Thymus pulegioides*, *Teucrium montanum*, *Medicago lupulina*) can only be expected in the fleece after wallowing (Fischer *et al.*, 1996).

Epizoochorous transport of material embedded in mud attached to hoofs was demonstrated by Fischer *et al.* (1995). In the hoofs of 30 sheep they found 382 diaspores belonging to 48 plant species. *Poa pratensis* was the most common species recorded. Entire seed-heads of *Scabiosa columbaria* and *Thymus pulegioides* were found. Diaspores of low-growing species (e.g. *Thymus pulegioides*, *Medicago lupulina*, *Linum catharticum*, *Plantago media*) were well represented, as compared with diaspores attached to the fleece, in which tall species dominated. Some low-growing species (*Potentilla verna*, *Euphrasia stricta* and *Gentiana verna*) were only recorded from hoofs and not in the fleece.

5.6.2 Dispersal distances for zoochorous transport

Most studies on zoochory do not allow an estimation of dispersal distances (Shmida and Ellner, 1983; Hillegers, 1985; Milton *et al.*, 1990). Dispersal distances may be estimated by attachment experiments (see also Bullock and Primack, 1977). Kiviniemi (1996) reported that seeds attached to the coats of fallow deer and cattle might be dispersed over 1 km. Species with hooks (*Agrimonia eupatoria*) revealed better adhesive properties than species with smooth seed surface (*Triglochin palustre*). For the first time, Fischer *et al.* (1996) investigated seed dispersal on the fleece during a whole vegetation period on a tame sheep in Germany. They marked seeds with waterproof paint and placed them on the fleece. Dispersal distances were derived by looking at the loss of marked seeds after distinct time periods (see also Kiviniemi, 1996). Seeds were dispersed over very long distances. At least 8% of artificially attached seeds with hooks (*Bromus erectus*) and 17% of seeds with smooth surface (*Helianthemum nummularium*) were transported for more than 40 days. During this time period, sheep could cover distances of over 100 km.

Dispersal distances as a result of endozoochorous transport may be estimated if the passage time through the gut and the distance travelled during this time period are known (e.g. 24 to more than 96 hours in sheep; Özer, 1979).

5.6.3 Effectiveness of dispersal

Dispersal can only be effective if the diaspores reach a safe site where they can germinate and establish (Bakker *et al.*, 1996b). This was nicely demonstrated in an experiment where excluding cattle led to a total absence of seedling emergence of the woody *Prosopis glandulosa*, whereas large amounts were found in the grazed area. It may be suggested that the rate of invasion of this species in grasslands of subtropical regions has substantially increased following the settlement and introduction of domestic ungulates in North America (Brown and Archer, 1987), giving an example of very large-scale dispersal.

The role of dispersal and safe sites in colonization at a smaller scale can be revealed by introduction experiments. The emergence of seedlings after introducing seeds from species that are present neither in the established vegetation nor in the soil seedbank demonstrated that dispersal was a limiting factor for establishment. As more seedlings survived in heavily grazed areas than in lightly grazed patches, it was shown that the availability of safe sites is another limiting factor (Bakker, 1989).

The role of grazing in relation to the dispersal of diaspores should be discussed in the framework of the species pool concept. Zobel *et al.* (1998) define the **community species pool** as the set of species present in the target community (e.g. the number of species in a relevé of 10 m^2 in a calcareous grassland). The **local species pool** is the set of species occurring in the landscape around a target community that is capable of coexisting in that community (e.g. the entire calcareous grassland on a hill). These species are able to immigrate into the community relatively rapidly – say, within a few years. The community species pool also includes species that are not present in the established vegetation of the pool, but do persist in the soil seedbank. The **regional species pool** is the set of species occurring in a certain region that is capable of coexisting in the target community. A **region** is a reasonably large area with a more or less uniform physiography and climate, from which species may be expected to reach the target community (e.g. all calcareous grasslands in southern Germany). Many species-rich calcareous grasslands are no longer exploited in a low-intensity farming system and tend to become abandoned or afforested. Subsequently, the number of calcareous grassland species has declined.

Restoration management including removal of the forest is supposed to proceed step by step (Poschlod *et al.*, 1997) (Figure 5.10). The first step is appropriate management of the community species pool and the emergence of species still present in the long-term persistent seedbank, which may be regarded as part of the local species pool. The second step is the invasion of wind-dispersed species from the local species pool. Many species are still lacking as long as the third step is not accomplished: this includes the reinstatement of the former grazing practice by large-scale movements of sheep flocks covering hundreds of kilometres. Hence, the third step is the colonization by herbivore-dispersed species from the local and regional species pools.

The implications for conservation management are as follows. Many plant species cannot rely on a long-term persistent seedbank for regeneration after their disappearance from the established vegetation. Dispersal of diaspores then becomes a serious bottle-neck in conservation management. We have to accept that wind dispersal does not reach as far as was believed until recently (Bakker *et al.*, 1996b). Therefore, wind dispersal cannot

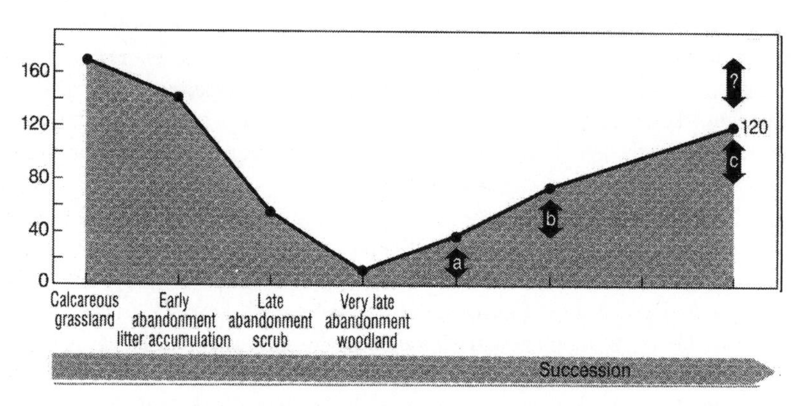

Figure 5.10 Changes in the number of species in the 'species pool' of the herb layer of limestone grasslands ($N = 170$) in the region of the Central Schwäbischen Alb, Germany, after abandonment and subsequent clearing, introduction of management measures, and sheep-grazing: (a) activation of the diaspore bank; (b) establishment of wind-dispersed species; (c) establishment of sheep-dispersed species. (After Poschlod *et al.*, 1997.)

replace the diversity of dispersal processes of the natural as well as the semi-natural landscape. The emergence of reintroduced species indicates that dispersal of diaspores may be an important limiting factor in restoration management. This implies that we do not need corridors as a static 'ecological infrastructure', so often proposed as an important step in the conservation of species from semi-natural landscapes. Knowing the mechanisms causing dispersal of diaspores, we need a dynamic 'moving ecological infrastructure' including livestock or wild herbivores (Poschlod *et al.*, 1996).

5.7 EFFECTS OF MULTI-SPECIES GRAZING ON THE VEGETATION

Multi-species grazing may have other effects on the vegetation than grazing by a single species, as herbivore species differ in food habits, terrain use and their potential to influence vegetation development. Multi-species grazing may thus be of value for the management of plant communities, yet too little is known to allow predictions for the impact of specific combinations of herbivore species at different stocking rates.

Some data are available on the terrain use of different herbivore species grazed together. The New Forest in Hampshire, England, is an area of about 37 500 ha of mixed vegetation which has been grazed for the past 900 years by deer, but most significantly by cattle and ponies. Their numbers have ranged between 2500 and 7000 during recent decades (Putman, 1986). The domestic herbivores were preferential grazers. Throughout the year both species spent over 50% of their total time on only 1140 ha of improved grassland consisting of an extremely short sward dominated by *Agrostis capillaris*, *Lolium perenne*, *Festuca rubra* and rosette plants like *Hypochaeris radicata*, *Bellis perennis*, *Leontodon autumnalis* and *Plantago lanceolata*. Cattle showed a relatively constant pattern of habitat use throughout the year with considerable emphasis on improved and streamside grassland and extensive use of heathland. Ponies showed marked seasonality in their use of plant communities. Improved grasslands also were important throughout the year, but wet heaths, bogs and natural acid grasslands showed a clearly seasonal pattern of use matching the growth of *Molinia caerulea*. Ponies browsed *Ulex europaeus* and deciduous woodland in wintertime (Putman *et al.*, 1987). Grazing pressure on the improved grasslands is very high. The grazing behaviour of ponies on the heavily grazed improved grasslands creates a mosaic of avoided latrine areas and grazed non-latrine areas. This generates an interaction between the two herbivore species: cattle are unable to feed on the short sward (10–20 mm height) cropped by ponies and hence concentrate on pony latrines (50 mm height) (Putman, 1986). Combined grazing of cattle and horses at lower stocking rates (1.6 animals/ha) on ancient commons near Oxford suggested that at any time the two species separate spatially. Both species established latrines, but cattle were not forced into the horse latrines by shortage of herbage, as in the New Forest improved grasslands. Due to the relatively low stocking rate, there was no difference in sward height between latrine and non-latrine areas (Putman *et al.*, 1991).

Another example concerns the extensive free-ranging system on the Isle of Rhum, Scotland, including sympatric cattle, red deer, goats and ponies (Gordon, 1989a). Cattle, red deer and ponies preferentially selected grassland communities, while goats preferred heathland. All species show marked seasonality in their use of the available plant communities. The majority of herbivore species selected the mesotrophic *Agrostis/Festuca* stands in spring and summer. During autumn and winter, cattle selected oligotrophic grasslands, *Molinia* and *Schoenus* fens, whereas red deer and ponies remained on mesotrophic communities. Female goats selected mesotrophic communities, whereas males selected oligotrophic communities. This pattern of selection reflects an interaction between the patterns of variation in the nutritive value, biomass and availability of plant material within and between communities and the characteristics of the ungulates, including body size and digestive system (hindgut fermenter or ruminant) (Gordon, 1989b). Where there is a diversity of plant communities there is also likely to be a diversity of niches for ungulate species to fill.

The implication for management given by Gordon (1989b) is that when developing mixed grazing systems for the economic development of rural ecosystems, care must be taken to create a system where competition is reduced by using animals that differ in either body size, digestive system or the use they make of different parts of the vegetation. The more different the herbivores, the less they will compete for the same resource. In fact, large herbivores may facilitate for smaller herbivores (Chapter 6). The same management recommendation is not necessarily advisable when applying multi-species grazing to the conservation of semi-natural and natural landscapes. Since many of the differences between animal species in foraging behaviour and diet selection can be attributed to variation in body size, it may be expected that species will differ in the spatial scale at which heterogeneity in the vegetation can be exploited, according to their body size and muzzle morphology. This should lead to species-specific effects on habitat heterogeneity. Two or more herbivore species then will potentially utilize a greater proportion of plant species within a community, which may result in a more uniform vegetation pattern at high stocking rates (Briske, 1996). Depending on the plant communities affected, the higher degree of utilization could, however, be associated with a greater species richness. At lower herbivore density species with complementary foraging strategies can be expected to have more complex effects on the vegetation, especially when facilitation occurs between species. With the present knowledge, it is unclear if, and especially how, multi-species grazing can contribute to increase variation in the structure and species composition of the vegetation. One of the challenges certainly would be to investigate the potential for controlling the encroachment of woody species and other unpalatable species by using different herbivore species.

5.8 GRAZING AND MANAGEMENT GOALS

To what extent did the introduction of grazing as a management tool meet the expectations? The main goal was to maintain or create conditions for as many species a possible. Grazing at different intensities is considered an advantage for reaching a greater vegetational differentiation (Klapp, 1965; Oosterveld, 1975; Harper, 1977). Grazing at a relatively low stocking rate leads to variation in, amongst others, soil compaction, dung deposition and, especially, heterogeneous removal of plant material due to selectivity by the herbivores. Extensive grazing creates a higher structural diversity as compared with haymaking. It is obvious that patterns in the structure of the vegetation, in particular the micro-patterns, cannot be established by haymaking, but grazing may be less successful than haymaking when selective grazing or excessive trampling threatens rare species, and when encroachment of unpalatable species takes place.

This section concludes by reviewing whether the management goals of the studies mentioned in section 5.2 were achieved. At the Schiermonnikoog salt marsh the aim of preventing the spread of tall grass communities by seasonal cattle grazing was realized. For the embanked sand flats, the aim of preventing bush encroachment by ponies was also achieved. Once bushes and *Calamagrostis epigejos* stands had developed, it proved to be difficult to remove them by combined cattle and pony grazing (Van Deursen *et al.*, 1993) but this was successfully accomplished for *Phragmites australis* stands (Drost *et al.*, 1990). The aim of slowing down the effects of desalinating soil and maintaining saline or brackish plant communities after embankment was fulfilled (Westhoff and Sykora, 1979).

Concerning heathland, the objective of maintaining heathland communities and preventing tree encroachment at the Westerholt by year-round sheep grazing was also accomplished (Figure 5.11), but at the same time grasses spread from the adjacent grassland area, which was certainly not a management goal (Bakker *et al.*, 1983). The impoverishment of the soil in the grassland area in order to re-establish heathland did not succeed, though after about 25 years an increase of *Erica tetralix* and *Calluna vulgaris* was found; other heathland species were still lacking (Bakker, 1989). The same was found in the dry parts of the Heest under seasonal cattle grazing, but here all indicators of heathland communities were lacking after 20 years, and tall herb communities were established at places where cattle rested. The aim of maintaining species-rich grassland communities on peat with seepage was not effectuated, and tall herb communities spread (Bakker and Grootjans, 1991). *Calluna* heathland overgrown by *Deschampsia flexuosa* at Wolfhezerheide could be restored and subsequently maintained by year-round cattle grazing, which fulfilled the management goal (Bokdam and Gleichman, 1989). Heathland with bryophytes and lichens overgrown by trees at Drouwenerzand could be maintained, after removal of the trees, by year-round sheep grazing, more or less according to the management goal (Van der Bilt and Nijland, 1993). Heather moorland required grazing in order to be maintained (Hester *et al.*, 1991), but at excessive stocking rates grasses and herbs caused 'pollution' (Welch, 1984b,c; Welch and Scott, 1995).

On calcareous soil, six years of secondary succession in a former arable field and the effect of controlled grazing by sheep was compared with the vegetation of established calcareous grasslands of 40, 50 and at least 100 years old near the experimental field. Despite the rate at which characteristic species of older calcareous grasslands arrived in the experimental field, the rate of community change was slow (Gibson and Brown, 1992). Succession towards species-rich calcareous grassland is expected to take at least a century. The results from this experiment show the perspective for succession, by showing the difference between the early stages of succession and the ancient semi-natural communities, considered as the target communities. The ancient grassland communities will not develop without appropriate management (e.g. sheep grazing), but there seems to be a limit to the extent at which community change can be accelerated by grazing. On the community scale, the most that natural restoration schemes can hope to do is to initiate and maintain the processes by which such grassland vegetation

Figure 5.11 Year-round sheep-grazing was effective in preventing the encroachment of *Betula pubescens* in Westerholt, The Netherlands.

arises. This community approach is in marked contrast with short-term achievements at the species level (Gibson and Brown, 1991a). Colonization of the experimental field appeared to be a random draw of species from the adjacent areas, including patches of ancient grassland. Consequently, nearly 75% of the local characteristic species of calcareous grasslands had established in the field seven years after the cessation of agricultural exploitation. Community change is apparently not limited by initial dispersal (Gibson *et al.*, 1987), if short distances are involved. With careful design – relatively poor soils next to surviving patches of ancient grassland and grazing heavily in accordance with site productivity – a restoration project can quickly produce an attractive species-rich grassland which contains newly established populations of many ancient grassland species. Such a project cannot be expected to recreate a community resembling the ancient grasslands in a short time: it can only initiate the long-term processes needed to do so. Only future results can show whether these processes will continue to operate in the experimental field. However, such patterns are indeed evident from observations of arable land abandoned between 1945 and the 1980s next to long-established calcareous grasslands in southern Britain (Gibson and Brown, 1991b).

From many references it can be concluded that the cessation of grazing results in the accumulation of litter, increase of tall herbs and eventually bush encroachment. The effect is always deterioration of the grazing-dependent plant communities from semi-natural landscapes. Such results stress the importance of grazing as a conservation tool in the sense of maintenance management. The rate of change after cessation of grazing seems to increase with the level of production of the plant communities concerned. This stresses the importance of abiotic conditions, as discussed in Chapter 11. In this line of thought, Milchunas and Lauenroth (1993) conclude in their review of a worldwide 236-site data set that the predominance of ecosystem-environmental variables rather than grazing variables in determining grazing impact suggest that 'where we graze' may be more important than 'how we graze'.

Relatively few references were found on the role of grazing in restoration management, i.e. counteracting the effects of previous strong impacts of drainage, input of artificial fertilizers or abandonment. Most effects seem to work at the level of changes in the structure of the vegetation, and hardly at all on higher levels in ecosystem hierarchy, such as abiotic conditions. The discussion on the effects of grazing and management goals is pursued further in Chapters 9 and 11 on the perspectives and limitations of grazing as a management tool.

5.9 CONCLUDING REMARKS

There is still no clear consensus about the importance of herbivores in plant ecology, presumably because their impact differs so much from biome to biome and from community to community within each biome. In order to reach consensus we need more well designed, properly replicated, long-term field studies, in which individual herbivore species are excluded in combination with seed-sowing and soil-disturbance treatments. Only then will we know the relative importance of seed limitation, micro-site limitation and herbivory in plant recruitment (Crawley, 1997).

Conservation management is interested in the question of how diversity, i.e. plant species and plant community richness, can be maintained or restored. Field experiments, especially in grasslands, show that herbivores often, but not always, increase species and community diversity. Moreover, their effects on diversity may vary at different spatial and temporal scales (Olff and Ritchie, 1998). These sometimes confusing results should be understood in order to manage plant diversity based on the knowledge of mechanisms, instead of by trial and error. Recent syntheses and models (see also Chapter 10) suggest that effects of herbivory on plant diversity should vary across environmental gradients of soil fertility and precipitation (Milchunas *et al.*, 1988). Herbivore effects may be better understood in the framework of variation in primary productivity, palatability of competitive plants species, and the evolutionary history of grazing, including the development of deterrents. These mechanisms should be combined with the processes of local colonization and local extinction of plant species (Olff and Ritchie, 1998).

REFERENCES

Al-Mufti, M.M., Sydes, C.L., Furness, S.B. *et al.* (1977) A quantitative analysis of shoot phenology and dominance in herbaceous vegetation. *Journal of Ecology*, **65**, 759–791.

Andresen, H., Bakker, J.P., Brongers, M. *et al.* (1990) Long-term changes of salt marsh communities by cattle grazing. *Vegetatio*, **89**, 137–148.

Bakker, J.P. (1985) The impact of grazing on plant communities, plant populations and soil conditions on salt marshes. *Vegetatio*, **62**, 391–398.

Bakker, J.P. (1989) *Nature Management by Grazing and Cutting*, Kluwer Academic Publishers, Dordrecht.

Bakker, J.P. and Grootjans, A.P. (1991) Potential for vegetation regeneration in the middle course of the Drentsche A brook valley (The Netherlands). *Verhandlungen Gesellschaft für Ökologie*, **20**, 249–263.

Bakker, J.P. and De Vries, Y. (1992) Germination and early establishment of lower salt-marsh species in grazed and mown salt marsh. *Journal of Vegetation Science*, **3**, 247–252.

Bakker, J.P., De Bie, S., Dallinga, J.H. *et al.* (1983) Sheep-grazing as a management tool for heathland conservation and regeneration in the Netherlands. *Journal of Applied Ecology*, **20**, 541–560.

Bakker, J.P., De Leeuw, J. and Van Wieren, S.E. (1984) Micro-patterns in grassland vegetation created and sustained by sheep-grazing. *Vegetatio*, **55**, 153–161.

Bakker, J.P., Bakker, E.S., Rosén, E. *et al.* (1996a) Soil seed bank composition along a gradient from dry alvar grassland to *Juniperus* shrubland. *Journal of Vegetation Science*, 7, 165–176.

Bakker, J.P., Bekker, R.M., Poschlod, P. *et al.* (1996b) Seed banks and seed dispersal: important topics in restoration ecology. *Acta Botanica Neerlandica*, **45**, 461–490.

Bakker, J.P., Esselink, P., Van Der Wal, R. and Dijkema, K.S. (1997) Options for restoration and management of coastal salt marshes in Europe, in *Restoration Ecology and Sustainable Development*, (eds K.M. Urbanska, N.R. Webb and P.J. Edwards), pp. 286–322, Cambridge University Press, Cambridge.

Bekker, D.L. and Bakker, J.P. (1989). Het Westerholt IX; veranderingen in vegetatiesamenstelling en-patronen na 15 jaar beweiden. *De Levende Natuur*, **90**, 114–119.

Berendse, F. (1985) The effect of grazing on the outcome of competition between plants species with different nutrient requirements. *Oikos*, **44**, 35–39.

Berg, G., Esselink, P., Groeneweg, M. and Kiehl, K. (1997) Micropatterns in *Festuca rubra*-dominated salt-marsh vegetation induced by sheep grazing. *Plant Ecology*, **132**, 1–14.

Bertness, M.D. and Ellison, A.M. (1987) Determinants of pattern in a New England salt marsh plant community. *Ecological Monographs*, 57, 129–147.

Black, J.L. and Kenney, P.A. (1984) Factors affecting diet selection by sheep. II. Height and density of pasture. *Australian Journal of Agricultural Research*, 35, 565–578.

Bobbink, R. and Willems, J.H. (1987) Increasing dominance of *Brachypodium pinnatum* in chalk grasslands: a threat to a species-rich ecosystem. *Biological Conservation*, **40**, 301–314.

Bobbink, R. and Willems, J.H. (1996) Herstelbeheer van kalkgrasland op de Bemelerberg. *Natuurhistorisch Manndblad*, **85**, 247–251.

Bobbink, R., Boxman, D., Fremstad, E. *et al.* (1992) Critical loads for nitrogen eutrophication of terrestrial and wetland ecosystems based upon changes in vegetation and fauna, in *Critical Loads for Nitrogen. Report from a workshop held at Lökeberg, Sweden, 1992*, (eds P. Grennfelt and E. Thörmelöf), pp. 111–159, Nordic Council of Ministers, Copenhagen.

Boeker, P. (1959) Samenauslauf aus Mist und Erde von Triebwegen und Ruheplätze. *Zeitschrift für Acker- und Pflanzenbau*, **108**, 77–92.

Bokdam, J. and Gleichman, J.M. (1989) Invloed van begrazing op Struikheide en Bochtige smele. *De Levende Natuur*, **90**, 6–14.

Briske, D.D. (1996) Strategies of plant survival in grazed systems: a functional interpretation, in *The Ecology and Management of Grazing Systems* (eds J. Hodgson and A.W. Illius), pp. 37–67, CAB International, Wallingford.

Brown, J.R. and Archer, S. (1987) Woody plant seed dispersal and gap formation in a North American subtropical savanna woodland: the role of domestic herbivores. *Vegetatio*, 73, 73–80.

Bullock, J.M., Clear Hill, B., Dale, M.P. and Silvertown, J. (1994) An experimental study of the effects of sheep grazing on vegetation change in a species-poor grassland and the role of seedling recruitment into gaps. *Journal of Ecology*, **31**, 493–507.

Bullock, S.H. and Primack, R.B. (1977) Comparative experimental study of seed dispersal on animals. *Ecology*, **58**, 681–686.

Bülow-Olsen, A. (1980a) Changes in the species composition in an area dominated by *Deschampsia flexuosa* as a result of cattle grazing. *Biological Conservation*, **18**, 257–270.

Bülow-Olsen, A. (1980b) Nutrient cycling in grassland dominated by *Deschampsia flexuosa* and grazed by nursing cows. *Agro-Ecosystems*, **6**, 209–220.

Cargill, S.M. and Jefferies, R.L. (1984) The effects of grazing by lesser snow geese on the vegetation of a sub-arctic salt marsh. *Journal of Applied Ecology*, **21**, 669–686.

Charles, W.N., Mc Cowan, D. and East, K. (1977) Selection of upland swards by Red Deer (*Cervus elaphus* L.) on Rhum. *Journal of Applied Ecology*, **14**, 55–64.

Crawley, M.J. (1997) Plant–herbivore dynamics, in *Plant Ecology* (ed. M.J. Crawley), pp. 401–474, Blackwell, Oxford.

Davidson, D.W. (1993) The effects of herbivory and granivory on terrestrial plant succession. *Oikos*, **68**, 23–35.

De Bonte, A.J. and Boosten, A. (1996) Changes in the vegetation of Meijendel influenced by grazing. *Acta Botanica Neerlandica*, **45**, 585.

Deinum, B. and Van Soest, P.J. (1969) Prediction of forage digestibility from some laboratory procedures. *Netherlands Journal of Agricultural Sciences*, **17**, 119–127.

De Leeuw, J. and Bakker, J.P. (1986) Sheep-grazing with different foraging efficiencies in a Dutch mixed grassland. *Journal of Applied Ecology*, **23**, 781–793.

De Leeuw, J., Olff, H. and Bakker, J.P. (1990) Year-to-year variation in peak above-ground biomass of six salt marsh angiosperm communities as related to rainfall deficit and inundation frequency. *Aquatic Botany*, **36**, 139–151.

Dierssen, K., Eischeid, I., Gettner, S. *et al.* (1996) *Abslchlussbericht Ökosystemforschung Wattenmeer TV A5.2, A5.3. Bioindikatoren im Supralitoral*, UBA-Forschungsbericht 10802085/01, Berlin.

Dijkema, K.S. (1983) The salt marsh vegetation of the mainland coast, estuaries and halligen, in *Flora and Vegetation of the Waddensea Islands and Coastal Areas. Report 9 of the Wadden Sea Working Group*, (eds K.S. Dijkema and W.J. Wolff), pp. 185–220, Balkema, Rotterdam.

Dore, W.G. and Raymond, L.C. (1942) Pasture studies XXIV. Viable seeds in pasture soil and manure. *Science Agriculture*, **23**, 69–79.

Drent, R.H. and Prins, H.H.T. (1987) The herbivore as prisoner of its food supply, in *Disturbance in Grasslands* (eds J. Van Andel, J.P. Bakker and R.W. Snaydon), pp. 131–147, Junk, Dordrecht.

Drent, R.H. and Van Der Wal, R. (1998). Cyclic grazing in vertebrates and the manipulation of the food resource, in *Herbivores, Between Plants and Predators*, (eds H. Olff, V.K. Brown and R.H. Drent), pp. 271–299, Blackwell Scientific Publications, Oxford.

Drost, H.J. and Muis, A. (1988) Begrazing van Duinriet op 'de Rug' in de Lauwersmeer. *De Levende Natuur*, **89**, 82–88.

Drost, H.J., Van Deursen, E.J.M. and Muis, A. (1990) Begrazing van Riet door runderen en paarden in de Lauwersmeer. *De Levende Natuur*, **91**, 68–74.

Dutoit, T. and Alard, D. (1995) Mécanisme d'une succession végétale secondaire en pelouse calcicole: une approche historique. *Comptes rendus de l'Académie des Sciences Série III, Sciences de la vie*, **318**, 897–907.

Ehrenburg, A., Van Til, M. and Mourik, J. (1995) Vegetatieontwikkeling en begrazingsbeheer van het zeedorpenlandschap bij Zandvoort. *De Levende Natuur*, **96**, 202–211.

Elkington, T.T. (1981) Effects of excluding grazing animals from grassland on sugar limestone in Teesdale, England. *Biological Conservation*, **20**, 25–35.

Ellenberg, H. (1978) *Vegetation Mitteleuropas mit den Alpen*, Ulmer, Stuttgart.

Erchinger, H.F., Coldewey, H.G. and Meyer, C. (1996) Interdisziplinäre Erforschung des Deichvorlandes im Forschungsvorhaben 'Erosionsfestigkeit von Hellern'. *Die Küste*, **58**, 1–45.

Erisman, J.W. (1990) *Acid Deposition in the Netherlands*, Report RIVM, Bilthoven.

Esselink, P., Schotel, J. and Van Gils, H. (1991) Nitrogen- and phosphorus-limited primary production of annuals. ITC Publication No. 11, ITC, Enschede.

Esselink, P., Dijkema, K.S., Reents, S. and Hageman, G. (1998) Vertical accretion and marsh-profile development in man-made tidal marshes after abandonment. *Journal of Coastal Research*, **14**, 570–582.

Fischer, A. (1995) Zum Verhalten von Rindern auf Moorgrünland. *Zeitschrift für Kulturtechnik und Landentwicklung*, **36**, 169–172.

Fischer, S., Poschlod, P. and Beinlich, B. (1995) Die Bedeutung der Wanderschäferei für den Artenaustausch zwischen isolierten Schaftriften. *Beihefte Veröfftlichungen Naturschutz und Landschaftspflege Baden–Württemberg*, **83**, 229–256.

Fischer, S., Poschlod, P. and Beinlich, B. (1996) Experimental studies on the dispersal of plants and animals by sheep in calcareous grasslands. *Journal of Applied Ecology*, **33**, 1206–1222.

Floate, M.J.S. (1970) Mineralization of nitrogen and phosphorus from organic materials and animal origin and its significance in the nutrient cycle in grazed uplands and hill soils. *Journal of the British Grassland Society*, **25**, 295–302.

Gardener, C.J., McIvor, J.G. and Jansen, A. (1993) Passage of legume and grass seeds through the digestive tract of cattle and their survival in faeces. *Journal of Applied Ecology*, **30**, 63–74.

Gibson, C.W.D. and Brown, V.K. (1991a) The effects of grazing on local colonisation and extinction during early succession. *Journal of Vegetation Science*, **2**, 291–300.

Gibson, C.W.D. and Brown, V.K. (1991b) The nature and rate of development of calcareous grasslands in southern England. *Biological Conservation*, **58**, 297–316.

Gibson, C.W.D. and Brown, V.K. (1992) Grazing and vegetation change: deflected or modified succession? *Journal of Applied Ecology*, **29**, 120–131.

Gibson, C.W.D., Watt, T.A. and Brown, V.K. (1987) The use of sheep grazing to recreate species-rich grassland from abandoned arable land. *Biological Conservation*, **42**, 165–183.

Gordon, I. (1989a) Vegetation community selection by ungulates on the Isle of Rhum. II. Vegetation community selection. *Journal of Applied Ecology*, **26**, 53–64.

Gordon, I. (1989b) Vegetation community selection by ungulates on the Isle of Rhum. III. Determinants of vegetation community selection. *Journal of Applied Ecology*, **26**, 65–79.

Grant, S.A., Hamilton, W.J. and Souter, C. (1981) The responses of heather-dominated vegetation in north-east Scotland to grazing by red deer. *Journal of Applied Ecology*, **69**, 189–204.

Grant, S.A., Suckling, D.E., Smith, H.K. *et al.* (1985) Comparative studies of diet selection by sheep and cattle: the hill grasslands. *Journal of Ecology*, **73**, 987–1004.

Grant, S.A., Torvell, L., Common, T.G. *et al.* (1996a) Controlled grazing studies on *Molinia* grassland: effects of different seasonal patterns and levels of defoliation on *Molinia* growth and responses of swards to controlled grazing by cattle. *Journal of Applied Ecology*, **33**, 1276–1280.

Grant, S.A., Torvell, L., Sim, E.M. *et al.* (1996b) Controlled grazing studies on Nardus grassland: effects of between-tussock sward height and species of grazer on *Nardus* utilization and floristic composition in two fields in Scotland. *Journal of Applied Ecology*, **33**, 1053–1064.

Gray, A.J. and Scott, R. (1977) Biological flora of the British Isles: *Puccinellia maritima. Journal of Ecology*, **65**, 699–716.

Grime, J.P. (1973) Control of species diversity in herbaceous vegetation. *Journal Environmental Management*, **1**, 151–167.

Grime, J.P. (1979) *Plant Strategies and Vegetation Processes*, Wiley, Chichester.

Hacker, J.B. and Minson, D.J. (1981) The digestibility of plant parts. *Herbage Abstracts*, **51**, 459–482.

Hansen, K. (1911) Weeds and their vitality. *Ugeskrift for Landmand*, **56**, 149.

Hansen, D. (1982) Entwicklung und Beeinflussung der Netto primär produktion auf Vorlandflächen und im Vogelschutzgebiet Hauke-Haien-Koog. *Schriftenreihe Institut für Wasserwirtschaft und Landschaftökologie*, **1**, 1–273. Christian-Albrechts-Universität, Kiel.

Harrison, A.F. (1978) Phosphorus cycles of forest and upland grassland ecosystem and some effects of land management practices, in *Phosphorus and the Environment: its chemistry and biochemistry*, (eds R. Potter and D.W. Fitzsimmon), pp. 175–199, Elsevier, Amsterdam.

Harper, J.L. (1977) *Population Biology of Plants*, Academic Press, London.

Hester, A.J., Miles, J. and Gimingham, C.H. (1991) Succession from heather moorland to birch woodland. I. Experimental alteration of specific environmental conditions in the field. *Journal of Applied Ecology*, **79**, 303–315.

Hewson, R. (1977) The effect on heather *Calluna vulgaris* of excluding sheep from moorland in north-east England. *Naturalist*, **102**, 133–136.

Hik, D.S. and Jefferies, R.L. (1990) Increases in the net above-ground primary production of a salt marsh forage grass: a test of the prediction of the herbivore optimization model. *Journal of Ecology*, **78**, 180–195.

Hill, M.O., Evans, D.F. and Bell, S.A. (1992) Long-term effects of excluding sheep from hill pastures in North Wales. *Journal of Ecology*, **80**, 1–13.

Hillegers, H.P.M. (1984) Begrazing met mergellandschapen in Zuid-Limburg. *De Levende Natuur*, **85**, 178–184.

Hillegers, H.P.M. (1985) Exozoöchoor transport van diasporen door Mergellandschapen. *Natuurhistorisch Maandblad*, **74**, 54–56.

Hodgson, J. (1979) Nomenclature and definitions in grazing studies. *Grass and Forage Science*, **14**, 11–18.

Hodgson, J., Herbes, T.D.A., Armstrong, R.H. *et al.* (1991) Comparative studies of the ingestive behaviour and herbage intake of sheep and cattle grazing indigenous hill communities. *Journal of Applied Ecology*, **28**, 205–227.

Huisman, J, Grover, J.P., Van der Wal, R. and Van Andel, J. (1998) Competition for light, plant species replacement, and herbivory along productivity gradients, in *Herbivores, Between Plants and Predators*, (eds H. Olff, V.K. Brown and R.H. Drent), pp. 23g–26g, Blackwell Scientific Publications, Oxford.

Hunter, R.F. (1962) Hill sheep and their pasture: a study of sheep grazing in south-east Scotland. *Journal of Ecology*, **50**, 651–680.

Huston, M. (1979) A general hypothesis of species diversity. *American Naturalist*, **13**, 81–101.

Hutchinson, K.J. and King, K.L. (1980) Management impacts on structure and function of sown grasslands, in *Grasslands, System Analysis and Man*, (eds A.I. Breymeyer and G.M. Van Dyne), pp. 823–852, Cambridge University Press, Cambridge.

Iacobelli, A. and Jefferies, R.L. (1991) Inverse salinity gradients in coastal marshes and the death of stands of *Salix*: the effects of grubbing by geese. *Journal of Ecology*, **97**, 61–73.

Jefferies, R.L. (1988) Pattern and process in arctic coastal vegetation in response to foraging by lesser snow geese. *Plant Form and Vegetation Structure*, (eds M.J.A. Werger, P.J.M. Van Der Aart, H.J. During and J.T.A. Verhoeven), pp. 281–300, SPB Academic Publishing, Den Haag.

Jensen, A., Skovhus, K. and Svendsen, A. (1990) Effects of grazing by domestic animals on saltmarsh vegetation and soils, a mechanistic approach, in *Saltmarsh management in the Wadden Sea region*, (ed. C.H. Ovesen), Ministry of the Environment/National Forest and Nature Agency, Horsholm.

Jeschke, L. (1987) Vegetationsdynamik des Salzgraslandes im Bereich der Ostseeküste der DDR unter dem Einfluss des Menschen. *Hercynia Neue Fassung*, **24**, 321–328.

Job, D.A. and Taylor, J.A. (1978) The production, utilization and management of upland grazing on Plynlimon, Wales. *Journal of Biogeography*, **5**, 173–191.

Joenje, W. (1985) The significance of grazing in the primary vegetation succession on embanked sandflats. *Vegetatio*, **62**, 399–406.

Kaiser, T. (1995) Vegetationskundliche Untersuchungen auf reliefiertem Niedermoor bei Extensivweide. *Zeitschrift für Kulturtechnik und Landentwicklung*, **36**, 175–177.

Kauppi, M. (1967) Über den Einfluss der Beweidung auf die Vegetation der Uferwiesen an der Bucht Liminganlahti im Nordteil des Bottnischen Meerbusens. *Aquilo, Series Botanica*, **6**, 347–369.

Kempski, E. (1906) *Über endozoische Samenverbreitung und speziell die Verbreitung von Unkräutern durch Tiere auf dem Wege des Darmkanals.* Inaugural-Diss., Univ. Rostock. Carl Georgi, Bonn.

Kenney, P.A. and Black, J.L. (1984) Factors affecting diet selection by sheep. I. Potential intake rate and acceptability of feed. *Australian Journal of Agricultural Research*, **35**, 551–563.

Kiehl, K., Eischeid, I., Gettner, S. and Walter, J. (1996) The impact of different sheep grazing intensities on salt marsh vegetation in Northern Germany. *Journal of Vegetation Science*, **7**, 99–106.

Kiviniemi, K. (1996) A study of adhesive seed dispersal of three species under natural conditions. *Acta Botanica Neerlandica*, **45**, 73–83.

Klapp, E. (1965) *Grünlandvegetation und ihre Standort*, Parey, Berlin.

Leonard, L.A., Hince, A.C. and Luther, M.E. (1995) Superficial sediment transport and deposition processes in a *Juncus roemerianus* marsh, West-Central Florida. *Journal of Coastal Research*, **1**, 322–336.

Malo, J.E. and Suárez, F. (1995) Establishment of pasture species on cattle dung: the role of endozoochorous seeds. *Journal of Vegetation Science*, **6**, 169–174.

McNaughton, S.J. and Banyikwa, F.F. (1995) Plant communities and herbivory, in *Serengeti II, Dynamics, Management and Conservation of an Ecosystem*, (eds A.R.E. Sinclair and P. Arcese), pp. 49–70, Chicago University Press, Chicago.

Milchunas, D.G., Sala, O.E. and Lauenroth, W.K. (1988) A generalized model of the effects of grazing by large herbivores on grassland community structure. *The American Naturalist*, **132**, 87–106.

Miles, J. (1987) Soil variation caused by plants: a mechanism of floristic change in grassland? in *Disturbance in Grasslands*, (eds J. Van Andel, J.P. Bakker and R.W. Snaydon), pp. 37–49, Junk, Dordrecht.

Milton, S.J., Siegfried, W.R. and Dean, W.R.J. (1990) The distribution of epizoochoric plant species: a clue to the prehistoric use of arid Karoo rangelands by large herbivores. *Journal of Biogeography*, **17**, 25–34.

Mitchley, J. (1988) Control of relative abundance of perennials in chalk grasslands in southern England. II. Vertical canopy structure. *Journal of Ecology*, **76**, 341–350.

Moore, D.R.J. and Keddy, P.A. (1989) The relationship between species richness and standing crop in wetlands: the importance of scale. *Vegetatio*, **74**, 99–106.

Mourik, J., Ehrenburg, A. and Van Til, M. (1995) Dune landscape near Zandvoort aan Zee, The Netherlands: use and management past and present, in *Directions in European Coastal Management*, (eds M.G. Healy and J.P. Doody), pp. 483–487, Samara Publishing Limited, Cardigan.

Müller, P. (1955) Verbreitungsbiologie der Blütenpflanzen. *Veröfftlichungen Geobotanisches Institut Eidgenossiche Technische Hochschule Stiftung Rübel Zürich*, **30**, 1–152.

Nicholson, I.A., Paterson, I.S. and Currie, A. (1970) A study of vegetational dynamics: selection by sheep and cattle in Nardus pasture, in *Animal Populations in Relation to their Food Resources*, (ed. A. Watson), pp. 129–143, British Ecological Society Symposium no. 10, Blackwell, Oxford.

Olff, H. and Ritchie, M. (1998) Effects of herbivores on grassland plant diversity. *Trends in Ecology and Evolution*, **13**, 261–265.

Olff, H., De Leeuw, J., Bakker, J.P. *et al.* (1997) Vegetation succession and herbivory on a salt marsh: changes induced by sea level rise and silt deposition along an elevational gradient. *Journal of Ecology*, **85**, 799–814.

Oosterveld, P. (1975) Beheer en ontwikkeling van natuurreservaten door begrazing. *Natuur en Landschap*, **29**, 161–171.

Osbourn, D.F. (1980) The feeding value of grass and grass products, in *Grass, its Production and Utilization*, (ed. W. Holmes), pp. 70–123, Blackwell, Oxford.

Özer, Z. (1979) Über die Beeinflussung der Keimfähigkeit der Samen mancher Grünlandpflanzen beim Durchgang durch den Verdauungstrakt des Schafes und nach Mistgärung. *Weed Research*, **19**, 247–254.

Parsons, A.J., Colett, B. and Lewis, J. (1984) Changes in the structure and physiology of a perennial ryegrass sward when released from a continuous stocking management: implications for the use of exclusion cages in continuously stocked swards. *Grass and Forage Science*, **39**, 1–9.

Peet, R.K., Glenn-Lewin, D.C. and Walker Wolf, J. (1983) Prediction of man's impact on plant species diversity, in *Man's Impact on Vegetation*, (eds W. Holzner, M.J.A. Werger and I. Ikusima), pp. 41–54, Junk, Den Haag.

Penning, P.D. (1986) Some effects of sward conditions on grazing behaviour and intake by sheep, in *Grazing at Northern Latitudes*, (ed O. Gudmundsson), pp. 219–226, Plenum, New York.

Perkins, D.F., Jones, V., Millar, R.O. and Neep, P. (1978) Primary production, mineral nutrients and litter decomposition in the grassland ecosystem, in *Production Ecology of British Moors and Montane Grasslands*, (eds O.W. Heal and D.F. Perkins), pp. 304–331, Springer, Berlin.

Persson, S. (1984) Vegetation development after the exclusion of grazing cattle in a meadow area in the south of Sweden. *Vegetatio*, **55**, 65–92.

Pigott, C.D. (1956) The vegetation of upper Teesdale in the North Pennines. *Journal of Ecology*, **44**, 545–586.

Poschlod, P., Deffner, A., Beier, B. and Grunicke, U. (1991) Untersuchungen zur Diasporenbank von Samenpflanzen auf beweideten, gemähten, brachgefallenen und aufgeforsteten Kalkmagerrasenstandorten. *Verhandlungen Gesellschaft für Ökologie*, **20**, 893–904.

Poschlod, P., Bakker, J.P., Bonn, S. and Fischer, S. (1996) Dispersal of plants in fragmented landscapes, in *Species Survival in Fragmented Landscapes*, (eds J. Settele, C. Margules, P. Poschlod and K. Henle), pp. 123–127, Kluwer Academic, Dordrecht.

Poschlod, P., Bonn, S., Kiefer, S. *et al.* (1997) Die Ausbreitung von Pflanzenarten und -populationen in Raum und Zeit am Beispiel der Kalkmagerrasen Mitteleuropas. *Berichte der Reinhold-Tüxen-Gesellschaft*, **9**, 139–157.

Prins, H.H.T. (1996) *Ecology and Behaviour of the African Buffalo: social inequality and decision making*, Chapman & Hall, London.

Prins, H.H.T. and Van der Jeugd, H.P. (1993) Herbivore crashes and woodland structure in East Africa. *Journal of Ecology*, **81**, 305–314.

Prins, H.H.T., Ydenberg, R.C. and Drent, R.H. (1980) The interaction of brent geese (*Branta bernicla*) and sea plantain (*Plantago maritima*) during spring staging: field observations and experiments. *Acta Botanica Neerlandica*, **29**, 585–596.

Putman, R.J. (1986) *Grazing in Temperate Ecosystems: Large Herbivores and the Ecology of the New Forest*, Croom Helm, Beckenham.

Putman, R.J., Pratt, R.M., Ekins, J.R. and Edwards, P.J. (1987) Food and feeding behaviour of cattle and ponies in the New Forest, Hampshire. *Journal of Applied Ecology*, **24**, 369–380.

Putman, R.J., Fowler, A.D. and Tout, S. (1991) Patterns of use of ancient grassland by cattle and horse and effects on vegetational composition and structure. *Biological Conservation*, **56**, 329–347.

Rawes, M. (1981) Further results of excluding sheep from high-level grasslands in the northern Pennines. *Journal of Ecology*, **69**, 651–669.

Regnéll, G. (1980) A numerical study of successions in an abandoned, damp calcareous meadow in S. Sweden. *Vegetatio*, **43**, 123–130.

Rejmánek, M. and Rosén, E. (1988) The effects of colonizing shrubs (*Juniperus communis* and *Potentilla fruticosa*) on species richness in the grasslands of Stora Alvaret, Öland (Sweden). *Acta Phytogeographica Suecica*, **76**, 67–72.

Ring, C.B. II, Nicholson, R.A. and Launchbaugh, J.L. (1985) Vegetational traits of patch-grazed rangeland in West-central Kansas. *Journal of Range Management*, **38**, 51–55.

Rosén, E. (1973) Sheep grazing and changes of vegetation on the limestone heath of Öland. *Zoon Supplement*, **1**, 137–151.

Rozé, F. (1993) Successions végétales après pâturage extensif par des chevaux dans une roselière. *Bulletin Ecologique*, **24**, 203–209.

Scherfose, V. (1993) Zum Einfluss der Beweidung auf das Gefässpflanzen-Artengefüge von Salz- und Brackmarschen. *Zeitschrift für Ökologie und Naturschutz*, **2**, 201–211.

Schiefer, J. (1981) Bracheversuche in Baden-Württemberg. Vegetations- und Standortsentwicklung auf 16 verschiedenen Versuchsflächen mit unterschiedlichen Behandlungen. *Beihefte Veröfftlichungen Naturschutz und Landschaftsplege, Baden-Württemberg*, **22**, 1–325.

Schmeisky, H. (1977) Der Einfluss von Weidetieren auf Salzpflanzengesellschaften an der Ostsee, in *Vegetation und Fauna*, (ed. R. Tüxen), pp. 481–498, Cramer, Vaduz.

Scholz, A. (1995) Vom Weidevieh gemiedene Pflanzen, Ausbreitung und Massnahmen zur Eindämmung. *Zeitschrift für Kulturtechnik und Landentwicklung*, **36**, 173–174.

Schreiber, K.F. (1997) Grundzüge der Sukzession in 20-jährigen Grünland-Bracheversuchen in Baden-Württemberg. *Forstwissenschaftlisches Centralblatt*, **116**, 243–258.

Schreiber, K.F. and Schiefer, J. (1985) Vegetations- und Stoffdynamik in Grünlandbrachen – 10 Jahre Bracheversuche in Baden-Württemberg, in *Sukzession auf Grünlandbrachen. Münstersche Geographische, Arbeiten 20*, (ed. K.F. Schreiber), pp. 111–153.

Shmida, A. and Ellner, S. (1983) Seed dispersal on pastoral grazers in open Mediterranean chaparral, Israel. *Israelian Journal of Botany*, **32**, 147–159.

Siira, J. (1970) Studies on the ecology of the seashore meadows of the Bothnian Bay with special reference to the Liminka area. *Aquilo Ser. Bot.*, **9**, 1–109.

Silvertown, J., Lines, C.E.M. and Dale, M.P. (1994) Spatial competition between grasses – rates of mutual invasion between four species and the interaction with grazing. *Journal of Ecology*, **82**, 31–38.

Slager, H., Groen, K. and Visser, H. (1993) Begrazing, betreding en ontzilting. *De Levende Natuur*, **94**, 106–110.

Slim, P.A. and Oosterveld, P. (1985) Vegetation development on newly embanked sandflats in the Grevelingen (The Netherlands) under different management practices. *Vegetatio*, **62**, 407–414.

Smith, R.S. and Rushton, S.P. (1994) The effects of grazing management on the vegetation of mesotrophic (meadow) grassland in Northern England. *Journal of Applied Ecology*, **31**, 13–24.

Snow, A.A. and Vince, S.W. (1984) Plant zonation in an Alaskan salt marsh. II. An experimental study of the role of edaphic conditions. *Journal of Ecology*, **72**, 669–684.

Srivastava, D.S. and Jefferies, R.L. (1996) A positive feedback: herbivory, plant growth, salinity, and the desertification of an Arctic salt marsh. *Journal of Ecology*, **84**, 31–42.

Ter Heerdt, G.N.J., Bakker, J.P. and De Leeuw, J. (1991) Seasonal and spatial variation in living and dead plant material in a grazed grassland as related to plant species diversity. *Journal of Applied Ecology*, **28**, 120–127.

Thalen, D.C.P., Poorter, H., Lotz, L.A. and Oosterveld, P. (1987) Modelling the structural changes in vegetation under different grazing regimes, in *Disturbance in Grasslands*, (eds J. Van Andel, J.P. Bakker and R.W. Snaydon), pp. 167–183, Junk, Dordrecht.

Tilman, D. and Pacala, S. (1993) The maintenance of species richness in plant communities, in *Species Diversity in Ecological Communities: Historical and geographical perspectives*, (eds R. Rickleffs and D. Schluter), pp. 13–25, University of Chicago Press, Chicago.

't Mannetje, L. (1978) Measuring quantity of grassland vegetation, in *Measurement of Grassland Vegetation and Animal Production*, (ed. L. 't Mannetje), pp. 63–95, Commonwealth Agricultural Bureaux, Hurley.

Van de Koppel, J., Rietkerk, M. and Weissing, F.J. (1997) Catastrophic vegetation shifts and soil degradation in terrestrial grazing systems. *Trends in Ecology and Evolution*, **12**, 352–355.

Van den Bos, J. and Bakker, J.P. (1990) The development of vegetation patterns by cattle grazing at low stocking density in the Netherlands. *Biological Conservation*, **51**, 263–272.

Van der Bilt, E.W.G. and Nijland, G. (1993) Tien jaar extensieve begrazing met heideschapen in het Drouwernerzand. *De Levende Natuur*, **94**, 164–169.

Van der Maarel, E. and Sykes, M.T. (1993) Small-scale plant species turnover in limestone grassland: the carousel model and some comments on the niche concept. *Journal of Vegetation Science*, **4**, 179–188.

Van der Wal, R., Van de Koppel, J. and Sagel, M. (1998) On the relationship between herbivore foraging efficiency and plant standing crop: an experiment with barnacle geese. *Oikos*, **82**, 123–130.

Van Deursen, E.J.M. and Drost, H.J. (1990) Defoliation and treading by cattle of reed *Phragmites australis*. *Journal of Applied Ecology*, **27**, 284–297.

Van Deursen, E.J.M., Cornelissen, P., Vulink, J.T. and Esselink, P. (1993) Jaarrondbegrazing in de Lauwersmeer: zelfredzaamheid van grote grazers en effecten op de vegetatie. *De Levende Natuur*, **94**, 196–204.

Van Dijk, H.J.W. (1992) Grazing domestic livestock in Dutch coastal dunes: experiments, experiences and perspectives, in *Coastal Dunes, Geomorphology, Ecology and Management for Conservation*, (eds R.W.G. Carter, T.G.F. Curtis and M.J. Sheehy-Skeffington), pp. 235–250, Balkema, Rotterdam.

Van Til, M. (1996) Vegetation development in relation to land use and management of the coastal village landscape near Zandvoort. *Acta Botanica Neerlandica*, **45**, 583–584.

Van Wijnen, H.J., Bakker, J.P. and De Vries, Y. (1997) Twenty years of salt marsh succession on the coastal barrier island of Schiermonnikoog (The Netherlands). *Journal of Coastal Conservation*, **3**, 9–18.

Van Wijnen, H.J., Van Der Wal, R. and Bakker, J.P. (1998) The impact of small herbivores on soil net nitrogen mineralization rate: consequences for salt-marsh succession. *Oecologia* (in press).

Vivier, J.P. (1997) Influence du Pâturage sur la disponibilité de l'Azote pour l'Exportation dans un Marais Salé. PhD thesis, University of Rennes.

Wahren, C.-H.A., Papst, W.A. and Williams, R.J. (1994) Long-term vegetation change in relation to cattle grazing in the subalpine grassland and heathland on the Bogong High Plains: an analysis of vegetation records from 1945 to 1994. *Australian Journal of Botany*, **42**, 607–639.

WallisDeVries, M.F. and Daleboudt, C. (1994) Foraging strategies of cattle in patchy grassland. *Oecologia*, **100**, 98–106.

Watt, A.S. (1962) The effect of excluding rabbits from Grassland A (Xerobrometum) in Breckland, 1936–60. *Journal of Ecology*, **50**, 181–198.

Welch, D. (1984a) Studies in the grazing of heather moorland in North-east Scotland. I. Site description and patterns of utilization. *Journal of Applied Ecology*, **21**, 179–195.

Welch, D. (1984b) Studies in the grazing of heather moorland in North-east Scotland. II. Response of heather. *Journal of Applied Ecology*, **21**, 197–207.

Welch, D. (1984c) Studies in the grazing of heather moorland in North-east Scotland. III. Floristics. *Journal of Applied Ecology*, **21**, 209–225.

Welch, D. (1985) Studies in the grazing of heather moorland in North-east Scotland. IV. Seed dispersal and plant establishment in dung. *Journal of Applied Ecology*, **22**, 461–472.

Welch, D. (1986) Studies in the grazing of heather moorland in North-east Scotland. V. Trends in *Nardus stricta* and other unpalatable graminoids. *Journal of Applied Ecology*, **23**, 1047–1058.

Welch, D. and Rawes, M. (1966) The intensity of sheep grazing on high-level bog in upper Teesdale. *Irish Journal of Agricultural Research*, **5**, 185–196.

Welch, D. and Scott, D. (1995) Studies in the grazing of heather moorland in North-east Scotland. VI. 20-year trends in botanical composition. *Journal of Applied Ecology*, **32**, 596–611.

Westhoff, V. and Sykora, K.V. (1979) A study on the influence of desalination on the *Juncetum gerardii*. *Acta Botanica Neerlandica*, **28**, 505–512.

Whittaker, R.H. (1965) Dominance and diversity in land plant communities. *Science*, **147**, 250–260.

Wielgolaski, F.E. (1975) Comparison of plant structure on grazed and ungrazed tundra meadows, in *Fennoscandian Tundra Ecosystems. I. Plants and micro-organisms* (ed. F.E. Wielgolaski), pp. 86–93, Springer, Berlin.

Willems, J.H. (1983) Species composition and above ground phytomass in chalk grassland with different management. *Vegetatio*, **52**, 171–180.

Wilmshurst, J.F., Fryxel, J.M. and Hudson, R.J. (1995) Forage quality and patch choice by wapiti (*Cervus elaphus*). *Behavioral Ecology*, **6**, 209–215.

Ydenberg, R.C. and Prins, H.H.T. (1981) Spring grazing and the manipulation of food quality by barnacle geese. *Journal of Applied Ecology*, **18**, 443–453.

Zeevalking, H.J. and Fresco, L.F.M. (1977) Rabbit grazing and species diversity in a dune area. *Vegetatio*, **35**, 193–196.

Zobel, M. (1997) The relative role of species pools in determining plant species richness: an alternative explanation of species coexistence? *Trends in Ecology and Evolution*, **12**, 266–269.

Zobel, M. Van der Maarel, E. and Dupré, C. (1998) Species pool: the concept, its determination and significance for community restoration. *Applied Vegetation Science*, **1**, 55–66.

6

Effects of large herbivores upon the animal community

Sipke E. Van Wieren
Tropical Nature Conservation and Vertebrate Ecology Group, Department of Environmental Sciences, Wageningen Agricultural University, Bornsesteeg 69, 6708 PD Wageningen, The Netherlands

6.1 INTRODUCTION

In any system large herbivores can interact either directly or indirectly with other species of animals. Through the activities that herbivores bring about (Figure 6.1), direct competition may take place for resources important to other fauna. Indirect effects are the result of changes in the structure and the species composition of the vegetation, mainly through the foraging process. The effects are mediated through species-specific characteristics of the herbivore involved (e.g. grazer or browser, small size or large size), and, probably most important, through density. To evaluate properly (and judge) the interactions between large herbivores and other fauna in a given system it is important to realize where the system ranks on a scale from natural to completely influenced by humans. In general this is quite easy as most systems are to some extent managed and are ranked as semi-natural.

Frequently other activities than grazing are applied too (e.g. mowing, logging, fertilizing, changing the watertable). Nevertheless, the issue is important in the light of the present debate on restoring ecosystems in temperate climates. In this debate the composition of the potential natural vegetation and that of the potential natural fauna play an important role (Figure 6.2). Equally important, however, is the question of whether large herbivores can act as keystone species in temperate ecosystems (Chapter 1). If they cannot, the interaction scheme can be represented by Figure 6.3; if they can, the system can be represented by Figure 6.4. The main differences are that density leads to differences in the proportion of macrostructural classes within the two systems: at low densities (Figure 6.3) woodland and scrub dominate,

Grazing and Conservation Management. Edited by M.F. WallisDeVries, J.P. Bakker and S.E. Van Wieren. Published in 1998 by Kluwer Academic Publishers, Dordrecht. ISBN 0 412 47520 0.

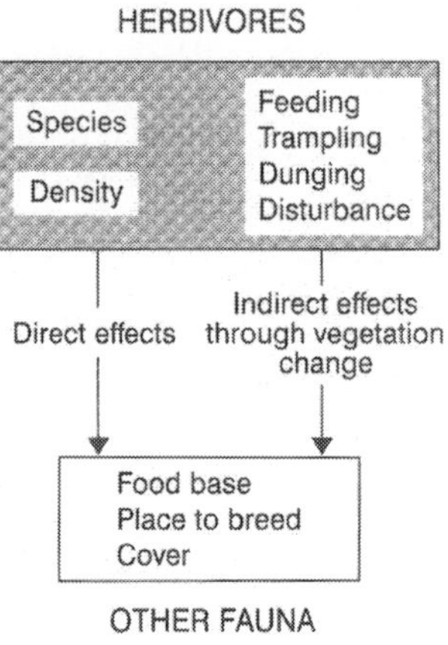

HERBIVORES

Species
Density

Feeding
Trampling
Dunging
Disturbance

Direct effects

Indirect effects through vegetation change

Food base
Place to breed
Cover

OTHER FAUNA

Figure 6.1 Potential relationships between processes generated by herbivores and other fauna.

while at moderate densities (Figure 6.4) the structural classes are more equally represented.

This in turn will have profound consequences for the animal community at large. The problem is that we do not yet know if large herbivores can act as keystone species in the temperate zone. Natural temperate ecosystems

Figure 6.2 Red deer have been introduced into a newly created polder in The Netherlands to complete the existing herbivore community.

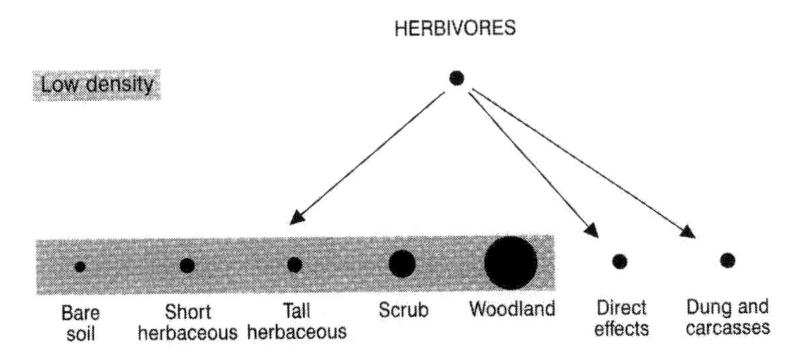

Figure 6.3 Effects of herbivory when herbivore density is low.

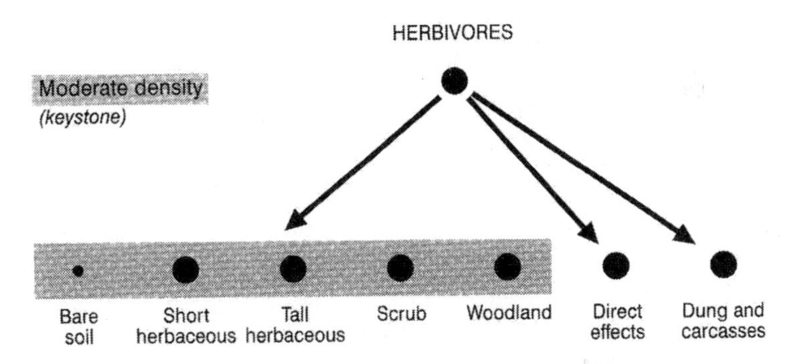

Figure 6.4 Effects of herbivory when herbivore density is moderate.

hardly exist any more, which makes the mechanism difficult to study. There is (some) evidence that large herbivores may have played a keystone role in temperate ecosystems from the Atlantic period until the human impact became dominant (Vera, 1997). When extrapolating from existing systems, however, many researchers feel that large herbivores cannot play a keystone role and that fairly low densities would be the rule under natural temperate conditions with most landscapes dominated by (closed) forests.

This chapter discusses the consequences of varying herbivore densities for the animal community. In part these consequences are inferred from the general relationships that exist between various vital resources supplied by the ecosystem and the presence of birds and mammals. Birds especially can demonstrate the effects because their ecology is well known and relationships with herbivores are well studied. Through the bird examples the consequences for many other animal species or animal guilds may be deduced by analogy.

6.2 RELATIONSHIPS BETWEEN MAMMALS AND BIRDS AND VEGETATION STRUCTURE

The dependence of 257 bird species and 57 mammal species on vegetation structure for nesting and foraging in western Europe is given in Tables 6.1 and 6.2, respectively.

Species of birds and mammals can be found in any of the major vegetation structures. Most mammal species (bats) are found in forests; very few species prefer short herbaceous vegetation. Old-growth forests are particularly important to insects (Warren and Key, 1991). For many species of birds, short and tall herbaceous vegetation is important for nesting and feeding sites.

It will be clear that when herbivory leads to large changes in the area of any of the vegetation structures, this will affect the animal species related to them. When areas get reduced, at first changes in abundance can be expected but later species richness will be affected as well and species will disappear from the system. As stated above and indicated in Figures 6.3 and 6.4, herbivores can directly affect other fauna through competition for food

Table 6.1 Distribution of western European birds (number of species) across various types of preferred nesting and feeding sites

	Number of species	
Habitat	Nesting site	Feeding site
Bare soil	18	40
Short herbaceous (open landscape)	62	57
Tall herbaceous (with and without scrub layer)	81	47
Scrub	20	12
Woodland	76	30

N = 257 species.
Unequal column totals result from many bird species feeding in habitats not represented here (e.g. aquatic).
Data compiled by Albert de Hoon and kindly placed at our disposal.

Table 6.2 Distribution of species of mammals (ungulates excluded) in western Europe across preferred habitat types; species can be related to more than one habitat type (*N* = 57 species; compiled after MacDonald and Barrett, 1993)

Habitat	Number of species
Short herbaceous	5
Tall herbaceous	22
Scrub	27
Woodland	41

or disturbance. Bird species that take advantage of large herbivores by following them to prey on insects and amphibians disturbed by their movements are the commensals. Examples are cattle egrets (*Bubulcus ibis*), starlings (*Sturnus vulgaris*), white wagtails (*Motacilla alba*) and white storks (*Ciconia ciconia*). Dung is the home of dung beetles (Scarabaeidae) which are favoured by predators like the great grey shrike (*Lanius excubitor*), red-backed shrike (*Lanius collurio*) and corvids. Carrion of large ungulates is of importance to the kite (*Milvus milvus*), black kite (*Milvus migrans*), golden eagle (*Aquila chrysaetos*), white-tailed eagle (*Haliaetus albicilla*) and corvids. The black vulture (*Aegypus monachus*) lived in western Europe until Roman times. Finally, some species of parasites (e.g. bot flies, warble flies) totally depend on the presence of ungulate hosts, and others such as ticks (Acarina), lice (Mallophaga) and horse flies (Tabanidae) are favoured by the presence of large ungulates.

6.3 HERBIVORE–HERBIVORE INTERACTIONS IN NATURAL AND SEMI-NATURAL SYSTEMS

Studies of interactions between large herbivores in natural multi-species systems are relatively scarce. For the most part, such studies have examined in detail the ecology of individual species and the partitioning of resources, but have not explicitly considered the actual mechanism of such partitioning and the scale of interactions between the species. Interpretation of this literature is thus hampered by not knowing whether lack of reported competition reflects a genuine lack of interaction or merely lack of appropriate investigation. Mathur (1991), for example, discusses ecological separation between sympatric populations of chital (*Axis axis*), sambar (*Cervus unicolor*) and nilgai (*Boselaphus tragocamelus*) in three national parks in India, in relation to use of both habitat and forage resources; in earlier studies in Nepal, clear separation in the use of habitat was recorded between chital, hog deer (*Axis porcinus*) and barasingha (*Cervus duvauceli*), and between chital, sambar, hog deer and muntjac (*Muntiacus reevesi*) in Chitwan National Park (Mishra, 1982). The individual habitat preferences of sable antelope (*Hippotragus niger*), waterbuck (*Kobus ellipsiprymnus*) and impala (*Aepyceros melampus*) in southern Africa are presented in an elegant analysis by Ben-Shahar and Skinner (1988); resource relationships of ungulates of the intensively studied Serengeti ecosystem are summarized by Jarman and Sinclair (1979). However, in none of these studies in tropical systems were the mechanisms or implications of such partitioning explored in detail.

In temperate systems, too, the majority of studies merely describe patterns of resource use and partitioning among sympatric species; e.g. Hudson (1976) for white-tailed deer (*Odocoileus virginianus*), mule deer (*Odocoileus hemionus*), wapiti (*Cervus elaphus*) and bighorn sheep (*Ovis canadensis*) in British Columbia; Cairns and Telfer (1980) for moose (*Alces*

alces), wapiti, white-tailed deer and American bison (*Bison bison*); Jenkins and Wright (1988) for white-tailed deer, wapiti and moose in Montana; Grodzinski (1975), summarizing the work of various authors on the ecology of roe deer (*Capreolus capreolus*), red deer (*Cervus elaphus*), fallow deer (*Dama dama*), moose and European bison (*Bison bonasus*) in Polish forests; and Chapman *et al.* (1985) for ecological separation between roe deer, fallow deer and Chinese muntjac sympatric within a commercial conifer forest in the UK.

What we cannot tell from any of these studies is whether or not the clear ecological separation expressed implies a lack of any potential for interaction, or whether separation is itself an explicit response to competition. Only where we may find evidence of a clear shift in resource use of a species in allopatry and sympatry may we suspect a competitive interaction – when clear overlap in resource use is accompanied by an inverse relationship in population sizes of a given species pair. In the absence of experimental manipulations, even this evidence for competition is circumstantial and based solely upon correlation, but it is at least suggestive of some degree of interaction.

In Kanha National Park in central India, for example, where a clear dietary overlap has been recorded between chital deer and barasingha in their use of open grassland plants, a recent decline in population numbers of barasingha coincided with a large increase in abundance within the park of chital (Martin, 1987). Chapman *et al.* (1985) recorded a high overlap in diet and winter habitat use for roe deer and Chinese muntjac in the King's Forest in Suffolk, UK, and there is increasing evidence of a decline in numbers of roe in areas of high muntjac density (Forde, 1989; Wray, 1992). By contrast, detailed analysis of population trends of roe deer, red deer, fallow and sika (*Cervus nippon*) in the New Forest of southern England over a 24-year period (Putman and Sharma, 1987) failed to show any correlation between population sizes of the four species, despite significant apparent overlap in resource use (Putman, 1986).

In fact, in general, evidence for actual competition between members of an established guild of large herbivores is hard to find. While this may in part reflect lack of appropriate investigation, the absence of reported competition is in any case not unexpected. Over an evolutionary time scale, natural selection would be expected to promote clear separation in resource use between regularly interacting sets of species specifically to minimize the loss of fitness incurred through competition. In effect we might expect that competitive interactions would become apparent only when an established system is perturbed from equilibrium in some way – perhaps challenged by a recent invader. In such context it is noteworthy that chital have only recently expanded their range and population size within the Kanha National Park; barasingha and chital have not had a long common evolutionary history (Martin, 1987). Likewise, Chinese muntjac are an introduced species in the UK and were first recorded in the New Forest in 1963. Sika deer, considered possible competitors for white-tailed deer on Assateague Island (Maryland, USA), were introduced to the island in 1925. The most widely quoted examples of competition within an ungulate community are reported from New Zealand, on interactions between red deer and sika (McKelvey, 1959) and red deer, fallow and white-tailed deer (Kean, 1959) – where the entire ungulate assemblage is introduced.

Analysis of the potential for competition among naturally occurring guilds of large ungulates implies that such interactions will be negative, or at best neutral. Yet equally there is potential for facilitation within such guilds, although the evidence here is also poor. It has been suggested that there is facilitation between a number of species of the African savanna – more specifically, the larger species facilitating for the smaller ones (Bell, 1971) – and within the guild of grazers (Prins and Olff, 1998). Cape buffalo (*Syncerus caffer*) are thought to facilitate for Egyptian geese (*Alopocheus aegyptiacus*) (H.H.T. Prins, personal communication). Most studies on facilitation have focused on relationships between livestock and other herbivores (e.g. geese, hares, deer). Examples will be given below in section 6.5, but here it can be noted that the possible facilitating effects of livestock on smaller herbivores may be considered to mimic relationships, in some situations, between herbivores in natural communities (e.g. bison–wapiti, aurochs–red deer, aurochs–geese).

In summary, very little real evidence is available for the existence of both competitive and facilitative relationships between herbivores in natural communities.

6.4 LOW AND MODERATE DENSITY

The density concept can only be used in the relative sense. Here, low and moderate density imply that the four major structural classes (Figures 6.3 and 6.4) are present, and that the plant species composition is (at most) only little altered. At high densities especially scrub and woodland are heavily affected (Figure 6.5), while frequently changes in plant species composition have occurred on a large scale.

At first sight the difference between low and moderate density is one of degree and not of kind (Figures 6.3 and 6.4). At low density, fewer herbivores are present; direct effects, and effects through dung and carrion, are less profound. More important is that woodland prevails at low density. If the woodland is closed, very little space may be left for open areas with short and tall herbaceous vegetation. As these vegetation structures are important for many species of birds, the latter will occur only in low numbers, have a patchy distribution, or may not be present at all. Also many insects rely on open areas in woodland, especially a number of butterflies (*Leptidea sinapsis, Photedes fluxa, Apamea scolopacina*) (Warren and Key,

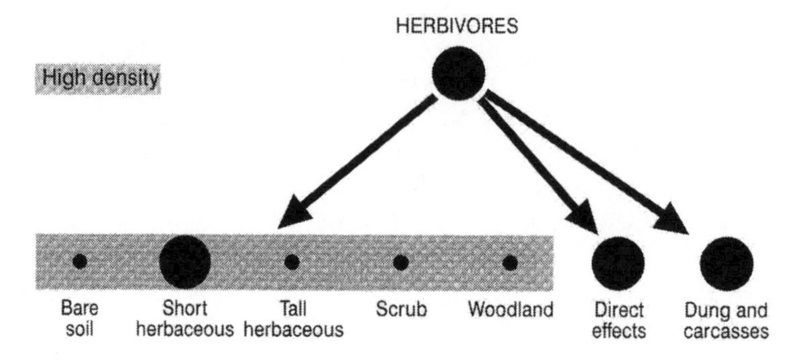

Figure 6.5 Effects of herbivory when herbivore density is high.

1991). The most important change when herbivore density becomes moderate is an increase in open areas (Figure 6.4). The landscapes thus created seem to provide optimal conditions for greatest species richness because all structural classes are present in a substantial proportion of the total area. The importance of a high structural diversity seems particularly important for a very large group of butterfly species. The majority of European butterflies depends on a combination of grassland, woodland and fringe communities (Bink, 1992) (Figure 6.6).

Figure 6.6 The small copper (*Lycaena phlaeas*), a species from grassland and heathland, benefits from extensive grazing.

6.5 HIGH DENSITY OF LARGE HERBIVORES

The examples discussed in this section are all taken from systems that are managed and influenced by humans. Some form of grazing with livestock is applied, generally for production but sometimes also for conservation purposes. Grazing can be year-round, seasonal (summer) or rotational. Stocking rate is frequently high. Generally, other management activities are applied as well (harvesting, logging, mowing, etc.) and the landscape is not shaped by herbivory alone.

As a result of heavy grazing, major changes occur (Figure 6.5). The impact of direct effects increases, as well as the amount of dung. The number of carcasses does not increase, because the herbivore population is harvested. Most striking is a large shift in the relative proportion of major structural classes. Woodland cover drastically decreases in favour of a large increase in short herbaceous vegetation. Next to these changes, plant species composition within structural classes may change as well. In extreme cases, grazing by large herbivores may cause actual transition from one entire community type to another; for example, variation in the level of grazing by cattle and horses has been shown to mediate a reversible shift between grassland and heathland vegetation (Bokdam and Gleichman, 1989). These changes, of course, can be expected to have knock-on effects on many animal species (Figure 6.7). These effects, however, frequently go unnoticed because the changes have occurred in the past, and the present landscape (including its present impoverished fauna) is taken for granted or even taken as the norm. Below, various examples will be given of grazing effects on other fauna within the different structural classes. Herbivore–herbivore interactions will be treated separately as well as consequences for the foodwebs.

6.5.1 Woodland

The effects of a substantial decrease in woodland cover upon the fauna have not been studied. It can be expected that large animals in particular – e.g. lynx (*Lynx lynx*), brown bear (*Ursus arctos*) – are among the first to suffer from a loss of woodland cover but more and more species will be affected as forest cover decreases further. As can be deduced from Tables 6.1 and 6.2, loss of forested area will have detrimental effects on very large numbers of species of animals, not to mention the huge number of species of insects that are related to old-growth forests (Warren and Key, 1991).

Heavy grazing, especially by browsing herbivores like deer, may change the species composition of forests. Preferred tree species (*Salix* spp., *Populus tremula*, *Quercus robur* and *Abies alba*) may decrease and give way to least preferred species (*Pinus sylvestris*, *Picea sitchensis* and *Pinus nigra*) (Gill, 1992). Within the bird community a change from deciduous forest to coniferous forest will lead to a decrease or disappearance of species like tawny owl (*Strix aluco*), nuthatch (*Sitta europaea*), hawfinch (*Coccothraustus coc-*

Figure 6.7 Sand lizard (*Lacerta agilis*) on a cow dung patch. This species needs bare patches for laying eggs and structurally diverse vegetation that may be provided by an extensive grazing regime.

cothraustus), wood warbler (*Phylloscopus bonelli*) and marsh tit (*Parus palustris*), and an increase or appearance of species like goldcrest (*Regulus regulus*), crested tit (*Parus cristatus*), coal tit (*Parus ater*), crossbill (*Loxia curvirostra*) and goshawk (*Accipiter gentilis*). Disappearance of deciduous forests will negatively affect bats.

6.5.2 Scrub

The effects on the scrub layer, too, have received little attention. Heavy browsing by deer can have detrimental effects on the scrub layer. In the United States, increasing white-tailed deer densities in intensively managed forests have led to a decline in species richness and abundance of woody and herbaceous vegetation (Behrend *et al.*, 1970; Warren, 1991; Miller *et al.*, 1992). Possible effects on songbirds were studied by deCelesta (1994) in enclosures with density varying from 3.7 to 24.9 deer/100 ha (Figure 6.8). Above a density of 7.9 deer/100 ha both species richness and abundance of songbirds nesting in the intermediate canopy (0.5–7.5 m above ground) declined significantly. No relationship with deer density was found in ground-nesting and upper canopy-nesting birds.

6.5.3 Tall herbaceous vegetation

As many insect species thrive in tall herbaceous vegetation, they can be seriously affected when heavy grazing leads to a sharp decrease of this structural

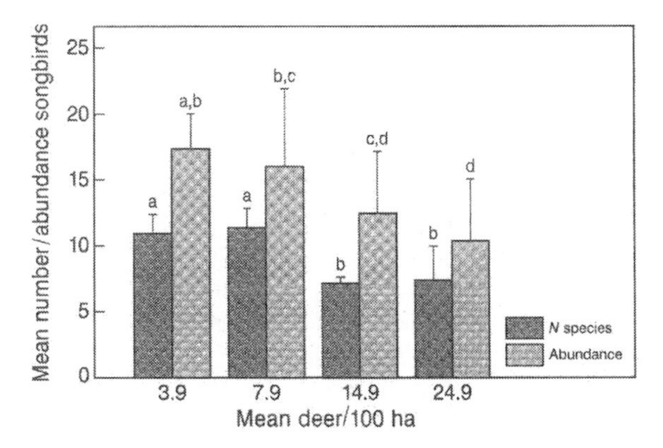

Figure 6.8 Mean (± SE) number of songbirds species and mean abundance of songbirds (numbers/100 ha ± SE), nesting in the intermediate canopy, by white-tailed deer density, in northwestern Pennsylvania (bars with dissimilar letters were different, $P < 0.05$). (After deCelesta, 1994.)

class. Red deer grazing reduced availability of *Vaccinium myrtillus* (Baines *et al.*, 1994). Numbers of lepidopterous larvae were almost fourfold higher, Hymenoptera threefold higher and Coleoptera, Aranaea, Diptera and Plecoptera all twofold higher in ungrazed exclosures than in grazed forest. Density ranged from 10 to 20 deer/100 ha. The results suggest that deer not only reduce available habitat for larvae but that competition occurs for the fresh green leaves of *Vaccinium*. Deer grazing may have subsequent knock-on effects on insectivorous forest birds like capercaillie (*Tetrao urogallus*) and black grouse (*Tetrao tetrix*) (Baines *et al.*, 1994).

Rahmann *et al.* (1987) compared different management systems in salt marshes in northern Germany. The number of insect species decreased with increasing stocking rate. A lightly grazed grassland (0.16 cows/ha) also had fewer species and biomass than an ungrazed area (Siepel *et al.*, 1987). The most important negative factors on the invertebrate communities in grazed systems are fewer flowering plant individuals, less litter, reduced standing crop, disturbance by treading, less prey for predators and a more extreme microclimate.

The effects of grazing on invertebrates are not always so unidirectionally negative. Macrofaunal invertebrates were studied after cessation of cattle grazing three years after the withdrawal of the grazers in the Leybucht area in Niedersachsen, Germany (Irmler and Heydemann, 1985). The number of species did not change, but populations of species sensitive to trampling, either directly or indirectly by damaging tall forage plants, increased in the abandoned area. They included springtails (Collembola), spiders (Aranea), butterflies (Lepidoptera), aphids (Aphinidae), bees (Apidae) and hoverflies (Syrphidae). In contrast, coprophagous species and those feeding on algae on the bare soil decreased. The cessation of grazing caused litter accumulation and hence the foodweb shifted from predominance of herbivorous animals to a foodweb dominated by detritivorous animals (Andresen *et al.*, 1990).

Some grasshopper species seem to benefit from heavy grazing (Fleischner, 1994). The presence of both short and tall vegetation can be a prerequisite for grasshoppers. In the egg stage, grasshoppers need short herbaceous for egg development, but high and dense vegetation is favoured by the nymphal and imaginal stages (Van Wingerden *et al.*, 1991). It seems that an optimum grazing level exists for maximum grasshopper survival.

Many bird species use the tall herbaceous vegetation for nesting (Table 6.1) and it is not surprising that they, too, will suffer negative effects from heavy grazing. In a complex rotational cattle grazing system, nest density of sharp-tailed grouse was twice as low when compared with ungrazed sites, because fewer suitable nest sites (tall dense vegetation) were available (Kirby and Grosz, 1995). Summer grazing by cattle (high density: 0.75 cattle/ha) reduced tall herbaceous vegetation on a salt marsh, thereby inversely affecting curlew (*Numenius arquata*), wood pigeon (*Columba palumbus*), linnet (*Carduelis cannabina*) and reed bunting (*Emberiza schoeniclus*) (Van Dijk and Bakker, 1980). Positive effects have also been reported in wetlands: here, grazers can increase the area of open water in reedbeds, thereby encouraging species which use this habitat for resting, feeding (ducks and coots) or nesting (grebes) (Gordon *et al.*, 1990).

6.5.4 Short herbaceous vegetation

This structural class is heavily favoured in many grazing systems and no doubt species related to it will benefit. For many species of mammals (Table 6.2) and insects, a very short sward is not a good habitat. Furthermore, the heavy grazing intensity needed to maintain the short swards increases the direct effects on other fauna. Only a limited number of species seem to benefit and wetland species have received most attention. Wintering wigeons (*Anas penelope*) in the Ouse Washes, for example, benefit from the effects of summer grazing with cattle at a stocking density of 2.5 cows/ha. This high grazing intensity prevents a succession towards dominance of the more unpalatable grasses *Glyceria maxima*, *Phalaris arundinacea* and *Elymus repens* (Owen and Thomas, 1979). The density of redshank breeding on the salt marshes of the Wash was positively correlated with the percentage cover of sea-couch grass (*Elymus athericus)* at a stocking rate of one cow/ha. This relatively heavy grazing intensity provided the most structurally diverse vegetation including tussocks for nest-building. Ungrazed salt marshes provide a uniform tall stand of sea-couch grass, which is less attractive for redshank (Norris *et al.*, 1997). Cattle grazing in summer (2.5–4.0 cattle/ha) increased the populations of mountain bluebird (*Sialia eurrocoides*) and of blue grouse (*Dendragapus obscurus*) from rare to common in Oregon grasslands (Anderson and Scherzinger, 1974). Summer grazing by cattle (0.75 cattle/ha)

Figure 6.9 Moderate summer grazing benefits meadow birds such as the redshank (*Tringa totanus*).

on a Dutch salt marsh increased nest densities of birds preferring short herbaceous vegetation: oystercatcher (*Haematopus ostralagus*), lapwing (*Vanellus vanellus*), redshank (*Tringa totanus*), black-tailed godwit (*Limosa limosa*) and skylark (*Alauda arvensis*) (Van Dijk and Bakker, 1980) (Figure 6.9).

Although many meadow bird species have profited from relatively intensive agricultural grassland exploitation, the recent intensification of agriculture (increasing density of livestock, early mowing, lowering watertable, increased fertilization levels) has led to increasingly negative effects on meadow birds – notably black-tailed godwit, ruff (*Philomachus pugnax*) and common snipe (*Gallinago gallinago*) (Beintema and Müskens, 1987; Green, 1988). For these species an optimum intensity of exploitation seems to be the most favourable. The black-tailed godwit, for example, took advantage of the transformation of wilderness into grasslands, expanding its range and increasing in numbers (Beintema, 1991). The species reached its greatest abundance around 1965 and then declined because of further intensification of agriculture.

6.5.5 Bare soil

Heavy grazing may lead to an increase in bare soil, positively affecting species depending on it. Some species of butterflies that require very short-cropped grasses surrounded by bare ground for egg-laying sites were favoured in areas heavily grazed by sheep or rabbits (Thomas, 1983; Thomas *et al.*, 1986). Numbers decline under light grazing, and species disappear when grazing stops (Thomas, 1991). The specialized grazing management needed to maintain these butterfly populations can be viewed as a generally poor substitute for more natural conditions that are now lost (Thomas, 1991).

In the United States, ground-breeding birds that are related to bare soil – e.g. mourning doves (*Zenaida macroura*) and quail (*Colinus virginianus*) – are positively affected by heavy grazing (Baker and Guthery, 1990; Guthery, 1986). Meadow lark (*Sturnella magna*) densities were higher at moderate grazing pressure.

6.5.6 Heavy grazing and consequences for foodwebs: the New Forest

Hill (1985) assessed the effects of vegetational change caused by centuries of heavy grazing in the New Forest (in Hampshire, southern England) on populations of mice, voles and shrews, by comparing the species diversity and population sizes of these small mammals within the Forest (grazed by deer, cattle and ponies) with those recorded in equivalent plant communities in areas outside the Forest boundary (grazed by deer but not by livestock). Data were collected separately in woodland, grassland and heathland communities.

While typical densities of bank voles (*Clethrionomys glareolus*), field voles (*Microtus agrestis*), woodmice (*Apodemus sylvaticus*) and shrews (*Sorex* spp.) were trapped in grasslands and heathlands outside the Forest boundary, only two captures were made (both of *A. sylvaticus*) during 2400 trap-nights in heathlands, and 25 individuals (20 *A. sylvaticus*; five *Microtus*) in acid grasslands within the Forest. These animals were transients and in effect no established populations of any rodents were recorded in such communities within the Forest. Within ungrazed sites, woodland areas throughout supported substantial populations of woodmice and bank voles, with lower densities recorded of yellow-necked mice (*Apodemus flavicollis*) and common shrew (*Sorex araneus*).

Hill has related the differences in small mammal densities in grazed and ungrazed sites to vegetational structure and it is clear that the differences recorded are linked to changes in vegetation resulting from heavy grazing. More direct evidence in support of such a conclusion may be adduced from comparison of rodent communities within experimental enclosures established within the New Forest (Putman *et al.*, 1989). Community structure and diversity of small mammals were assessed in two adjacent 5.5 ha woodland pens – one grazed and the other protected from grazing for a period of 25 years. Three species of small mammals (*Apodemus sylvaticus*, *Clethrionomys glareolus* and *Sorex araneus*) were regularly recorded throughout a two-year trapping period in the ungrazed pen, with a further two species recorded occasionally (*Apodemus flavicollis*, *Sorex minutus*). In the grazed area, *A. sylvaticus* was the only species ever recorded – and at consistently lower population density than in the ungrazed pen.

Changes in the abundance of prey species will probably have a wider effect through the community web. Petty and Avery (1990), for example, have shown that immediately after the cessation of grazing in upland areas about to be afforested, populations of field voles show an initial surge in population density. This increased food supply is exploited by predators such as short-eared owl (*Asio flammeus*), long-eared owl (*Asio otus*), kestrel (*Falco tinnunculus*) and barn owl (*Tyto alba*) whose breeding success and population sizes also increase until canopy closure within the new plantations.

Perhaps the most complete analysis to date of the effects of heavy grazing by domestic herbivores upon the dynamics of the community as a whole comes from the series of detailed studies published on the ecology of the New Forest. Reduced diversity and overall abundance of small mammals in the heavily grazed communities of the Forest, demonstrated to be the direct result of the sustained heavy grazing pressure (Hill, 1985), can be shown in turn to have had an effect on the foraging behaviour, diet, population density and breeding success of a diverse array of predators, such as fox (*Vulpes vulpes*), badger (*Meles meles*), buzzard (*Buteo buteo*), kestrel and tawny owl (Tubbs and Tubbs, 1985; Putman, 1986).

Studies of the diets of foxes in the New Forest reveal that while New Forest animals did consume small rodents when available, the frequency and relative proportion in the diet was lower than that recorded in other areas; the bulk of rodent prey consumed consisted of rats (*Rattus norvegicus*) and field voles (Putman, 1986). Few birds were taken and it was noted that the foxes clearly relied heavily on invertebrate material (particularly earthworms and beetles) and carrion. Diets of foxes within the Forest were shown to contain a higher proportion of beetle and other insect prey than was apparent in samples from nearby areas with much lower grazing densities; they also consumed considerable quantities of fruits and fungi, particularly in autumn. Perhaps in response to scarcity of suitable prey, the overall density of foxes within the Forest is itself somewhat low and estimated as 2.18/100 ha, with a minimum adult density of 0.75/100 ha.

Badger densities within the Forest are also lower than recorded for the same species in less heavily grazed areas elsewhere in Hampshire (Packham, 1983). Despite the low density of their preferred prey of earthworms within the Forest, badgers do not adjust their diet as do the foxes to make opportunistic use of alternative foodstuffs that are more readily available, but remain earthworm specialists. Since worm densities are so low, individual badgers are forced to forage over extremely large areas – too large to be maintained as the typical exclusive territories of an individual or a social group. Overall population densities are low, and low prey density further transforms the entire social and spatial organization of those populations (Packham, 1983).

Avian predators are also affected by the low rodent abundance. Tubbs (1974, 1982) first noted that there seemed to be a close correlation between the breeding success of buzzards in the New Forest and concurrent population densities of grazing cattle and ponies. Tubbs noted that in the New Forest, as in other parts of England where rabbits are not readily available, buzzards appear to rely very heavily upon rodent prey and breeding success is directly related to the abundance of such rodent prey. Although he had no direct data on the changing abundance of small mammals within the Forest over the years, he nonetheless showed a clear correlation between the number of buzzard pairs attempting to breed in any year and the numbers of livestock grazed on the Forest over the preceding three years (see also Putman, 1986). Hirons (1984) examined diet and breeding success of other raptors within the Forest whose diets would normally be expected to contain high numbers of rodents. Tawny owls, essentially woodland predators, continued to maintain a high proportion of rodent prey within the diet, though this was composed almost exclusively of *Apodemus sylvaticus*. Rodents contributed 42% of all prey taken (as against 60%–70% recorded from studies elsewhere: Hirons 1984); New Forest owls compensated with an increased reliance on invertebrate prey – particularly dung beetles. Territory sizes of tawny owls in the New Forest were similar to those recorded elsewhere, but Hirons noted a significant reduction in the proportion of pairs breeding in any one year (25% in the Forest as against 65% of pairs in areas outside). Owls are woodland birds and, of all the Forest communities, woodlands showed least difference in rodent densities from ungrazed areas outside the Forest (Hill, 1985); in the more open habitats of grassland or heathland, small rodents were virtually absent. Recorded density of kestrels (Hirons, 1984) was 0.0625 pairs/100 ha within the Forest, compared with 0.25 pairs/100 ha on farmland outside the Forest. In an analysis of the diet of these essentially open-country predators, Hirons noted that mammalian prey constituted only 30% by weight of vertebrate prey taken and the kestrels relied heavily on small birds and on common lizards (*Lacerta vivipara*), which occurred in 45% of all pellets analysed (Hirons, 1984). No kestrels were recorded as breeding successfully within the Forest boundary during the two years of the study.

This extensive series of studies within the New Forest is explored here in some detail to illustrate the whole suite of subtle changes that may be expected within any ecological system following the introduction of a regime of heavy grazing. It emphasizes once more that the ecological effects of grazing are not restricted to the immediate and obvious response of the vegetation itself.

6.5.7 Heavy grazing and vegetation dynamics: the bird community of the Slikken van Flakkee

In 1971 an estuary in the southwest Netherlands (Grevelingen) was embanked and tidal movements ceased. The sandflats and salt marshes became permanently dry and a new vegetation developed after desalination. Vegetation succession and the development of the bird community were studied in a part of the area: the Slikken van Flakkee (Slob, 1989; Van Schaik and De Jong, 1989). About half of the area (650 ha) is sown with grasses and grazed with 50 Heck cattle and 25 horses; the other half (600 ha) is left ungrazed. The development of the breeding bird populations is closely related to the vegetational succession (Figure 6.10). The number of bird species is higher in the ungrazed area because of the occurrence of scrub and woodland. The number of breeding pairs, however, is much greater in the grazed area. Although the number of species in the grassland remains similar, the grazed grasslands have become increasingly important breeding areas for the typical meadow species: oystercatcher, lapwing, black-tailed godwit and redshank. Each area is thus important for a different group of species. The ungrazed area is important to a limited number of species which are absent in the grazed area. Although there are no species that breed exclusively in the grazed area, the total number of breeding pairs is much higher. The number of breeding pairs decreases in the ungrazed area but increases in the grazed area. The presence of uniform short herbaceous vegetation in the grazed area also attracts wintering geese (Table 6.3).

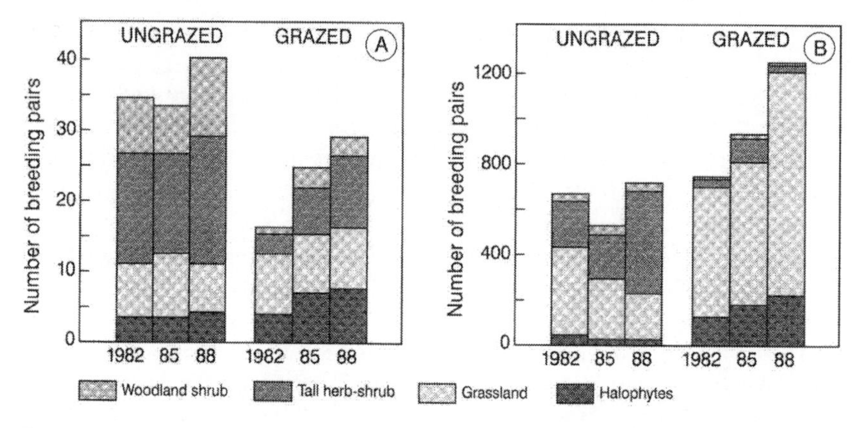

Figure 6.10 The number of (A) bird species and (B) breeding pairs of the breeding bird community, across macrostructural classes of vegetation in the grazed and ungrazed Slikken van Flakkee in 1982, 1985 and 1988. (After Van Wieren, 1991.)

6.5.8 Herbivore–herbivore interactions

It would appear that levels of competition within natural communities of large herbivores may generally be rather low, but that an increase in the level of interaction may be expected following introductions of novel species into the system. In most cases, imposition of a regime of grazing for conservation purposes will be by introduction of livestock; this introduction of 'exotic' species, typically at high densities, may then be expected to have a profound effect on native herbivores. The focus is generally the food-base but effects on cover and direct effects through interference/disturbance play a role as well. Interaction through food can be competitive or facilitative,

Table 6.3 Total number of barnacle geese (*Branta leucopsis*) counted in the grazed and ungrazed Slikken van Flakkee in the period November–February (one count/month) (from Van Wieren, 1991)

Year	Grazed	Ungrazed
1981	7 180	1 200
1982	12 500	0
1983	36 000	370
1984	11 000	0
1985	10 400	2 600
1986	11 200	1 945
1987	6 000	0
1988	22 400	5 930

very much depending on density and the characteristics of the species involved.

In the United States, the effects of livestock grazing on native browsing ungulates have been widely studied. The effects are generally negative. The introduction of cattle to native rangelands in the Sierra Nevada mountains in California had a profound impact on mule deer. Moderate and heavy grazing by cattle reduced hiding cover for mule deer in aspen and riparian habitats (Loft *et al.*, 1987). Mule deer preferred ungrazed to grazed pastures (Ragotzkie and Bailey, 1991). Heavy cattle-grazing triggered habitat shifts in mule deer, especially at the end of the summer: female mule deer occurred relatively less often on preferred habitats and more often on less preferred habitats as competition for resources increased (Loft *et al.*, 1991). Deer spent more time feeding and less time resting, with increasing stocking rates (Kie *et al.*, 1991), and female mule deer avoided cattle when travelling through their home range (Loft *et al.*, 1993). Cattle seem to affect mule deer mostly through effects on cover and because of interference competition. The species have little dietary overlap.

Sheep grazing of shrubs (*Artemisia*) in late spring and fall in the United States possibly competes with white-tailed deer and wapiti, leaving less food for the latter. The effect was less with moderate grazing than with heavy grazing. The effects of cattle on deer do not seem to be only negative: cattle-grazing has been reported to enhance forage availability for mule deer and white-tailed deer (Willms *et al.*, 1979; Wallace and Krausman, 1987; Gavin *et al*, 1984).

In the New Forest of southern England, the four native or introduced species of wild ungulates (red, sika, roe and fallow deer) share the complex vegetation mosaic of woodlands, heathlands and grassland with large populations of domestic cattle and ponies, turned out onto the Forest under ancient rights of common (Putman, 1986). Calculated levels of overlap in use of resources presented by Putman (1986) for all pairwise combinations of species in all seasons revealed a high level of overlap in diet between the two domestic species (cattle and ponies) and between them and both sika and fallow deer. While diets of cattle, ponies and fallow deer are equivalent to those reported in other published studies, diets described for sika deer in the New Forest differ significantly from what is described as typical for the species elsewhere (Mann, 1983; Mann and Putman, 1989). Sika in Dorset and in five forest areas in Scotland all have much the same diet, composed primarily of heather (30–40%) and grasses (50–70%); only in the New Forest does the diet appear to change, with a marked increase in intake of browse and reduced intake of grasses. Nor is the direction of the change what might be expected in terms of availability of habitat; in principle, the diversity of vegetation within the New Forest offers better opportunities for grazing. Such an unexpected shift in diet may well be the result of competition and it has been suggested elsewhere (Putman, 1986) that for sika in the

New Forest there may indeed be real competition from grazing ponies. In this and other cases, such demonstrated shifts in resource use are of course not necessarily damaging. They reveal a change in patterns of resource use in response to the presence of another species, but need not deleteriously affect population performance. Indeed, populations of sika deer in the New Forest are expanding in both numbers and range; changes in population numbers since the early 1960s show no correlation with densities of either cattle or ponies. By contrast, numbers of both roe deer (Putman and Sharma, 1987) and fallow deer (S.K. Sharma, unpublished) within the Forest in any year show a significant inverse correlation with the average number of ponies grazed in the Forest over the preceding three-year period.

Facilitative interactions can also occur. Introduced cattle grazing on the Isle of Rhum resulted in greater biomass and a greater availability of green grass in a *Molinia*-dominated grassland than in ungrazed sites (Gordon, 1988). The areas grazed by cattle were preferentially grazed by red deer and the reproductive performance of deer in these areas was improved: the calf–hind ratio was found to be higher in grazed sites (Gordon, 1988). The effect on elk (wapiti) of cattle-grazing intensity was demonstrated by Anderson and Scherzinger (1974) (Figure 6.11): when uncontrolled heavy grazing in wintering areas for wapiti was replaced by a system of controlled moderate grazing, the number of wintering wapiti markedly increased.

Grazing effects on smaller (grass-eating) herbivores are generally positive (Figure 6.12). In an abandoned grass-heath grazed with Icelandic ponies, the rabbit density was positively correlated with the grazing pressure of the ponies. On their own, rabbits could only maintain a short sward when at very high density (Oosterveld, 1983). Frylestam (1976) reported a negative association between hares (*Lepus europeus*) and cattle density (Figure 6.13).

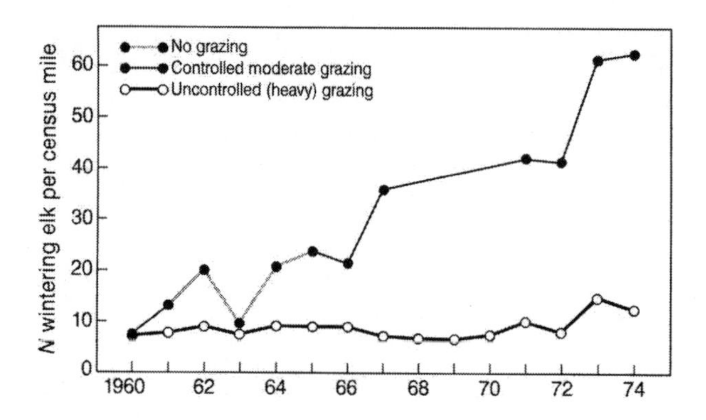

Figure 6.11 Number of wintering elk (wapiti) in wildlife management areas with uncontrolled and controlled grazing in eastern Oregon. (After Anderson and Scherzinger, 1974.)

This negative association is mainly based on direct avoidance behaviour because cattle grazing also provides short fresh grazing for the hares. Hares avoid dense and high vegetation when feeding. The cattle generally occupied only a part of the enclosure and the hares occurred mainly in the other part of the plot.

Figure 6.12 Small grazers with a preference for very short swards, such as hare (*Lepus europeus*) and barnacle goose (*Branta leucopsis*), are facilitated by the grazing activities of large herbivores.

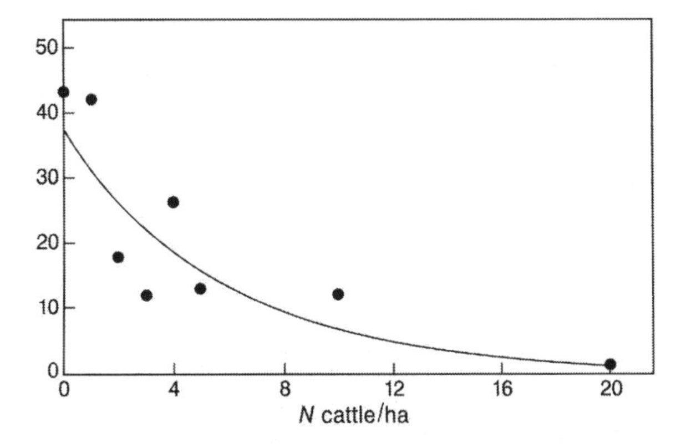

Figure 6.13 Relationship between occurrence of hares and density of cattle in various pastures in Sweden ($y = 36.8x^{-0.17}$; $r = 0.94$, $P < 0.05$). (After Frylestam, 1976.)

In a productivity gradient related to the age of a salt marsh not exploited by cattle, the small herbivores (hares, rabbits and geese) showed the highest density at intermediate standing crop (Figure 6.14) (Van de Koppel *et al.*, 1996). The small herbivores appeared unable to keep the standing crop at a level that was favourable for themselves. This suggests that a regression in succession caused by large herbivores could facilitate the smaller species. The impact of cattle grazing on the older, more productive part of the salt marsh indeed coincided with the occurrence of geese (Olff *et al.*, 1997). The

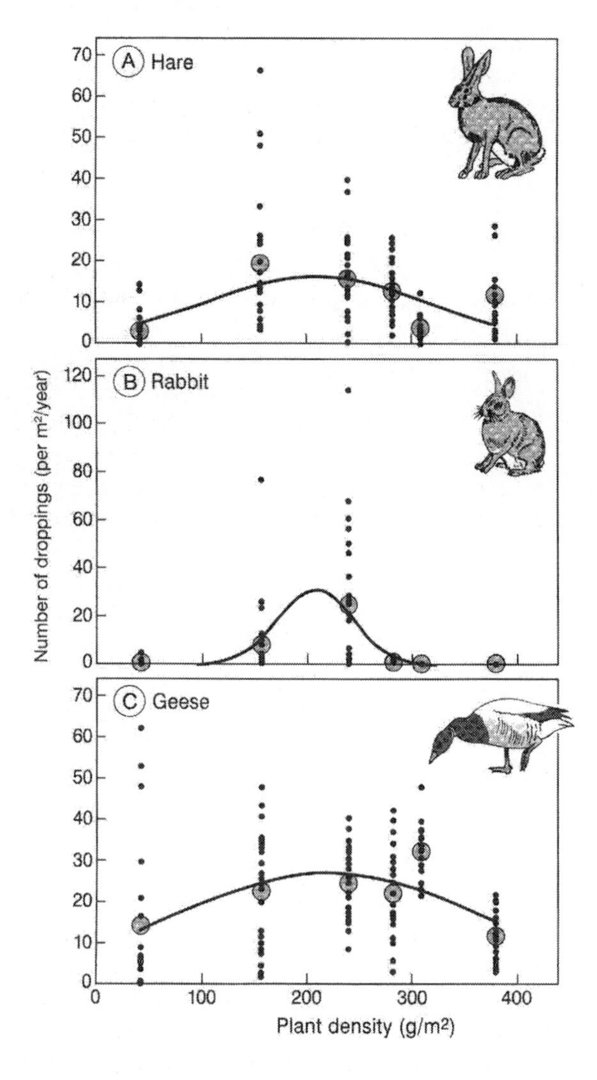

Figure 6.14 Relationships between plant density and the occurrence of hares, rabbits and geese on the salt marsh of the isle of Schiermonnikoog. (After Van de Koppel *et al.*, 1996.)

facilitation of large herbivores for small herbivores is currently monitored in a large-scale experiment.

A very rapid response of barnacle geese (*Branta leucopsis*) to management changes was observed at the Hamburg–Hallig marshes, Germany. Until the summer of 1991, this entire marsh area of 763 ha was heavily grazed by sheep during summer at a stocking density of 10 sheep/ha. An experiment to study the effect of management change on goose visitation was started in 1991, when 25% of the area became moderately grazed at a density of 3 sheep/ha; 40% of the area was left abandoned; and in the remaining 35% the stocking rate was kept at 10 sheep/ha. Figure 6.15B shows the abrupt change in goose visitation with changing management against a continuing overall increase of the Baltic-flyway population of barnacle geese (Figure 6.15A) (Bakker *et al.*, 1997). The geese who continued to use the marsh showed a shift from their traditional area, where grazing had decreased or been abandoned, to the intensively sheep-grazed area.

In summary, grazing by livestock generally seems to have negative effects on the browsing ungulates. Part of the explanation may be that the studied species (mule deer, white-tailed deer, roe deer) are specialized browsers who cannot switch easily to non-browse forage if browse is disappearing through the activities of the more catholic large grazers. Further, being small and cryptic species, they are sensitive to a decrease in cover. Facilitation is to some extent expected with species that are either grazers or intermediate feeders (red deer, wapiti), but here the density of the domestic species is of paramount importance. Small herbivores are mostly facilitated because they themselves have difficulty in maintaining the vegetation in a favourable condition.

6.6 GRAZING AND ANIMAL SPECIES RICHNESS: POTENTIAL AND LIMITS OF A MANAGEMENT TOOL

The relationships between large herbivores and other fauna can be summarized as follows. If large herbivores do not play a keystone role in natural ecosystems, or when the density is low, then the effects on the vegetation and other species of animals will be limited. Because of the low density, the herbivores do not structure the vegetation, while further direct effects and effects through dung and carcasses will be of minor importance (Figure 6.3).

If large herbivores are keystone species in natural ecosystems, or if densities are moderate in managed systems, then the effects on vegetation structure, and other effects, will be more substantial (Figure 6.4). The major difference with the low-density situation is an increase in open areas with short and tall herbaceous vegetation. Species related to these structural classes will be positively affected and it is likely that animal species richness will be higher when compared with the low-density situation (Figure 6.16). If large herbivores should happen not to play a keystone role, then the question arises what the natural habitat of those species that now depend on some form of grazing management would be or would have been, e.g. some spe-

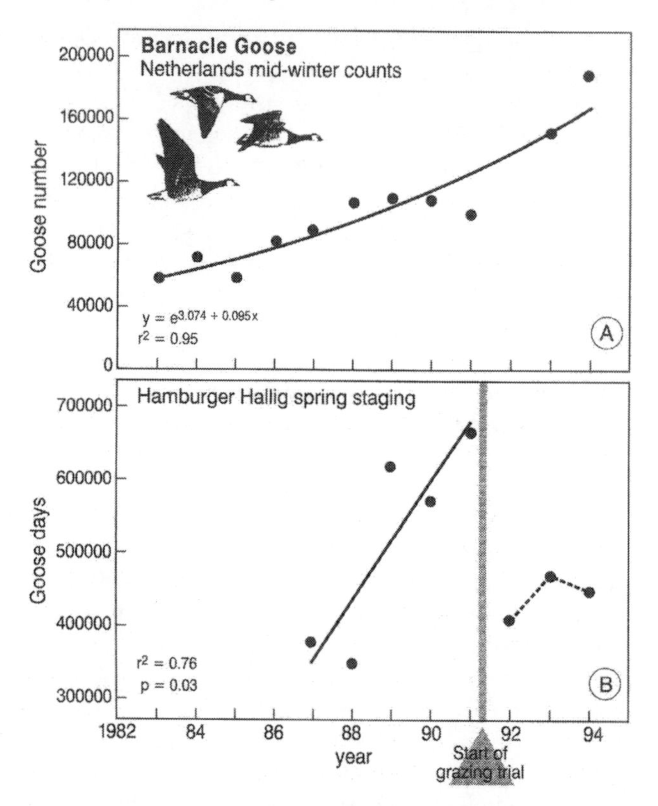

Figure 6.15 (A) Recent change in the wintering population of the Baltic-flyway population of barnacle geese in the Netherlands, and (B) number of goose days in the Hamburger-Hallig marshes. (From Bakker *et al.*, 1997.)

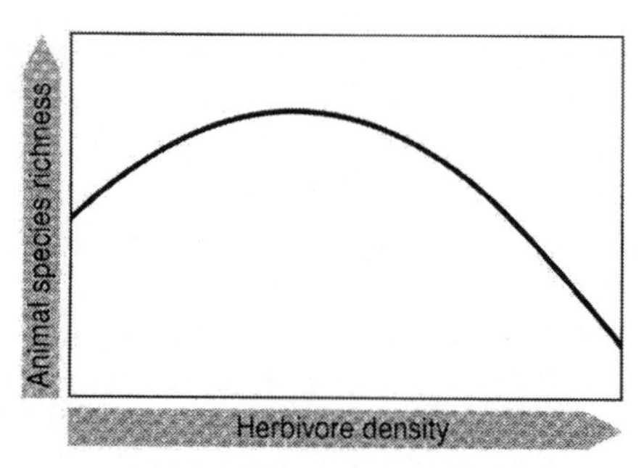

Figure 6.16 Generalized relationship between herbivore density and animal species richness.

cialized species of butterflies or some species of wetland birds. It is possible that the natural habitat of some of these species has disappeared (Thomas, 1991), which may justify the sometimes limited (aimed at only a few species) management goal that is pursued.

If large herbivores can have keystone effects, some managed grazing systems can be viewed as ecological analogues of natural systems because livestock then play the role of wild herbivores that have become extinct (aurochs, tarpan) or have disappeared from former occupied areas (e.g. bison, European bison). It has to be stressed that care must be taken with this approach. Densities should not be too high and it is likely that the mechanism only works properly if large herbivores have been part of the system in the past. The dramatic and detrimental effects of livestock grazing on animal species richness in western North America (Fleischner, 1994) can in part be attributed to bison never having been a significant part of the system. In other areas this may be different – for example, in large parts of Europe and less arid regions of the United States.

If the density of large herbivores becomes high or very high, animal species richness will decrease (Figure 6.16). This is mainly because important vegetation structures (forest, scrub) substantially decrease or disappear and with them many species who depend on them. A series of knock-on effects working its way through the community web can be expected. Only a limited number of species strongly related to open landscapes will profit and show an increase in numbers. The grazing regime applied will not affect all species of open landscapes in the same way. Much depends on density, as illustrated in Figure 6.17 for two species of wetland birds. It is thus unlikely that one grazing system will provide suitable conditions for all species. Heavy grazing tends to level heterogeneity in open areas (with tall herbaceous vegetation disappearing) and then only very few species will profit. The prevalence of short swards may benefit other herbivores, notably grass-eating small ungulates, and the small species (geese, teals, rabbits and hares). The conclusion is that with high grazing densities no rules can be given and all depends on the specific requirements of the species one wants to conserve.

When grazing is applied as a management tool it should be realized that some species will benefit but others will not. A choice has to be made. If ani-

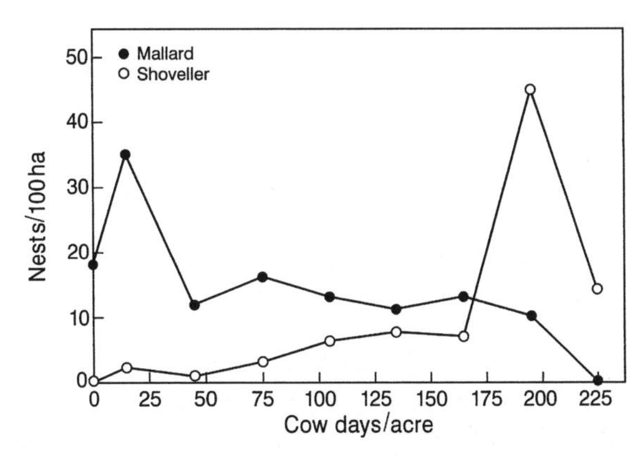

Figure 6.17 Density of nests (per 100 ha) of mallard and shoveller in relation to cattle density in the previous year in the Ouse Washes. (After Thomas, 1980.)

mal species richness is the goal, a moderate grazing intensity is likely to give the best results as it ensures the presence of substantial areas of the most important structural classes. The monitoring of these structural classes can be a useful tool in adjusting grazing intensity in the right direction.

ACKNOWLEDGEMENT

Dr R.J. Putman kindly provided part of the text on the New Forest.

REFERENCES

Anderson, E.W. and Scherzinger, R.J. (1974) Improving quality of winter forage for elk by grazing cattle. *Journal of Range Management*, **28**, 120–125.

Andresen, H., Bakker, J.P., Brongers, M. *et al.* (1990) Long-term changes of salt marsh communities by cattle grazing. *Vegetatio*, **89**, 137–148.

Baines, D., Sage, R.B. and Baines, M.M. (1994) The implications of red deer grazing to ground vegetation and invertebrate communities of Scottish native pinewoods. *Journal of Applied Ecology*, **31**, 776–783.

Baker, D.L. and Guthery, F.S. (1990) Effects of continuous grazing on habitat and density of ground-foraging birds in south Texas. *Journal of Range Management*, **43**, 2–5.

Bakker, J.P., Esselink, P., Van Der Wal, R. and Dijkema, K.S. (1997) Options for restoration and management of coastal salt marshes in Europe, in *Restoration Ecology and Sustainable Development*, (eds. K.M Urbanska, N.R. Webb and P.J. Edwards), pp. 286–322, Cambridge University Press, Cambridge.

Behrend, D.F., Mattfield, G.F., Tierson, W.C. and Wiley, J.E. (1970) Deer density control for comprehensive forest management. *Journal of Forestry*, **68**, 695–700.

Beintema, A.J. (1991) Status and conservation of meadow birds in the Netherlands. *Wader Study Group Bulletin*, **61** (Suppl.), 3–5.

Beintema, A.J. and Müskens, G.J.D.M. (1987) Nesting success of birds breeding in Dutch agricultural grasslands. *Journal of Applied Ecology*, **24**, 743–759.

Bell, R.H.V. (1971) A grazing system in the Serengeti. *Scientific American*, **225**, 86–93.

Ben-Shahar, R. and Skinner, J. (1988) Habitat preferences of African ungulates derived by uni- and multi-variate analyses. *Ecology*, **69**, 1479–1485.

Bink, F.A. (1992) *Ecologische atlas van de dagvlinders van Noord-West Europa*, Instituut voor Bos- en Natuuronderzoek en Unie van Provinciale Landschappen, Schuyt and Co, Haarlem.

Bokdam, J. and Gleichman, J.M. (1989) De invloed van runderbegrazing op de ontwikkeling van struikheide en bochtige smele. *De Levende Natuur*, **90**, 6–14.

Cairns, A.L. and Telfer, E.S. (1980) Habitat use by 4 sympatric ungulates in boreal mixed forest. *Journal of Wildlife Management*, **44**, 849–857.

Chapman, N.G., Claydon, K., Claydon, M. and Harris, S. (1985) Distribution and habitat selection by muntjac and other species of deer in a coniferous forest. *Acta Theriologica*, **30**, 287–303.

deCelesta, D.S. (1994) Effects of white-tailed deer on songbirds within managed forests in Pennsylvania. *Journal of Wildlife Management*, **58**, 711–718.

Fleischner, T.L. (1994) Ecological costs of livestock grazing in western North America. *Conservation Biology*, **8**, 629–644.

Forde, P.G. (1989) The comparative ecology of muntjac (*Muntiacus revesi*) and roe deer (*Capreolus capreolus*) in a commercial coniferous forest. PhD thesis, University of Bristol.

Frylestam, B. (1976) Effects of cattle grazing and harvesting of hay on density and distribution of an European hare population, in *Ecology and Management of European Hare Populations* (eds Z. Pielowski and Z. Pucek), pp. 199–203, Mammals Research Institute, Bialowieza, Warsaw.

Gavin, T.A., Suring, L.H., Vobs, P.A. and Meslow, E.C. (1981) Population characteristics, spatial organisation, and natural mortality in the Columbian white-tailed deer. *Wildlife Monographs* **91**.

Gill, R.M.A. (1992) A review of damage by mammals in north temperate forests: 1. Deer. *Forestry*, **65**, 145–169.

Gordon, I.J. (1988) Facilitation of red deer grazing by cattle and its impact on red deer performance. *Journal of Applied Ecology*, **25**, 1–10.

Gordon, I.J., Duncan, P., Grillas, P. and Lecomte, T. (1990) The use of domestic herbivores in the conservation of the biological richness of European wetlands. *Bulletin Ecologique*, **21**, 49–60.

Green, R.E. (1988) Effects of environmental factors on the timing and success of breeding of common snipe *Gallinago gallinago* (Aves: Scolopacidae). *Journal of Applied Ecology*, **25**, 79–93.

Grodzinski, W. (1975) The role of large herbivorous mammals in the functioning of forest ecosystems – a general model. *Polish Ecological Studies*, **1/2**, 5–15.

Guthery, F.S. (1986) *Beef, Brush and Bobwhites: quail management in cattle country*, Caesar Kleberg Wildlife Research Institute, Kingsville, Texas.

Hill, S.D. (1985) Influences of large herbivores on small rodents in the New Forest, Hampshire. PhD thesis, University of Southampton.

Hirons, G.J.M. (1984) The diet of tawny owls (*Strix aluco*) and kestrels (*Falco tinnunculus*) in the New Forest, Hampshire. *Proceedings of the Hampshire Field Club and Archeological Society*, **40**, 21-26.

Hudson, R.J. (1976) Resource division within a community of large herbivores. *Le Naturaliste Canadien*, **103**, 153–167.

Irmler, U. and Heydemann, B. (1985) Der Einfluss der Rinderbeweidung auf die Struktur der Salzwiesen-Biozönose. *Verhandlungen Gesellschaft fur Ökologie*, **13**, 71–76.

Jarman, P.J. and Sinclair, A.R.E. (1979) Feeding strategy and the pattern of resource partitioning in ungulates, in *Serengeti: Dynamics of an Ecosystem*, (eds A.R.E Sinclair and M. Norton-Griffiths), pp. 130–163, University of Chicago Press, Chicago.

Jenkins, K.J. and Wright, R.G. (1988) Resource partitioning and competition among cervids in the northern Rocky Mountains. *Journal of Applied Ecology*, **25**, 11–24.

Kean, R.I. (1959) Ecology of the larger wildlife mammals of New Zealand. *New Zealand Science Reviews*, **17**, 35–37.

Kie, J.G., Evans, C.J., Loft, E.R. and Menke, J.W. (1991) Foraging behaviour by mule deer: the influence of cattle grazing. *Journal of Wildlife Management*, **55**, 665–674.

Kirby, D.R. and Grosz, K.L. (1995) Cattle grazing and sharp-tailed grouse nesting success. *Rangelands*, **17**, 124–126.

Loft, E.R., Menke, J.W., Kie, J.G. and Bertram, R.C. (1987) Influence of cattle stocking rate on the structural profile of deer hiding cover. *Journal of Wildlife Management*, **51**, 655–664.

Loft, E.R., Menke, J.W. and Kie, J.G. (1991) Habitat shifts by mule deer: the influence of cattle grazing. *Journal of Wildlife Management*, **55**, 16–26.

Loft, E.R., Kie, J.G. and Menke, J.W. (1993) Grazing in the Sierra Nevada: home range and space use patterns of mule deer as influenced by cattle. *Californian Fish and Game*, **79**, 145–166.

Macdonald, D. and Barrett, P. (1993) *Mammals of Britain and Europe*, HarperCollins, London.

Mann, J.C.E. (1983) The social organisation and ecology of the Japanese sika deer (*Cervus nipon*) in southern England. PhD thesis, University of Southampton.

Mann, J.C.E. and Putman, R.J. (1989) Diet of British sika deer in contrasting environments. *Acta Theriologica*, **34**, 97–109.

Martin, C. (1987) Interspecific relationships between barasingha (*Cervus duvauceli*) and axis deer (*Axis axis*) in Kanha National Park, India, and relevance to management, in *Biology and Management of the Cervidae*, (ed. C.M. Wemmer), pp. 299–306, Smithsonian, Washington.

Mathur, V.B. (1991) Ecological interaction between habitat composition, habitat quality and abundance of some wild ungulates in India. PhD thesis, University of Oxford.

McKelvey, P.J. (1959) Animal damage in North Island protection forests. *New Zealand Science Reviews*, **17**, 28–34.

Miller, S.G., Bratton, S.P. and Hadigan, J. (1992) Impacts of white-tailed deer on endangered and threatened vascular plants. *Natural Areas Journal*, **12**, 67–75.

Mishra, H.R. (1982) The ecology and behaviour of chital (*Axis axis*) in the Royal Chitwan National Park, Nepal. PhD thesis, University of Edinburgh.

Norris, K., Cook, T., O'Dowd, B. and Durdin, C. (1997) The density of redshank *Tringa totanus* breeding on the salt marshes of the Wash in relation to habitat and its grazing management. *Journal of Applied Ecology*, **34**, 999–1013.

Olff, H., De Leeuw, J., Bakker, J.P. *et al.* (1997) Vegetation succession and herbivory on a salt marsh: changes induced by sea level rise and silt deposition along an elevational gradient. *Journal of Ecology*, **85**, 799–814.

Oosterveld, P. (1983) Eight years of monitoring of rabbits and vegetation development on abandoned arable fields grazed by ponies. *Acta Zoologica Fennica*, **174**, 71–74.

Owen, M. and Thomas, G.J. (1979) The feeding ecology and conservation of wigeon wintering at the Ouse Washes, England. *Journal of Applied Ecology*, **16**, 795–809.

Packham, C.G. (1983) The influence of food supply on the ecology of the badger. BSc Honours thesis, University of Southampton.

Petty, S.J. and Avery, M.J. (1990) *Forest Bird Communities*, Occasional Paper 26, Forestry Commission, Edinburgh.

Prins, H.H.T. and Olff, H. (1998) Species richness in African grazer assemblages: towards a functional explanation, in *Dynamics of Tropical Ecosystems*, (eds D.N. Newberry, H.H.T. Prins and N. Brown), British Ecological Society Symposium, Vol. 37, pp. 449–490, Blackwell, Oxford.

Putman, R.J. (1986) *Large Herbivores and the Ecology of the New Forest*, Croom Helm, London.

Putman, R.J. and Sharma, S.K. (1987) Long-term changes in New Forest deer populations and correlated environmental change. *Zoological Symposium*, **58**, 167–179.

Putman, R.J., Edwards, P.J., Mann, J.C.E. *et al.* (1989) Vegetational and faunal changes in an area of heavily grazed woodland following relief of grazing. *Biological Conservation*, **47**, 13–32.

Ragotzkie, K.E. and Bailey, J.A. (1991) Desert mule deer use of grazed and ungrazed habitats. *Journal of Range Management*, **44**, 487–490.

Rahmann, M., Rahmann, H., Kempf, N. *et al.* (1987) Auswirkungen unterschiedlicher landwirtschaftlicher Nutzung auf die Flora und Fauna der Salzwiesen an der ostfriesischen Wattenmeerkuste. *Senkenbergia Maritima*, **19**, 163–193.

Siepel, H., Van de Bund, C.F., Van Wingerden, W.K.R. *et al.* (1987) *Beheer van graslanden in relatie tot de ongewervelde fauna: ontwikkeling van een monitorsysteem*, Report 87/29, Research Institute for Nature Management, Arnhem.

Slob, G.J. (1989) *15 jaar vogelontwikkelingen in het afgesloten Grevelingen bekken*, Report Staatsbosbeheer, Utrecht.

Thomas, G.J. (1980) The ecology of breeding waterfowl at the Ouse Washes, England. *Wildfowl*, **31**, 73–88.

Thomas, J.A. (1983) The ecology and conservation of *Lysandra bellargus* (Lepidoptera: Lycaenidea) in Britain. *Journal of Applied Ecology*, **20**, 59–83.

Thomas, J.A. (1991) Rare species conservation: case studies of European butterflies, in *The Scientific Management of Temperate Communities for Conservation*, (eds I.F. Spellerberg, F.B. Goldsmith and M.G. Morris), pp. 149–197, Blackwell, Oxford.

Thomas, J.A., Thomas, C.D., Simcox, D.J. and Clarke, R.T. (1986) Ecology and declining status of the silver-spotted skipper butterfly (*Hespera comma*) in Britain. *Journal of Applied Ecology*, **23**, 365–380.

Tubbs, C.R. (1974) *The Buzzard*, David and Charles, Newton Abbott.

Tubbs, C.R. (1982) The New Forest: conflict and symbiosis. *New Scientist*, **1**, 1–10.

Tubbs, C.R. and Tubbs, J.M. (1985) Buzzards, *Buteo buteo*, and land use in the New Forest, Hampshire, England. *Biological Conservation*, **31**, 41–65.

Van de Koppel, J., Huisman, J., Van der Wal, C.F.R. and Olff, H. (1996) Patterns of herbivory along a productivity gradient: an empirical and theoretical investigation. *Ecology*, 77, 736–745.

Van Dijk, A.J. and Bakker, J.P. (1980) Beweiding en broedvogels op de Oosterkwelder van Schiermonnikoog. *Waddenbulletin*, **15**, 134–140.

Van Schaik, A.W.J. and De Jong, D.J. (1989) *Vegetatieontwikkelingen Slikken van Flakkee 1972–1987*, Rijkswaterstaat, Dienst Getijdewateren.

Van Wieren, S.E. (1991) The management of populations of large mammals, in *The Scientific Management of Temperate Communities for Conservation*, (eds I.F. Spellerberg, F.B. Goldsmith and M.G. Morris), pp. 103–127, Blackwell, Oxford.

Van Wingerden, W.K.R.E., Musters, J.C.M., Kleukers, R.M.J.C. *et al.* (1991) The influence of cattle grazing intensity on grasshopper abundance (Orthoptera: Acrididae). *Proceedings of Experimental and Applied Entomology*, NEV, Amsterdam, **2**, 28–34.

Vera, F.W.M. (1997) Metaforen voor de wildernis: Eik, Hazelaar, Rund en Paard. Doctoral thesis, Wageningen Agricultural University. Wageningen.

Wallace, M.C. and Krausman, P.R. (1987) Elk, mule deer, and cattle habitats in central Arizona. *Journal of Range Management*, **40**, 80–83.

Warren, M.S. and Key, R.S. (1991) Woodlands: past, present and potential, in *The Conservation of Insects and their Habitats*, (eds N.M. Collins and J.A. Thomas), pp. 155–211, Academic Press, London.

Warren, R.J. (1991) Ecological justification for controlling deer populations in eastern national parks. *Transactions of North American Wildlife Natural Resource Conference*, **56**, 56–66.

Wilms, W., McLean, A., Tucker, R. and Ritcey, R. (1979) Interactions between mule deer and cattle on big sagebrush range in British Columbia. *Journal of Range Management*, **32**, 299–304.

Wray, S. (1992) The ecology and management of European hares (*Lepus europaeus*) in commercial forestry. PhD thesis, University of Bristol.

Part Three

Management Applications

7

Hydrological conditions and herbivory as key operators for ecosystem development in Dutch artificial wetlands

J. Theo Vulink and Mennobart R. Van Eerden
Ministry of Transport, Public Works and Water Management, Institute for Integral Freshwater Management and Waste Water Treatment (RIZA), PO Box 17, 8200 AA Lelystad, The Netherlands

7.1 INTRODUCTION

In western Europe marshes and estuaries have been disappearing in recent centuries due to human activities (e.g. drainage and reclamation for agricultural purposes: Finlayson and Moser, 1991; Duncan, 1992; Schultz, 1992). Wetlands are considered to be vulnerable habitats that deserve special attention for nature conservation (Finlayson and Moser, 1991). In The Netherlands some new wetlands, in total about 30 000 ha, have been created as a result of land reclamation in the past few decades. The international importance of the newly developed wetlands as breeding, wintering and stop-over sites for many bird species is closely related to the pioneer stage of these young areas. Nowadays, these areas are mostly enclosed by dikes and, although of large physical size, are no longer subject to the local water level fluctuation that is typical for natural wetlands.

Species richness was an important conservation goal of these new artificial wetlands. As high species richness in wetlands is associated with the early successional stages of the vegetation, the imposition of some level of herbivory was a logical management device right from the beginning of the development of the new wetland areas in order to maintain habitats containing these early succession stages. The option for no intervention in a hands-off scenario was not considered favourable in the management programme that was developed (see Chapter 2 for the discussion on the ecological frame of reference).

Grazing and Conservation Management. Edited by M.F. WallisDeVries, J.P. Bakker and S.E. Van Wieren. Published in 1998 by Kluwer Academic Publishers, Dordrecht. ISBN 0 412 47520 0.

In keeping with the strong empirical traditions of the authority in charge (originally the water management authority, Rijkswaterstaat), research was started to explore the primary system processes as well as to monitor the characteristic groups of wetland bird species that can be regarded as indicators for the functioning of these new systems. With respect to the effect of herbivory within the dynamics of such wetland systems our database consists of information at three levels:

- fixed densities of cattle and horses, stocked in summer in restricted areas;
- free-ranging cattle, horses, and red deer on a year-round basis;
- avian herbivores affecting marsh vegetation during summer.

The data are used to formulate and evaluate the management practices of these reserves in the wider perspective of nature development and ecosystem management in The Netherlands. As the complex interplay between mammalian and avian herbivores in combination with hydrological conditions is most intensively studied in the Oostvaardersplassen, we mainly refer to this system but include our experience in other artificial wetlands where appropriate.

In this chapter we will evaluate how the management goal – maintaining early successional stages and related species richness with respect to wetland birds – can be achieved in a freshwater marsh system. We will argue that:

- the inundated reed area can only be maintained by manipulation of the hydrological conditions;
- wet grasslands can be maintained in an open condition by large herbivores on a year-round basis if they, in turn, can use well drained grasslands in the winter period;
- greylag geese (*Anser anser*) play a key role in this system by preventing encroachment by emergent vegetation (Figure 7.1).

Inundated reed areas used for moulting, and wet grasslands which are used for grazing before and after moult, are linked by these greylag geese. The spin-off of the aforementioned management practices for other animal species will be discussed in relation to a species-based versus an ecosystem approach.

7.2 DESCRIPTION OF THE OOSTVAARDERSPLASSEN SYSTEM AND MANAGEMENT PRACTICES

7.2.1 The Oostvaardersplassen system

Oostvaardersplassen (52° 26′ N, 5° 19′ E), the largest freshwater marsh of northwest Europe, became established by a combination of human activity and natural developments. When Zuidelijk Flevoland was embanked in 1968, the new polder consisted initially of vast reed marshes (*Phragmites aus-*

Figure 7.1 Moulting greylag geese constitute habitat modifiers in the marsh zone.

tralis) totalling 40 000 ha, initially sown from aeroplanes (Van der Toorn and Hemminga, 1994). As a result of changing views and attitudes towards nature and environment, Oostvaardersplassen was allowed to develop as a conservation area in the lower wet part of the polder, comprising 3600 ha of marshland (Figure 7.2). The construction of a bank around the marsh, and of a pumping station as well as a water outlet in 1975, made it possible to regulate the water level. The marsh is situated on a mineral clay soil with a high soil fertility (clay content *c.* 30%, calcium content *c.* 10%, C : N ratio > 10). Some relevant climate characteristics are: mean winter and summer temperature 2°C and 18°C, respectively; mean annual precipitation *c.* 800 mm and evaporation *c.* 600 mm. The water depth in the marsh zone is 0–50 cm and the maximum variation in elevation in the nature reserve is *c.* 1.2 m.

In order to extend the array of habitats in a hydrological gradient from wet to dry, a peripheral zone of 2000 ha dry, partly cultivated land (hereafter referred to as the border zone) was added to the nature reserve in 1982. At the time the border zone was added, about 900 ha had already been converted into arable fields. Various habitats were created within the border zone. These consisted mainly of extensive wet grasslands with shallow pools and ditches, extensive well drained grasslands (hereafter called dry grasslands) and half-open mosaic vegetation. The rationale for this management policy was to satisfy the habitat requirements of characteristic wetland bird species, such as geese, spoonbills, herons, egrets and harriers, to attain a more diverse wetland system. To create these habitats the largest

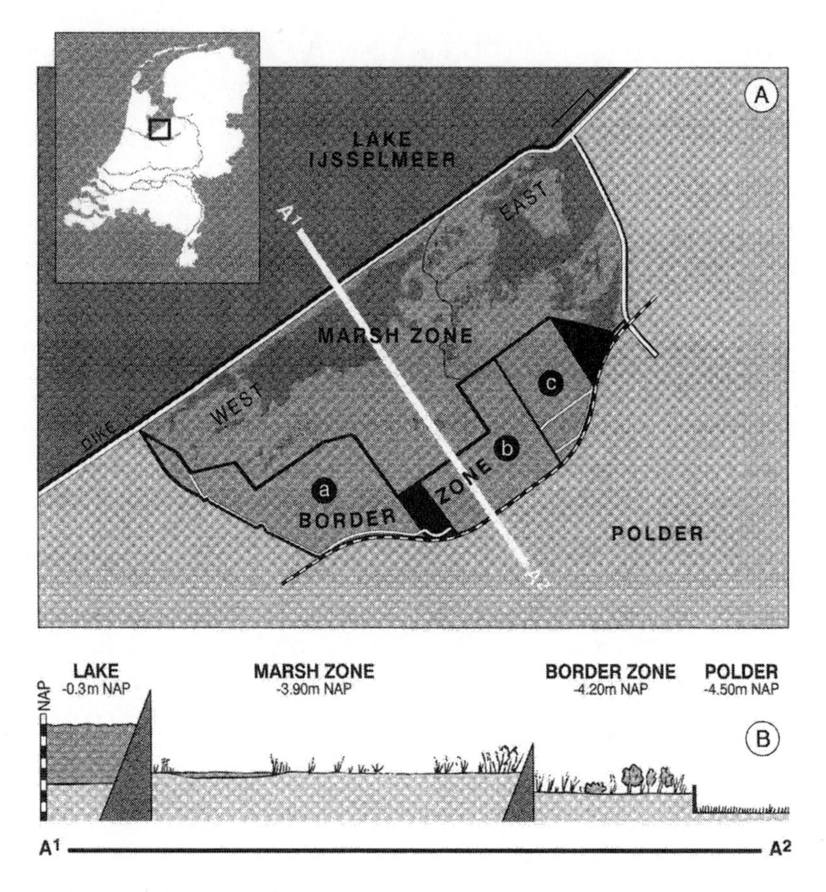

Figure 7.2 (A) Location of Oostvaardersplassen nature reserve in the Netherlands. Main units for nature management: (1) marsh zone (West and East); (2) border zone. Initial units for grazing management are shown as: (a) year-round grazing; (b) summer grazing; (c) summer grazing and mowing. (B) Cross-section of Oostvaardersplassen and its surroundings showing reversed level of soil surface due to differences in subsidence after reclamation (NAP = Dutch Ordnance Level).

part of the former arable fields and a small part of the remaining reed and tall herbaceous vegetation were turned into grassland by burning reed vegetation and sowing grass mixtures.

In the Oostvaardersplassen three habitat types are of paramount importance for birds. Shallow open water with freshwater mudflats, inundated half-open reed vegetation and wet grassland, inundated for the greater part of winter and spring, harbour the highest number of characteristic species. The drier habitats in the marsh (dry homogeneous reedbeds) and in the border zone (dry grassland, reed, tall herbaceous vegetation, shrubs), have a species richness that is much lower compared with the more heterogeneous and wetter areas mentioned above (Figure 7.3).

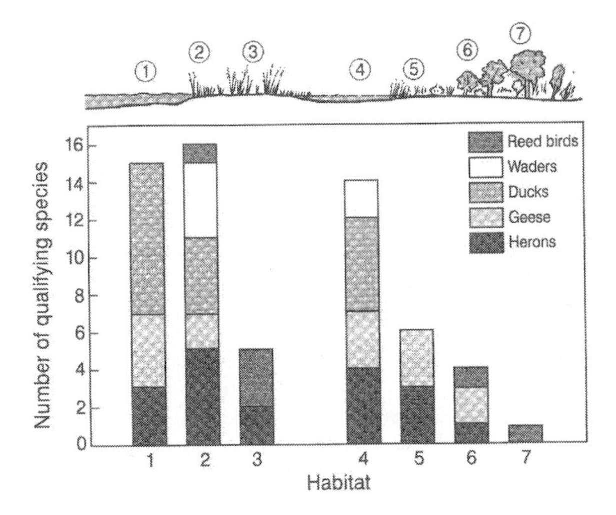

Figure 7.3 Number of wetland bird species in the different habitats in Oostvaardersplassen. Only those species which met the criterion that at least 1% of the biogeographical population of that species (or subspecies) must be present on a regular basis are taken into account, using population sizes from Rose and Scott (1994) and Meininger *et al.* (1995). For non-breeding birds, this 1% threshold is the most quantitative, bird-specific criterion of the 12 criteria for identifying internationally important wetlands proposed by the Ramsar Convention (Van den Tempel and Osieck, 1994).

Reed birds: *Circus aeruginosus, Porzana porzana, Cyanosylvia cyanecula, Locustella luscinioides, Panurus biarmicus*
Waders: *Recurvirostra avosetta, Philomachus pugnax, Limosa limosa, Chlidonias niger*
Ducks: *Anas penelope, A. strepera, A. crecca, A. acuta, A. querquedula, A. clypeata, Aythya ferina, Aythya fuligula, Mergus albellus, M. merganser*
Geese: *Cygnus cygnus, Anser fabalis, Anser albifrons, Anser anser, Branta leucopsis*
Herons: *Phalacrocorax carbo, Botaurus stellaris, Egretta garzetta, Casmerodius alba, Platalea leucorodia*
Habitat: 1, extensive shallow open water; 2, mud flats and inundated marsh vegetation; 3, not-inundated marsh vegetation; 4, extensive wet grassland; 5, extensive dry grassland; 6, mosaic vegetation of dry grassland, reed, tall herbs and shrubs; 7, young woodland, dominated by *Salix* species and *Sambucus* (after Van Eerden *et al.*, 1995).

Without additional management measures this diversity of wetland habitats will decrease in time. Spontaneous succession in the wet parts of freshwater wetlands like Oostvaardersplassen will culminate in a late successional stage of either open water or dense emergent vegetation. Which of these two late successional stages will occur depends on the water depth, fluctuations in the water level and activities of herbivorous wetland birds (Van der Valk, 1994). In the border zone the short vegetation will soon be overgrown by tall grasses (*Phragmites*, *Calamagrostis epigejos*), tall herbs (*Cirsium arvense*, *Urtica dioica*) and shrubs (*Salix* spp., *Sambucus nigra*) (nomenclature of plant species follows Van der Meijden *et al.*, 1990).

For the development and management of this newly created marshland a conceptual diagram concerning management goals and biological processes was constructed (Figure 7.4). The maintenance of habitat diversity, including habitats in the marsh zone and wet grassland areas in the border zone, is the most important management goal for the Oostvaardersplassen system. In order to maintain the early successional stages, water level management and grazing with cattle and horses were applied.

7.2.2 Hydrological conditions

The hydrological conditions of artificial wetlands are different from those of most other wetlands. Oostvaardersplassen, for example, has a lower soil level outside the marsh zone than inside it, as an effect of further reclamation of the adjacent polder and continuous subsidence of the drained soil (Figure 7.2b). Paradoxically, the wetland itself is perched above the surrounding landscape and exists only thanks to the encircling bank.

Hydrological factors were considered important to supplement and enforce the diversification initiated by the avian herbivores. The hydrological conditions were managed by water-level fluctuations (small-scale variation by outlet and pump) and by forced draw-down of the water level with subsequent raising after about four years (marsh-wide effect). This cyclic water-level management, inducing a cyclic vegetation succession, was tried out as an experiment, based on knowledge and experience in Oostvaardersplassen and in 'prairie glacial marshes' in Iowa, North America (Van der Valk and Davies, 1976, 1978; Van der Valk, 1994; Ter Heerdt and Drost, 1994). Herbivory by wild geese is also closely related to the water level (see below).

7.2.3 Grazing in the border zone

Because of a lack of experience with grazing in young wetlands in The Netherlands, it was decided to apply different grazing treatments (Polman and Schmidt-Ter Neuzen, 1987). The extremes in grazing regimes ranged from year-round grazing with free-ranging cattle and horses at relatively low stocking rate to intense summer grazing with cattle and horses in combination with mowing. The latter management was specifically tuned to maintain large-scale short vegetation communities with related plant and bird species. For the duration of the experiment the area was divided provisionally into three grazing sections (Figure 7.2A). This division was maintained until 1993, whereupon it was decided to apply year-round grazing with free-

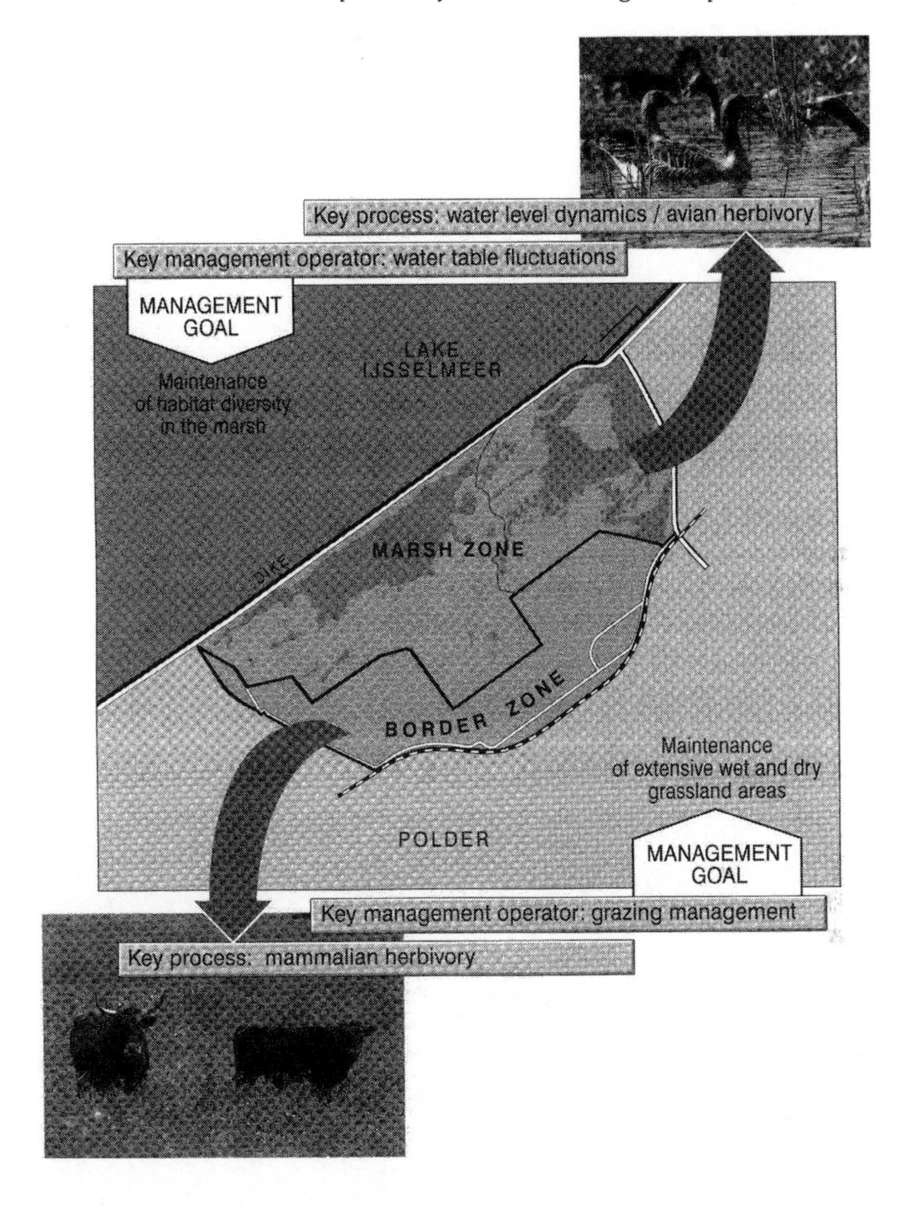

Figure 7.4 A conceptual diagram illustrating the ecological key processes and management operators for each subsystem and the relationships between subsystems within the Oostvaardersplassen as well as the relationship between the reserve and the neighbouring lake and polder. The total species richness of the Oostvaardersplassen, the lake and the polder together is supposed to be higher than the sum of the species of the single parts.

ranging cattle and horses over the entire area. From then on the area with year-round grazing expanded gradually in accordance with the increase of the herbivore populations.

In Oostvaardersplassen only roe deer (*Capreolus capreolus*) were present as wild ungulates from the onset. The other large herbivores were introduced: Heck cattle (*Bos taurus*; cross-bred from primitive races) (Figure 7.5) in 1983 and Konik horses (*Equus ferus*; a primitive breed of horses, originating from Poland) (Figure 7.6) in 1984 (Vera, 1988). Mixed grazing with Heck cattle and Konik horses took place from April 1986 onwards within a gradually extended fenced area. Red deer (*Cervus elaphus*) were introduced in 1992 to complete the herbivore assemblage with an intermediate feeder (terminology after Hofmann, 1973) from which a special impact on the development of shrubs was expected. The sex ratio of all introduced populations was about 1 : 1. Just before the introduction of red deer, a fence was placed around the whole nature reserve to minimize risk of accidents along the adjacent railway. Roe deer and red deer could make use of the entire nature reserve, and measures were taken to allow free entry and exit of smaller species.

7.2.4 Monitoring

The broad-scale vegetation development was monitored by five integral aerial surveys: in 1974, 1980, 1985, 1988 and 1992 (Jans and Drost, 1995). Effects of large herbivores on the structure of grassland vegetation were

Figure 7.5 Heck cattle, originally bred from various cattle breeds to resemble the extinct aurochs (*Bos primigenius*), are used as large grazers in restoration projects.

Figure 7.6 Konik horses are considered to be close relatives of the extinct European wild horse, the tarpan (*Equus ferus ferus*).

monitored by keeping continuous records of reed development. In Oostvaardersplassen the spread of reed is a direct consequence of low grazing intensity combined with high soil fertility and the presence of rhizome fragments of reed. There was a close positive relationship between reed height and reed cover in the reed height range 0–1.5 m (Huijser *et al.*, 1996). On wet and dry grassland three types of vegetation structure were distinguished in relation to reed cover: vegetation with low reed cover (reed height 0–50 cm), moderate reed cover with grassy undergrowth (reed height 50–100 cm) and high reed cover without other grasses (reed height > 100 cm). Reed height measurements in permanent transects were used to quantify the spatial distribution of reed cover. The reed cover classes were also applied to describe habitat use by geese. Birds were counted each month from a high-winged Cessna 172 aeroplane and from the ground. Geese and ducks were counted over the complete area, as were the colony-breeding spoonbills, herons, egrets and cormorants. Transects and plots were used to monitor scattered and rare species outside the breeding season as well as most breeding birds (Beemster, 1993).

7.3 VEGETATION AND BIRDS WITHIN THE MARSH ZONE IN RELATION TO WATER-LEVEL MANAGEMENT

In order to illustrate the response by birds to the long-term changes in water level (Figure 7.7), we used data for four characteristic bird species which depend on different habitats within the marsh: teal (*Anas crecca*, a small seed-eating duck typical for pioneer situations in shallow water of 0–10 cm), greylag goose (*Anser anser*, an obligate herbivore able to graze tall marsh plants), bittern (*Botaurus stellaris*, a fish-eater of clear shallow water) and bearded tit (*Panurus biarmicus*, a reed-dwelling songbird feeding on insects in summer and reed seeds in winter). Several stages can be discerned in the succession of the marsh zone in the period 1968–1995 (Figure 7.8).

First drying stage: 1968–1975

As a consequence of drainage and a continuous subsidence, the field level in the agriculturally exploited polder sank below that of the nature reserve. This led to a water flow from Oostvaardersplassen to the adjacent polder and a slow desiccation of the marsh. The vegetation expanded by a succession from annuals and biennials to emergent perennials like *Typha latifolia* and *Phragmites*. The most important pioneer species were *Senecio congestus*, *Rumex maritimus*, *Ranunculus sceleratus*, *Chenopodium rubrum*, *Aster tripolium* and *Juncus bufonius*. These species produced enormous amounts of seeds that were consumed by dabbling ducks like teal, pintail (*Anas acuta*) and mallard (*A. platyrhynchos*). Teal were present in peak numbers in 1970 when about 150 000 were counted at maximum (Figure 7.9). Bearded tit increased in response to the expanding reed vegetation and reached their maximum numbers in the period 1974–1975. The moulting greylag geese population and breeding pairs of bittern started to settle in this period.

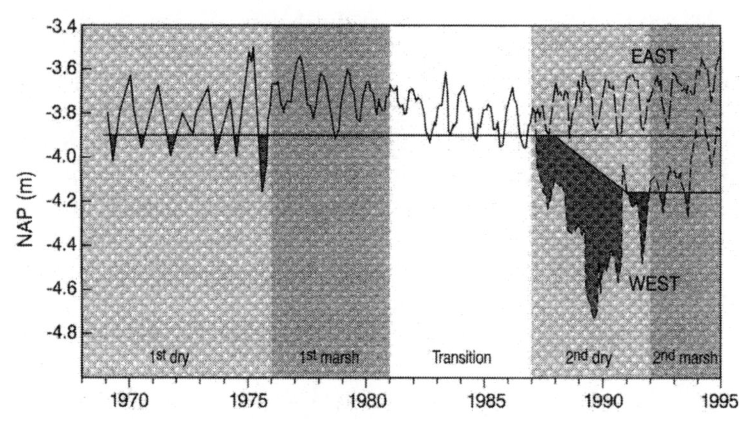

Figure 7.7 Water level in Oostvaardersplassen (NAP = Dutch Ordnance Level) in the period 1975–1995 for two compartments, one of which experienced a period of draw-down in order to restore the vegetation in the western part. Depending on the water level, different stages can be distinguished: first drying stage, 1968–1975; first marsh stage, 1976–1980; transition stage, 1981–1986; second drying stage in western part only, 1987–1991; second marsh stage in western part only, 1992–1995. In the western part, soil subsidence resulted from the draw-down. Periods when the water level was lower than soil surface are shaded in black.

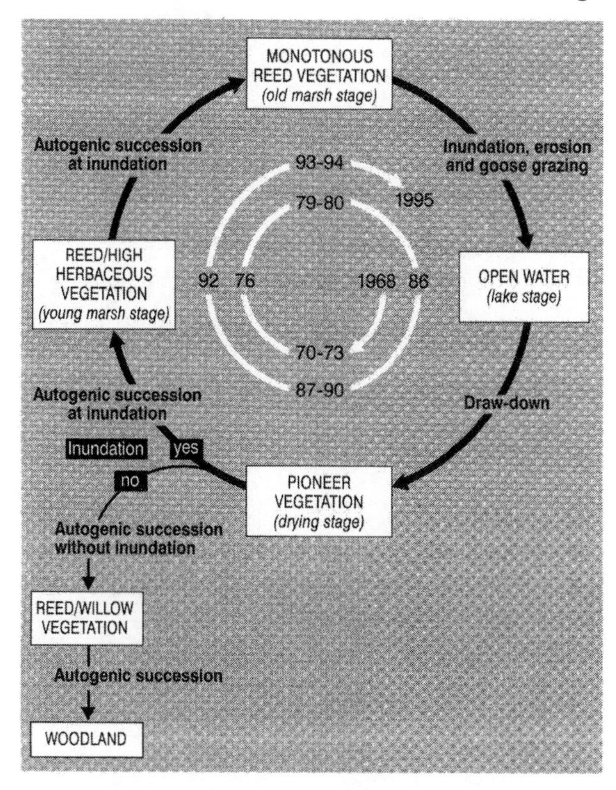

Figure 7.8 Schematic representation of the cyclic vegetation development between 1968 and 1995. The first cyclic vegetation development (1968–1986) took place in both parts, the second one (1986–1995) only in the western part following draw-down. The vegetation in the eastern part did not change much during the period 1986–1995. (After Jans and Drost, 1995.)

Until 1975 the marsh area was also an important stopover site for waders like avocet (*Recurvirostra avosetta*), black-tailed godwit (*Limosa limosa*) and ruff (*Philomachus pugnax*). These species benefited from large quantities of midge larvae (Chironomidae) in the mud. Rails – water rail (*Rallus aquaticus*) and spotted crake (*Porzana porzana*) – were common breeding birds during this period.

First marsh stage: 1976–1980
From 1976 onwards the water level was raised about 20 cm during the summer period with the help of a pumping station. The area with pioneer and reed vegetation declined. A large part of the reed vegetation became inundated. The number of ducks decreased due to the strong reduction of seed-producing annuals and biennials. For example, the number of teal decreased

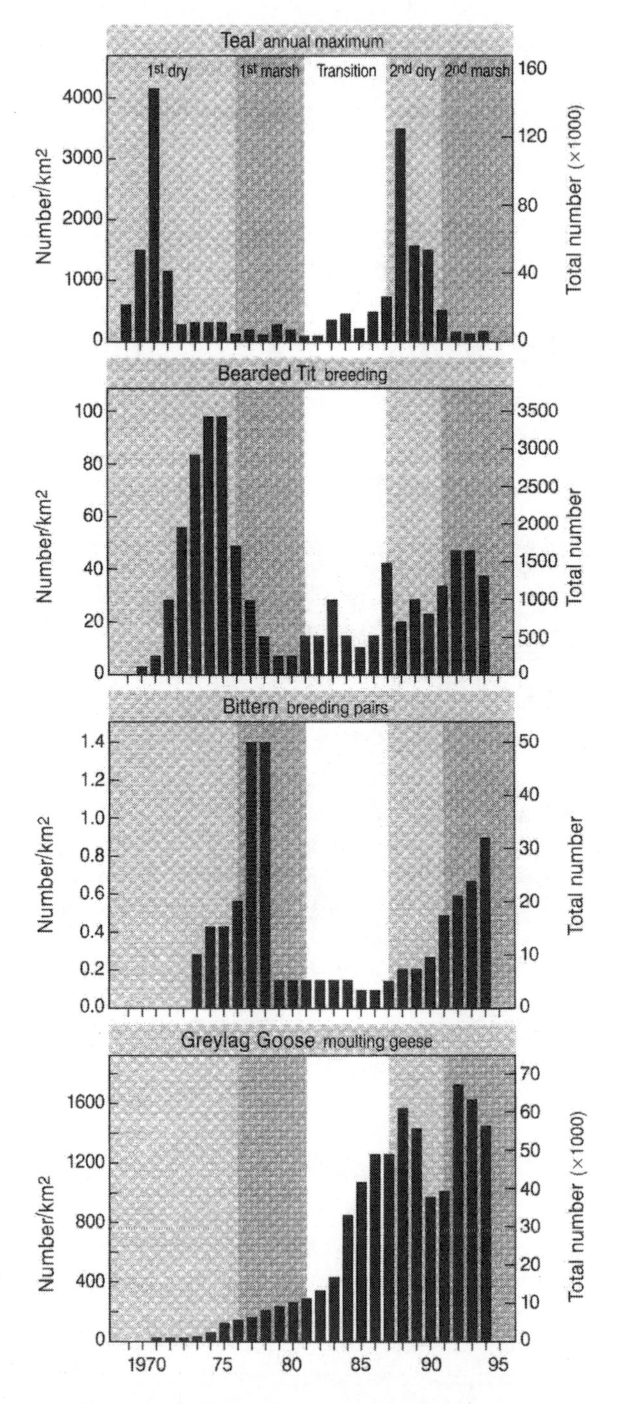

Figure 7.9 The development in number of the characteristic species teal, bearded tit, bittern and greylag goose in relation to changes in the water level in Oostvaardersplassen, 1975–1995.

from over 150 000 in 1970 to fewer than 15 000 in 1977 (Figure 7.9). The moulting greylag geese increasingly became attracted to the marsh, especially the inundated reed vegetation, where reed leaves were the most important food source (Zijlstra *et al.*, 1991). As the bearded tit occupies both the canopy of the reed vegetation and the ground, its numbers declined due to inundation of reed stands.

The number of waders decreased sharply due to the decreased availability of benthic food as a consequence of the raised water level. For some other bird species the food situation improved after the rise of the water level, especially for fish-eating birds such as spoonbill (*Platalea leucorodia*) and grey heron (*Ardea cinerea*), illustrated here by the relatively high number of breeding pairs of bittern (Figure 7.9). The strong decrease in breeding pairs of bittern in 1979 was probably due to the severe winter of 1978/79. The number of breeding pairs of some ducks and rails peaked shortly after 1975. Coot (*Fulica atra*) rose from about 300 in the drying stage to some 25 000 breeding birds in 1976–1978. Occasionally rare species like purple heron (*Ardea purpurea*), little egret (*Egretta garzetta*) and the great white egret (*Egretta alba*) were recorded breeding during this stage. Expansion of marsh vegetation, which after inundation became exploitable to the herbivores, as well as an explosion of young fish were the cause for these changes.

Transition stage: 1981–1986

Between 1980 and 1982 there was little variation in water level between summer and winter. From 1982 onwards the difference (about 20 cm) between relatively high winter and low summer water levels was re-established. As a consequence of erosion and grazing by wetland birds the mosaic pattern of the vegetation with pioneers and inundated helophytes decreased and was replaced by open shallow water. The food situation for fish-eating birds also deteriorated, as illustrated by a relatively low number of breeding bittern. The exponential growth in number of greylag geese continued, partly due to an overall increase of the northwest European population (Madsen, 1991) and to a new tradition for geese from a larger part of central Europe to use Oostvaardersplassen as their moulting place (Zijlstra *et al.*, 1991).

Second drying stage: 1987–1991

The cyclic water-level management started with an artificial draw-down in the western part (2100 ha) of the marsh. To provide a refuge for bird species that depend on the wet stages, the dry stage was not implemented in the whole marsh area.

During this second drying stage the pioneer and emergent species became re-established through a succession comparable to that in the period 1968–1975. The area covered with mature reed increased, with about 750 ha after four years. In the wet eastern part (1500 ha) the development towards an open shallow lake continued. All moulting geese, the nesting spoonbills and the majority of other marsh birds concentrated in the wet eastern part and total numbers dropped. Dabbling ducks like teal increased again in response to the peak seed production in the western part. In 1988 no fewer than 126 000 were present after many years of low numbers (Figure 7.9). These changes in bird populations and distribution within the marsh confirmed the hypothesis that inundated reed vegetation is essential for moulting greylag geese and most other marsh birds, but also that the reed vegetation is only grazed when it is inundated.

Second (induced) marsh stage: 1992–1995

From the winter of 1991/92 onwards the water level in the western marsh part was raised again by retaining the surplus precipitation. This raising of the water level turned the western part into marsh again – and although temporary, this was the phase with the highest species richness in the system. Mineralization of annual and biennial plants induced an increase in phytoplankton, zooplankton, macrofauna and young fish. At this stage the area offered most food and protection for breeding birds as well as herbivorous and fish-eating wetland birds.

During this period the area with pioneer vegetation declined and the number of dabbling ducks, like teal, decreased again in response to the lowered seed production (Figure 7.9). The reed vegetation changed into dense stands. The young *Phragmites* and *Typha* plants in the western part were heavily grazed by wetland birds, and the number of moulting greylag geese increased again (Figure 7.9). The number of fish-eaters such as bittern increased in the western part. Bearded tits took advantage of the reed seeds in winter and total numbers increased again, but not to the levels of the period 1970–1975. In the wet eastern part, the process of opening up of the reed vegetation by moulting greylag geese was consolidated.

7.4 HERBIVORY IN THE MARSH ZONE MEDIATED THROUGH WATER-LEVEL MANAGEMENT: GREYLAG GEESE AS HABITAT MODIFIERS OF THE MARSH SYSTEM

7.4.1 Population development and habitat use of greylag geese

Greylag geese have performed a moult migration to Oostvaardersplassen from central Europe (Poland, former German Democratic Republic) and southern Scandinavia (Denmark, Sweden), starting to use the area immediately after embankment in 1968 (Zijlstra *et al.*, 1991). The number of moulting geese started to increase strongly from 1976 onwards (Figure 7.9), when a large part of the reed vegetation was inundated. After a period of exponential increase the numbers dropped during the experimental drawdown of the water level in part of the marsh. Thereafter numbers increased again but at present (1997) they seem to have stabilized. The moulting population in Oostvaardersplassen is the largest concentration in Europe today (Fox *et al.*, 1995).

Greylag geese remain in the marsh for about five weeks during the moult. About 95% of their diet consists of *Phragmites*, the rest being *Urtica*, *Salix viminalis* and young leaves of *Typha*. During moult the geese can lose up to 950 g of body mass (M.R. Van Eerden *et al.*, unpublished data). Just before and after moult they exploit young grasses in the immediate neighbourhood of the marsh for fattening. Due to the relatively high quality as a consequence of the delayed grass growth under wet circumstances, they explicitly prefer wet grassland within the nature reserve. Outside the reserve leaves from cereals are also selected (Figure 7.10). Due to the severe winters of 1985 and 1986, growth and maturity of cereals were delayed and they were exploited intensively in these years by pre-moulting geese in late spring.

7.4.2 Grazing effects

Geese that use the area during the moult have a marked impact on the marsh system (Figure 7.11). By their intense grazing they have prevented an encroachment of emergent vegetation in the marsh, which would have caused loss of habitat diversity. By grazing, numerous passages were made and altered through the years. The closed reed vegetation of 1968–1975 was changed into a mosaic structure. This process lasted 10–15 years for most areas (Figure 7.12). As can be seen, the percentage cover by reed decreased, but the overall pattern of occurrence remained the same. This vegetation mosaic was also used by breeding greylag geese, nesting spoonbills and moulting shovellers (*Anas clypeata*) – a duck that lives on zooplankton in the eutrophic water of the marsh system. Geese droppings increased the phyto- and zooplankton production in the water (M.R. Van

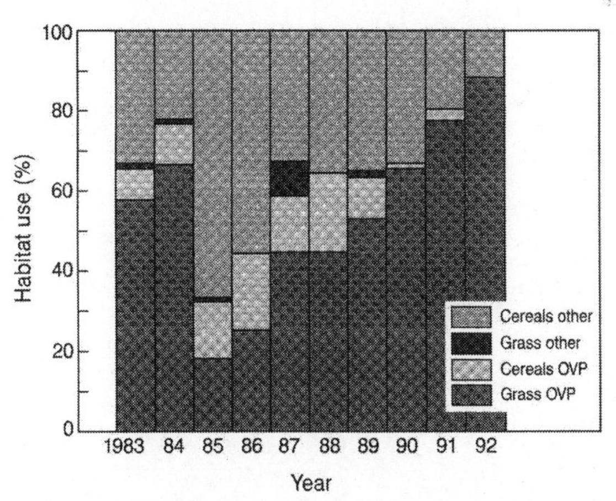

Figure 7.10 Habitat use of moulting greylag geese, within and outside the Oostvaardersplassen nature reserve around the period of wing moult. From 1985 onward, reed and tall herbaceous vegetation, and arable fields in the reserve, were converted to grassland, which improved the carrying capacity for geese in the border zone.

Eerden and R. Henkes, unpublished data). The shallow water between the reed tussocks was also used as a spawning place by various fish species. The fish production in turn was exploited by spoonbills, herons and egrets later in the season, when the fish concentrated in the more open areas after the water level had fallen.

To sum up, two interactions between greylag geese and their habitat were recorded: one involving redistribution of nutrients, which affects the food web, and one resulting in habitat modification. In the case of the fish-eating spoonbill, a target species for nature management in The Netherlands, both types of effect influence their habitat. The same holds for planktivorous ducks such as shoveller. Because of their crucial role with respect to the functioning of the system, the habitat requirements of geese were assessed outside the marsh as well. The use that large mammalian herbivores make of this zone is important for the geese, because of their facilitating effect on the grass sward.

7.5 HERBIVORY IN THE BORDER ZONE: INTERRELATIONS BETWEEN GRAZING PRESSURE BY LARGE HERBIVORES AND BIRDS

7.5.1 Population development and habitat use of large herbivores

The area grazed by Heck cattle and Konik horses increased, with the growth of their populations, from about 375 ha (including 30 ha dry grassland) in

Figure 7.11 Habitat diversity in the marsh as a result of grazing geese and water level management.

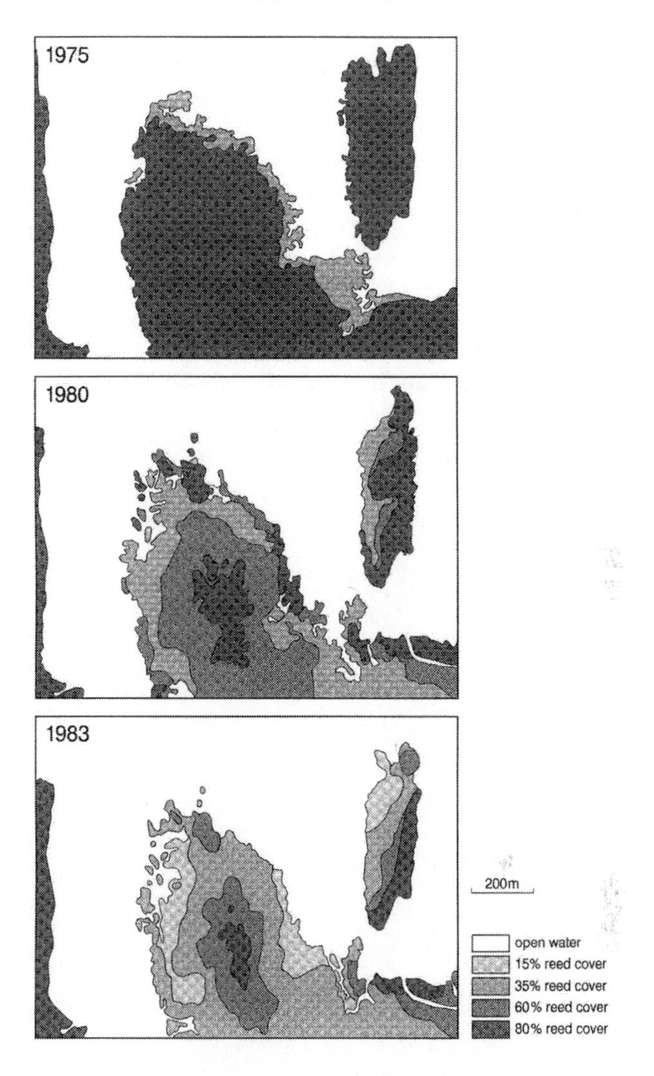

Figure 7.12 Decrease of reed cover in the marsh zone according to aerial photographs (after Van Eerden *et al.*, personal observation). Percentage cover decreased over the years, but the general outline of the area covered by reed only did so to a minor extent.

1986 to about 1300 ha (including 750 ha dry grassland) in 1995. Just after their introduction, the area of dry grassland available was relatively small, leading to a very high density in this habitat. From 1988 onwards the stocking rate was maintained at 0.25–0.50 animals/ha for the whole area (equivalent to a density of 0.5–1.0 animals/ha for the preferred dry grassland alone). After 1987 the herds were in an exponential growth phase. The slope of the regression lines (Figure 7.13) can be considered as intrinsic growth rates of the populations (Caughley and Krebs, 1983). In 1985 and 1986 the mortality was relatively high (Table 7.1), probably due to the relatively small area of dry grassland available, in conjunction with severe winters.

Habitat use data were collected in a year-round grazed study plot (650 ha) in the border zone (Figure 7.2) during the years 1991–1993. Three vegetation types were distinguished in this study area: dry grassland (100 ha); mosaic vegetation of dry grassland with tall reed (100 ha); and vegetation with tall reed, tall herbs, shrubs and some grasses (*Poa trivialis* and *Calamagrostis epigejos*) (450 ha). Cattle and horses showed a high preference for dry grassland during the whole year (Figure 7.14). As grasses represent the most important forage for cattle and horses, the forage supply on dry grassland was an important factor throughout the year (Cornelissen and Vulink, 1995). In summer, *Phragmites* was also an important food plant for cattle and horses.

When the supply of fresh grasses became depleted during the winter, cattle supplemented their diet with twigs and bark of *Salix* species. Horses, additionally, took stolons of *Phragmites*, roots and dead stems of *Urtica* and some bark and twigs of *Salix* species (Cornelissen and Vulink, 1995). These alternative food sources have a relatively low content of digestible organic matter as compared with fresh grasses (Vulink and Drost, 1991; Cornelissen *et al.*, 1995). In cattle, the deterioration in food digestibility caused a reduced rate of food passage and thus a decline of metabolizable energy intake rate (Chapter 9). Therefore the herbivores had to rely on their fat

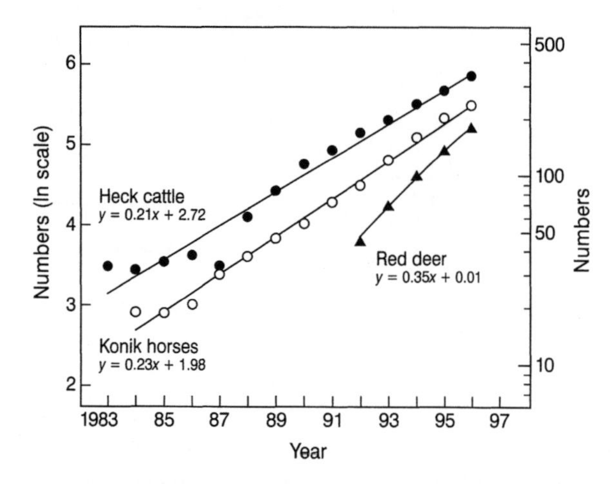

Figure 7.13 Population development of Heck cattle, Konik horses and red deer. The total number of introduced Heck cattle was 34 in 1983, 12 in 1987, and 15 in 1989. The total number of introduced Konik horses was 18 in 1984, 2 in 1986, 2 in 1987, and 2 in 1988. The total number of introduced red deer was 44 in 1992 and 13 in 1993. The slope of the function gives the average annual growth factor of the population.

Table 7.1 Birth and death rate (as percentage of the population) of Heck cattle, Konik horses and red deer in Oostvaardersplassen (from 1995 onwards no data were available about birth and mortality of young calves and foals)

	Heck cattle		Konik horses		Red deer	
Year	Birth	Mortality	Birth	Mortality	Birth	Mortality
1984	16	6	–	–	–	–
1985	32	24	17	6	–	–
1986	32	46	60	25	–	–
1987	47	9	28	10	–	–
1988	46	8	31	6	–	–
1989	31	10	31	4	–	–
1990	30	11	35	4	–	–
1991	31	7	30	6	–	–
1992	26	10	41	11	43	7
1993	30	7	37	6	47	6
1994	28	11	37	11	36	1
1995	–	–	–	–	37	5

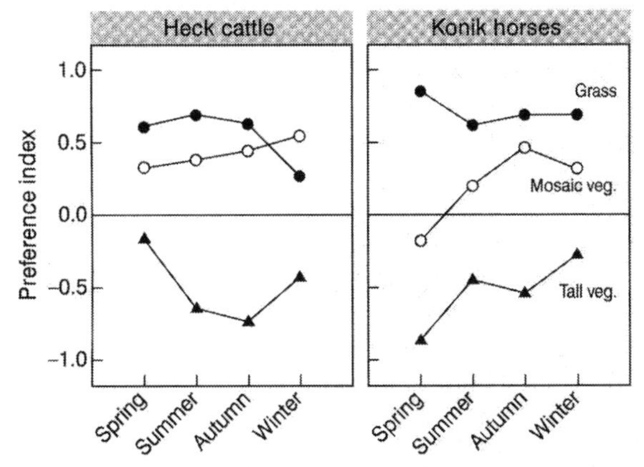

Figure 7.14 Preference indices according to Jacobs (1974) of Heck cattle and Konik horses for different habitats during spring (April–June), summer (July–September 14), autumn (September 15–December) and winter (January–March). The use that cattle and horses made of the different vegetation type was corrected for the available area of different vegetation types: dry grassland (Grass); mosaic of grassland and reed (Mosaic veg.); tall vegetation with reed, tall herbs and shrubs (Tall veg.).

reserves (Cornelissen and Vulink, 1995). Horses are able to maintain a higher passage rate than cattle and can compensate for this low diet quality to some extent by maintaining a higher daily intake than cattle (Duncan *et al.*, 1990). When depletion of grasses in winter occurs too early, cattle lose over 25% of their peak body mass. Without supplementary food, starvation will lead to increased mortality, especially in cattle. This indeed happened in the severe winters of 1985 and 1986 (Table 7.1).

Preliminary analysis of habitat use and diet composition showed that red deer and roe deer both used the marsh zone and the border zone during spring and summer (Cornelissen and Vulink, 1996). During autumn and winter they were concentrated in the border zone. Dry grassland was an important foraging area for red deer in the winter period. In winter, grasses were also important forage for roe deer. The diets of red deer and roe deer in the winter period consisted of 60–80% and 50–70% grasses, respectively (De Jong *et al.*, 1997). In spring and summer, dicots and browse were preferred: twigs with young leaves from *Salix* species and *Sambucus* were the most important forage. In contrast with the diet of cattle and horses, the reed content in the diet of red deer and roe deer was negligible.

Like cattle and horses, red deer also had a preference for dry grassland; during winter they preferred to forage on the seasonal cattle pastures (Cornelissen and Vulink, 1996). It appeared that red deer avoided the Heck cattle and Konik horses (see also Loft *et al.*, 1991, 1993). However, cattle grazing can also be facilitating to red deer (Gordon, 1988). In Oostvaardersplassen intensive grazing by cattle and horses maintains a low reed cover on dry grassland, which is expected to improve the quality of the winter foraging area of red deer.

In summary, Oostvaardersplassen offers good summer habitat for large herbivores but winter habitat is marginal, particularly for cattle. Wintergreen grasses and dwarf shrubs are missing, leading to a marginal performance of Heck cattle during severe winters followed by a wet and cold spring. In such years supplementary food (hay) has been supplied so far. Due to the strong preference of cattle and horses for dry grassland vegetation, the potential impact of year-round grazing on dry grassland vegetation would be relatively high compared with the impact on wet grassland and tall vegetation dominated by reed and tall herbs.

7.5.2 Grazing effects in relation to density of large herbivores

Because of the importance of short vegetation as habitat for wetland birds, grazing research in Oostvaardersplassen has focused on the dynamics between short grassy vegetation, tall reed with some herbs and shrubs (hereafter called tall vegetation) and herbivores. The discussion of grazing effects will concentrate on the vegetation development and habitat use of bird species in three main habitats: wet grassland, dry grassland, and tall vegetation.

(a) *Wet grassland*

Wet grassland was created on former arable fields by sowing a mixture of cultivars of *Lolium perenne*, *Phleum pratense* and *Trifolium repens* in 1980.

After 10 years of grazing and water-level management (summer grazing and inundation from November until April) the sown grasses were largely replaced, through spontaneous succession, by species like *Agrostis stolonifera*, *Juncus bufonius*, *Alopecurus geniculatus* and *Poa trivialis* (Jans and Drost, 1995).

A grazing pressure of about 1.0 animals/ha during summer (June–October) is required to stop encroachment of wet grassland by reed (based on 10 years of management experience). This is supported by data from another artificial wetland (Lauwersmeer) in The Netherlands. A lowering of the stocking rate from 1.0 to 0.4 animals/ha resulted in an invasion of wet grassland by reed (E.J.M. Van Deursen *et al.*, unpublished data). Intensive grazing maintained the extensive wet grassland as a suitable foraging area for greylag geese (Vulink *et al.*, personal observation). Many other species of ducks, waders, herons and spoonbills also benefited from this management (Figures 7.3, 7.15).

(b) Dry grassland

Dry grasslands were created in well drained reed stands and former arable fields by burning the reed vegetation and sowing a mixture of *Lolium*, *Phleum* and *Trifolium* in 1982 or a mixture of *Dactylis glomerata* and *Festuca rubra* during the years 1988–1990. In dry grasslands *Lolium* and *Trifolium* were for a large part (50%) replaced by *Poa trivialis* within six years through spontaneous succession (Jans and Drost, 1995). In the area where the mixture of *Dactylis* and *Festuca* was used, the species composi-

Figure 7.15 Extensive wet grassland is an important habitat for wetland birds.

tion remained remarkably constant. After six years the vegetation still consisted of 80–90% of the two species originally sown (Jans and Drost, 1995).

The summer-grazed area of dry grassland (1.7 animals/ha) had a short vegetation with a low reed cover (reed height 0–50 cm) most of the year. During July–August only 5% of the area had a moderate reed cover (Vulink *et al.*, personal observation).

In the year-round grazed area on dry grassland additional mowing took place until August 1988, resulting in a sharp boundary between grassland and reed vegetation. Due to the low stocking rate (0.45 animals/ha, based on the original 270 ha dry grassland), about half of the year-round grazed dry grassland became overgrown by reed after mowing stopped. This resulted in vegetation with a mosaic pattern of different reed height classes (Figure 7.16). Subsequently the area of open dry grassland (reed height < 50 cm) increased again due to the population increase of Heck cattle and Konik horses. In 1991 the stocking rate was about 0.75 animals/ha (based on the original 270 ha dry grassland).

The area dominated by reed (reed height > 50 cm) varied within years as well between years. In summer, the area dominated by reed (reed height > 50 cm) was highest. During autumn and winter the area with a reed height between 50 and 100 cm reverted partly to short grassland again through trampling and decay of litter (Vulink *et al.*, personal observation).

Greylag geese, white-fronted geese (*Anser albifrons*) and barnacle geese (*Branta leucopsis*) preferred open grassland (reed height < 50 cm). Wintering white-fronted and barnacle geese made use of the intensively grazed parts of dry grassland as well the summer-grazed and year-round grazed areas (Vulink *et al.*, personal observation).

The increase in reed cover on the year-round grazed dry grassland led to a temporary high density of voles in comparison with the summer-grazed dry grassland during the years 1988–1991. The vole density index (the number of voles caught per 100 catch-nights) in the year-round grazed dry grassland peaked at around 40 just after the mowing stopped and gradually decreased to the same constant low level (index < 2) as in the summer-grazed dry grassland after about five years (N. Beemster and Vulink, personal observation). As there is a significant relationship between vole density and density of vole-eating raptors like hen harrier (*Circus cyaneus*) and kestrel (*Falco tinnunculus*) (Dijkstra *et al.*, 1996), the density of these raptors showed the same trend: a temporary high density on the year-round grazed dry grassland just after cessation of mowing. This transition may be seen as an analogue for a sudden reduction in grazing intensity. In another Dutch artificial wetland (Lauwersmeer), the densities of hen harrier and kestrel peaked just after lowering the stocking rate for summer grazing, and decreased gradually thereafter (Beemster and Vulink, personal observation).

(c) *Tall vegetation*

In the non-cultivated core area, three vegetation types were distinguished: well drained reedbed vegetation; vegetation with tall herbs; and shrub veg-

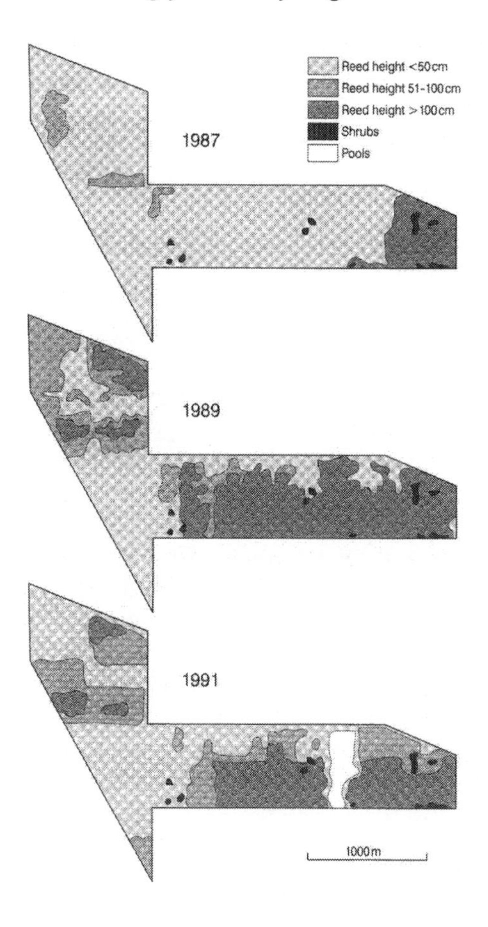

Figure 7.16 The development of reed cover (November situation) on the year-round grazed dry grassland (total 270 ha originally). The sharp boundary between grassland (reed height < 50 cm) and well drained reedbed vegetation (reed height > 100 cm) in 1987 was a result of additional mowing. In 1990 some pools were dug. At the start of the study period about 125 ha of dry grassland area had already changed into a vegetation of reed and tall herbs as a consequence of cessation of mowing together with a low grazing intensity. In 1991 the area was grazed by a herd of about 170 Heck cattle and about 90 Konik horses. (After Vulink *et al.*, personal observation.)

etation. The shrubs consisted of *Sambucus* and *Salix* species, the latter mainly *S. alba*, *S. triandra* and *S. viminalis*. These *Salix* species had germinated from wind-dispersed seeds. Solitary *S. cinerea* and *S. caprea* trees also became established in later years. *Phragmites* was dominant in the reedbed vegetation. The tall herbaceous vegetation is a mosaic of vegetation types dominated by tall herbs like *Cirsium* spp. and *Urtica*; it occurs on the higher grounds. At present (1997) most parts of the tall herbaceous vegetation are in the *Urtica–Sambucus* stage.

The grazing intensity by cattle and horses on well drained reedbed vegetation with some tall herbs in the year-round grazing area was highest in winter and early spring, i.e. outside the growing season (Figure 7.14). Vegetation development in the grazed and ungrazed vegetation, therefore, was largely the same: *Cirsium* and *Phragmites* were slowly replaced by *Urtica*. However, trampling of the litter layer in the grazed reedbed created germination gaps for other species, especially *Sambucus*, which can be suppressed by intensive cattle grazing (Vulink *et al.*, personal observation). In contrast, summer grazing with a relatively high animal density can turn dense stands of reed into a grassy vegetation. Study of an experimental area, where cattle were enclosed in tall vegetation, showed that cattle grazing (0.4 heifers/ha during May–November) could suppress reed dominance within four years and instead favour *Poa trivialis* (Vulink *et al.*, personal observation). The replacement of reed by grasses led to higher forage quality for herbivores in this habitat, especially during autumn and winter (Vulink and Drost, 1991).

Changes in populations of breeding birds were also related to vegetation changes in the grazed areas. Because the density of single species was too low to correlate with vegetation development, species were grouped in three categories: species related to vegetation dominated by reed, by tall herbs or by shrubs. Density of species (number of breeding pairs/10 ha) related to reed vegetation decreased (Spearman rank correlation, $Rs = -0.750$, $P < 0.05$); density of species related to tall herbs was rather stable ($Rs = 0.429$, N.S.); and density related to shrubs increased ($Rs = 0.857$, $P < 0.05$) (after Beemster, 1993; Koffijberg, 1995).

7.6 RETROSPECT AND PERSPECTIVES

7.6.1 Cyclic succession in the marsh and the need for draw-downs

With respect to management at the level of the ecosystem, the extent of human interference is an important point of discussion. In the marsh zone of Oostvaardersplassen, water-level management was found to be crucial for habitat diversity and related species richness. Artificial draw-down of water level is commonly used to re-establish emergent vegetation in open water areas in North America (Harris and Marshall, 1963; Smith and Kadlec, 1983), where the muskrat (*Ondatra zibethicus*) has an important additional grazing impact (Van der Valk and Davies, 1978). In Oostvaardersplassen an artificial draw-down period of about four years resulted in the re-establishment of emergent marsh vegetation dominated by *Typha* and *Phragmites* (Huijser *et al.*, 1995) and hence can be said to re-set the successional clock.

The application and frequency of artificial draw-downs in Oostvaardersplassen was further evaluated by development of a vegetation model. With this model the vegetation development in relation to water-level management can be predicted in terms of the proportional area of the main habitat types (Van Deursen, 1994). As shown by bird census data over many years, the proportion of reed vegetation that was inundated during spring and early summer was of crucial importance for wetland birds in the marsh system (Figure 7.3). Re-establishment of reed vegetation is greatly dependent on water level and herbivory by geese. The area of inundated reed was taken as a direct function of the water level, but it is surely also a function of greylag geese density. Experience in Oostvaardersplassen showed that, in spring, large areas of inundated reed vegetation would be present only after a dry period of about four years. Reed vegetation younger than about four years will be lost almost completely within a year, predominantly due to grazing of the rhizomes of inundated reed vegetation by geese during autumn and winter. Hence a dry period of at least four years is an important prerequisite for the re-establishment of inundated reed vegetation in Oostvaardersplassen. In northwestern Europe such dry periods (about four years) are scarce. Based on natural precipitation and evaporation in the period 1939–1989, inundated reed vegetation during springtime is predicted to be absent in most years (Figure 7.17A). The area of inundated reed vegetation with regular draw-down of water level in the entire area is shown in Figure 7.17B. In this case alternating of dry stages, marsh stages and lake stages occurs, but the entire area is in the same stage. Draw-downs alternating between the eastern and western compartments lead to a permanent presence of spring-inundated reed vegetation (Figure 7.17C). As long as marshes with habitats comparable to those of Oostvaardersplassen are absent in The Netherlands or in the surrounding area, an alternating draw-down approach as shown in Figure 7.17C is preferable to the other two approaches. This strategy of diversification can be viewed as a risk-minimizing approach for safeguarding vulnerable wetland birds and other marshland-related species during dry periods.

7.6.2 The role of herbivory in the subsystems

Herbivory by greylag geese was strongly related to the area of inundated reed vegetation and constituted an important process in the dynamics of the marsh by itself. Because of its impact on the habitat pattern in the marsh system, benefiting many other species, this avian herbivore can be considered as a keystone species of the marsh system (Meffe *et al.*, 1994; Bond, 1993; Mills *et al.*, 1993). The term keystone species (introduced by Paine, 1969), although originally applied to a top carnivore in the rocky intertidal zone, has since been used in a broader context. The term is also applied to important non-carnivorous species and habitat-modifying species that do not necessarily have direct trophic effects on other species (see also Chapter 1; Bond, 1993; Mills *et al.*, 1993).

Just after moult, wet grassland is an important foraging area for greylag geese. Inundation of the marsh, the impact of the geese themselves and graz-

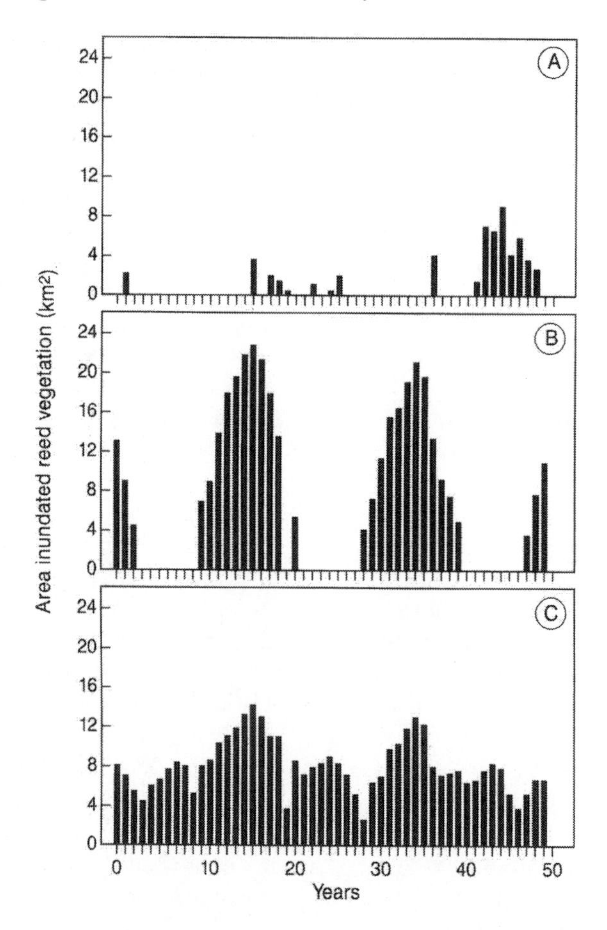

Figure 7.17 Calculated area of reed inundated during springtime in Oostvaardersplassen for different strategies of water-level management, calculated for 50 years of weather data: (A) based on natural precipitation and evaporation; (B) with a regular draw-down of water level in the entire area (dry stage of four years with intervals of 15 years); (C) as (B) but with draw-down alternating between two compartments. (After Van Eerden *et al.*, 1995.)

ing by large herbivores in the border zone are important factors in maintaining this habitat. Grazing affects vegetation structure as well as forage quality. Under repeated or continuous grazing pressure grass maturation is inhibited and regrowth stimulated, resulting in a relatively high forage quality. Both geese and large herbivores tend to concentrate on intensively grazed short swards (see also Ydenberg and Prins, 1981; McNaughton, 1984; Groot Bruinderink, 1989; Summers and Critchley, 1990; WallisDeVries and Daleboudt, 1994). Intensely grazed swards in wetlands and salt marshes support a higher species richness than ungrazed situations (e.g. plant species in salt marshes: Bakker, 1985; number and density of bird species: Larsson, 1969; Soikkeli and Salo, 1979; Duncan, 1992; see also Chapter 5). A grazing regime with a low stocking rate over a longer period is at the expense of extensive short grass vegetation, which is vital for geese as well as for large herbivores.

The density of grazers affects the condition of the sward and the extent to which other colonizers like *Phragmites* and shrubs can develop. Intensive grazing with cattle and horses kept wet grassland free from reed vegetation and facilitated birds such as herons, spoonbills, geese and waders. Dry grassland was maintained by intense grazing and functioned as an important winter foraging area for geese. For the large herbivores, too, dry grassland is the most important habitat. At a lowered grazing intensity, grassland quickly changed into a mosaic of short swards interspersed with reed and tall herbs. The transition stage of open dry grassland to a closed vegetation with reed and tall herbaceous vegetation is an especially important habitat for voles and foraging area for predators, e.g. vole-eating raptors (section 7.5.2). Thus, a different group of species may also benefit from a lesser grazing impact or a periodic reduction in herbivory (Chapter 7).

7.6.3 Interactions between avian and mammalian herbivores and integrated management

As mentioned before, greylag geese have a large impact on the marsh system. This means that habitat requirements of this species before and after moult should be considered with care. Wet grassland is of special importance to geese around the moulting period, because of its high forage quality in that period compared with dry grassland.

Grazing by cattle and horses at rather high stocking rates (*c.* 1.0 animals/ha) during summer can prevent wet grassland from being overgrown with reed and shrubs. It is not known, however, to what extent a year-round grazing regime, with lower large herbivore density, can maintain large-scale wet grassland. Until the present time, no area of wet grassland has been included within the year-round grazed area in Oostvaardersplassen, but information on the use of wet grassland by cattle and horses in year-round grazed areas is available from two other artificial wetlands in The Netherlands – Lauwersmeer and Grevelingen. These data show a high preference of free-ranging cattle and horses for dry grassland vegetation during the whole year and an underutilization of wet grasslands (Cornelissen and Vulink, 1995).

Dry grassland is a main habitat for large herbivores: the productivity of dry grassland determines the maximum sustainable density of large herbivores. On the other hand, extensive wet grassland area is an important habitat for wetland-related bird species (Figure 7.3) and plays a key role in the functioning of the marsh system. The area of wet grassland can be manipulated by adjustment of the summer grazing pressure of cattle and horses. It is therefore related to the area of dry grassland, which determines year-round carrying capacity for these herbivores. The maximum density of animals that can survive without supplementary forage in most winters is about 0.5 animals/ha on dry grassland and zero on wet grassland. The estimated ratio between the area of wet and dry grassland to maintain wet grassland, then, is about 1 : 8 (after Van Eerden *et al.*, 1995). For example, 500 grazers will maintain about 125 ha wet grassland and need about 1000 ha dry area during winter. This 1 : 8 ratio results in a maximum substantiated area in Oostvaardersplassen of about 200 ha open wet grassland, given a total size of the border zone of 2000 ha (wet plus dry grassland).

Taking into account the habitat use of moulting geese in the marsh zone, we estimate that a minimum of 40 000 geese is necessary to prevent encroachment of the marsh by reed. Assuming that at any one time 10 000 geese occupy the marsh, about 30 000 geese have to concentrate their foraging activities on fields adjacent to the marsh (especially wet grassland) for 10 days on average before and after moult (a total of 300 000 goose days) (M.R. Van Eerden *et al.*, personal observation). From field studies, it was assessed that wet grassland yields food for 400 goose days/ha and dry grassland for 100 goose days/ha (M.R. Van Eerden, unpublished data). Using these figures, about 300 ha wet grassland and 160 ha dry grassland is needed for about 30 000 moulting geese (Table 7.2). If more than about 200 ha wet grassland is desired to receive all moulting greylag geese, the dry area has to be expanded to accommodate the large herbivores or, for instance, reed on wet grassland has to be mown.

Inundated grassland, accompanied by pools and ditches, was also important as foraging area for spoonbills, herons and egrets in early spring. Considering the scale of Oostvaardersplasssen and the possible function as a breeding place for these rare species in northwestern Europe, an estimate was made of the requirements of the existing colonies of spoonbill (200

Table 7.2 The area of wet grassland required to sustain moulting greylag geese in relation to number of cattle and horses, and the area of dry grassland needed for these larger herbivores in winter

Area of wet grassland (ha)	Grazers (*N*)	Area of dry grassland (ha)	Goose days spent on grassland in OVP (*N*)	Goose days spent outside OVP (*N*)	Maintained in OVP (%)
50	200	400	60 000	240 000	20
100	400	800	120 000	180 000	40
200	800	1600	240 000	60 000	80
300 (100 ha mown)	800	1600	280 000	20 000	93
400 (200 ha mown)	800	1600	320 000	0	100

pairs) and grey heron (100 pairs) and to develop conditions for minimum numbers of breeding pairs of great white egret (10–15), little egret (10–15), night heron (*Nycticorax nycticorax*) (25) and bittern (10). Based on field studies (1987–1991) the food availability and consumption in shallow water by these wading birds was estimated (Voslamber, 1994; Voslamber and Buijse, 1996). Improved conditions for these species require an increase of 500–1000 ha of the area with pools, ditches and marsh with clear water, associated with wet grassland. For maintenance of the desired size of wet grassland for geese and these other wetland birds with year-round grazing and without additional management measures, Oostvaardersplassen should be enlarged by 2000–6500 ha to about 8000–12 500 ha (based on the ratio 1 : 8 for the surface area wet and dry grassland).

Interestingly, this calculated area coincides with the lower range for self-sustaining populations of large herbivores according to Belovsky (1987) and Soulé (1987) (see also Van Wieren, 1991). Not only is scale important but also the completeness of the system. Natural lacustrine and riverine marshes are surrounded by higher grounds which can function as wintering areas for large herbivores. Almost all nature reserves in The Netherlands, like the fenced Oostvaardersplassen, are too small to include complete year-round habitats for cattle. Close linkage with complementary landscapes is essential to create adequate year-round habitats for cattle and other large herbivores (Bokdam and WallisDeVries, 1992; WallisDeVries, 1995).

Large herbivores are able to retard vegetation succession (McNaughton, 1979, 1985; Bakker, 1985). The impact of grazing on vegetation pattern will be largely related to abiotic environmental conditions. In areas with low to intermediate productivity, the impact of natural grazing is predicted to be relatively high, because moderate herbivore densities are sufficient to consume the relatively low production of the vegetation (Oksanen, 1990). It is questionable, however, whether in the longer term large herbivores could significantly alter the vegetation succession in productive temperate areas without additional management measures (Chapter 3) (Thalen, 1984; Van Wieren, 1991). Presumably, in a productive environment with large seasonal variation in forage quality, large herbivores in their numerical response are more 'followers' of vegetation dynamics than 'determinants' (Chapter 10).

We now can hypothesize that after introduction of cattle and horses, a series of cycles both in the herbivore population and in the vegetation assemblage will ultimately lead to the disappearance of the herbivores (Table 7.3). The different stages of herbivore population development can be described by the phases of a generalized population growth curve (Stanier *et al.*, 1976), with an initial phase followed by an exponential growth phase, saturation and equilibrium. The cycle is completed with a sudden decline to low numbers after resource depletion or other causes inducing mass mortality. The length of the sequential cycles will decrease over time, since it can be expected that during each phase of low numbers the area of open dry grassland will dwindle as a consequence of the encroachment by reed and shrubs. A crucial assumption is that the large herbivores in their numerical response are not able to offset this encroachment during the next phases of population build-up. Thus, with a hands-off scenario, the carrying capacity of an isolated wetland reserve for large herbivores will decrease in time. This prediction is supported by insights from recent plant–herbivore models (Van de Koppel *et al.*, 1996; Chapter 10). It differs from classic views on plant–herbivore relationships, which predict a steady state after some oscillations (e.g. Caughley 1976, 1979).

In artificial wetlands, management based on an ecosystem approach depends on measures that mimic a setback of succession in natural systems. Several options can be considered, such as inundation with salt water for artificial wetlands near marine and estuarine ecosystems, periodically intense stocking, manipulation of year-round grazing herds by limiting areas made available on the analogy of partial migration, or mowing as an equivalent to short intensive 'rotational' grazing during migration. Sudden changes, involving either water level or stocking rate, can be seen as disturbances. Disturbance–recovery cycles can be considered as key ecological processes in ecosystems and do have profound effects on species richness (Hobbs and Huenneke, 1992; Noss, 1992).

Regulation of the populations of large herbivores also requires attention as a potential measure in the context of an ecosystem approach. Populations of large herbivores, which are enclosed in this semi-natural nature reserve,

Table 7.3 Expected phases and most important characteristics of herbivore population and vegetation after introduction of cattle and horse populations in isolated, eutrophic artificial wetlands

Phase	Herbivore population	Vegetation development	Vegetation pattern
1	Lag phase; population starts to increase	Encroachment of short vegetation with reed, tall herbs and shrubs	Small-scale mosaic vegetation with gradual transition between vegetation types
2	Exponential phase; maximum growth rate	Change of reed, tall herbs and shrubs into short and rough grassy vegetation	Large-scale mosaic vegetation, relatively sharp boundaries between vegetation types arise
3	Stationary phase; growth delayed as consequence of food shortage and intrinsic factors, e.g. stress	A large part of the vegetation types change into grassy vegetation	Relatively large-scale short and tall grassy vegetation with some reed, tall herbs and shrubs
4	Crash phase; population crash can take place during a severe winter in combination with a cold and wet spring, high risk for (parasitic) diseases	Overgrazed short grassy vegetation with bare ground as potential germination place for herbs and shrubs	Relatively large-scale short grassy vegetation

will eventually be food-limited as in more natural undisturbed ecosystems like the Serengeti (Sinclair, 1985) or Lake Manyara National Park (Prins, 1996). Social mechanisms will have some effect on population control but there is no evidence that these mechanisms are able to fine-tune the population size (Caughley and Krebs, 1983; Putman *et al.*, 1996). If ethical or practical considerations make it undesirable to have introduced mammalian herbivores regulated by their resources, this presents a limitation for the implementation of year-round grazing in artificial wetlands like Oostvaardersplassen. Accepting the reality of these limitations implies some measure in order to regulate the populations when saturation densities are reached.

7.6.4 Area of tension between species-based and ecosystem approaches in management of artificial wetlands

In a period of 25 years, a wide range of habitats and high species richness developed in Oostvaardersplassen despite the rather monotonous start on a former sea bottom. The natural factors of hydrological conditions and herbivory, together with measures such as artificial draw-down, inundation of wet grassland, intense stocking and mowing in a diversified regime, had an important influence on vegetation succession. This management resulted in temporal and spatial heterogeneity in habitat patterns and related species richness, characteristic for ecological reference systems (riverine and lacustrine systems). We support the theory that ecosystems fluctuate between different states following the non-equilibrium paradigm (Vitousek and White, 1981; Westoby *et al.*, 1989; Pickett *et al.*, 1993; Rietkerk *et al.*, 1996; Rietkerk and Van de Koppel, 1997) instead of fluctuating around a steady state. A management aiming to preserve a fixed patterning of communities is a contradiction of the properties of the system. Instead, the conservation of large areas at non-equilibrium is better served by managing the processes rather than the patterns. The processes of primary importance for management are those that determine vegetation dynamics. These should be manipulated so as to retain a dynamic patchwork of successional stages. This contrasts with a more static and traditional approach of selecting a target vegetation type or a habitat type for a single species. Such a species-based approach can be a feasible option for relatively small-scale areas. In practice it often involves a relatively high degree of management effort.

An ecosystem approach should be based on knowledge of functioning of the system and of so-called ecological reference systems, which can serve as a model for a particular area. A deep understanding of the role of keystone species and processes is the basis of this approach. However, systems ecology and its application to conservation problems is less developed than the ecology of single species (Meffe *et al.*, 1994). Despite the complexity of ecosystem management and the uncertainty of principles underlying systems ecology, the present challenge is to implement management of large-scale nature reserves by managing crucial processes.

In enclosed artificial wetlands a hands-off scenario – a superficially 'natural' way out of the uncertainty – could very well have the opposite effect than was intended. In these wetlands an ecosystem approach without any human intervention is likely to result in the loss of early successional stages and associated species richness – the very values for which the areas were designated. To prevent this degradation some specific measures, which imitate natural re-sets of succession in ecological reference systems, are needed. Both practical experience with long-term management of processes and improvement of integrated knowledge about components of the system will be necessary for a wise management based on understanding of the system.

ACKNOWLEDGEMENTS

Thanks are due to Wouter Dubbeldam and Perry Cornelissen for their effort in compiling data and the preparation of figures. Dick Visser carefully drew the final figures. Menno Zijlstra, Nico Beemster, Marcel Huijser, Bram Smit, Fré van de Klei, Niels Kooiman and Aaldert Muis contributed considerably to the database which formed the basis for this chapter. The long-term monitoring of vegetation development in the study areas was made possible by Hans Drost, who was responsible for aerial photographs and interpretation of vegetation maps. The wardens of Oostvaardersplassen (V.L. Wigbels, J. Griekspoor, N. Dijkshoorn, R. van Baarle) gave us the opportunity to work in the areas and provided miscellaneous technical assistance. Rudolf Drent, Herbert Prins, Michiel WallisDeVries, Jan Bakker, Sip van Wieren and Rory Putman are acknowledged for constructive comments on earlier drafts of this chapter.

REFERENCES

Bakker, J.P. (1985) The impact of grazing on plant communities, plant populations and soil conditions on salt marshes. *Vegetatio*, **62**, 391-398.

Beemster, N. (1993) Broedvogels in de Oostvaardersplassen: natuurlijke successie en effecten van begrazing in het onontgonnen deel van het buitenkaadse gebied in de periode 1987-1992. Intern rapport 1993-5 Lio. Rijkswaterstaat Directie Flevoland, Lelystad.

Belovsky, G.E. (1987) Extinction models and mammalian persistence, in *Viable Populations for Conservation*, (ed. M.E. Soulé), pp. 35–57, Cambridge University Press, Cambridge.

Bokdam, J. and WallisdeVries, M.F. (1992) Forage quality as a limiting factor for cattle grazing in isolated Dutch nature reserves. *Conservation Biology*, **6**, 399–408.

Bond, W.J. (1993) Keystone species, in *Biodiversity and Ecosystem Function*, (eds E.D. Schulze and H.A. Mooney), pp. 237–253, Springer Verlag, New York.

Caughley, G. (1976) Plant–herbivore systems, in *Theoretical Ecology: Principles and Applications*, (ed. R.M. May), pp. 94–113, Blackwell Scientific Publications, Oxford.

Caughley, G. (1979) What is this thing called carrying capacity? in *North American Elk: ecology, behaviour and management*, (eds M.S. Boyce and L.D. Hayden-Wing), pp. 2–8, University of Wyoming, Laramie, Wyoming.

Caughley, G. and Krebs, C.J. (1983) Are big mammals simply little mammals writ large? *Oecologia*, **59**, 7–17.

Cornelissen, P. and Vulink, J.T. (1995) *Begrazing in jonge wetlands*, Flevobericht nr 367, Ministerie van Verkeer en Waterstaat, Rijkswaterstaat Directie IJsselmeergebied, Lelystad.

Cornelissen, P. and Vulink, J.T. (1996) *Edelherten en reeën in de Oostvaardersplassen: demografie, terreingebruik en dieet*, Flevobericht nr 397, Ministerie van Verkeer en Waterstaat, Rijkswaterstaat Directie IJsselmeergebied, Lelystad.

Cornelissen, P., Van Deursen, E.J.M. and Vulink, J.T. (1995) *Jaarrondbegrazing op de Zoutkamperplaat in het Lauwersmeergebied; effecten op de vegetatie en zelfredzaamheid van runderen en paarden*, Flevobericht nr 379, Ministerie van Verkeer en Waterstaat, Rijkswaterstaat Directie IJsselmeergebied, Lelystad.

De Jong, C.B., Cornelissen, P. and Vulink, J.T. (1997) Grote Grazers in de Oostvaardersplassen: dieetsamenstelling op basis van faecesanalyse. RIZA werkdocument 97.114X, Lelystad

Dijkstra, C., Beemster, N., Zijlstra, M. *et al.* (1996) *Roofvogels in de Nederlandse Wetlands*, Flevobericht nr 381, Ministerie van Verkeer en Waterstaat, Rijkswaterstaat Directie IJsselmeergebied, Lelystad.

Duncan, P. (1992) *Horses and Grasses: the nutritional ecology of equids and their impact on the Camargue*, Ecological Studies 87, Springer Verlag, New York.

Duncan, P., Foose, T.J., Gordon, I.J. *et al.* (1990) Comparative nutrient extraction from forages by grazing bovids and equids: a test of the nutritional model of equid/bovid competition and coexistence. *Oecologia*, **84**, 411–418.

Finlayson, M. and Moser, M. (1991) *Wetlands*, International Waterfowl and Wetlands Research Bureau, Slimbridge, UK.

Fox, A.D., Kahlert, J. Ettrup, H. *et al.* (1995) Moulting greylag geese *Anser anser* on the Danish island of Saltholm; numbers, phenology, status and origins. *Wildfowl*, **46**, 16–30.

Gordon, I.J. (1988) Facilitation of red deer grazing by cattle and its impact on red deer performance. *Journal of Applied Ecology*, **25**, 1–10.

Groot Bruinderink, G.W.T.A. (1989) The impact of wild geese visiting improved grasslands in the Netherlands. *Journal of Applied Ecology*, **26**, 131–146.

Harris, S.W. and Marshall, H.W. (1963) Ecology of water-level manipulations on northern marsh. *Ecology*, **44**, 331–343.

Hobbs, R.J. and Huenneke, L.F. (1992) Disturbance, diversity, and invasion: implications for conservation. *Conservation Biology*, **6**, 324–337.

Hofmann, R.R. (1973) *The Ruminant Stomach Structure and Feeding Habitats of East African Game Ruminants*, East African Literature Bureau, Nairobi.

Huijser, M.P., Drost, H.J. and Röling, Y.J.B. (1995) Vegetatieontwikkeling en cyclisch waterpeilbeheer in de Oostvaardersplassen. *De Levende Natuur*, **96**, 213–222.

Huijser, M.P., Vulink, J.T. and Zijlstra, M. (1996) *Begrazing in de Oostvaardersplassen; vegetatiestructuur en terreingebruik door grote herbivoren en ganzen*. Intern rapport nr 1996-5 Lio, Rijkswaterstaat Directie IJsselmeergebied, Lelystad.

Jacobs, J. (1974) Quantitative measurement of food selection. A modification of the forage ratio and Ivlev's selectivity index. *Oecologia*, **14**, 413–417.

Jans, L. and Drost, H.J. (1995) *De Oostvaardersplassen; 25 jaar vegetatie onderzoek*, Flevobericht nr 382, Ministerie van Verkeer en Waterstaat, Rijkswaterstaat Directie IJsselmeergebied, Lelystad.

Koffijberg, K. (1995) Broedvogels en vegetatie op het Stort, Oostvaardersplassen, in 1994. Intern rapport 1995-25 Lio, Rijkswaterstaat Directie IJsselmeergebied, Lelystad.

Larsson, T. (1969) Land use and bird fauna on shore meadows in southern Sweden. *Oikos*, **20**, 136–155.

Loft, E.R., Menke, J.W. and Kie, J.G. (1991) Habitat shifts by mule deer: the influence of cattle grazing. *Journal of Wildlife Management*, **55**, 16–26.

Loft, E.R., Kie, J.G. and Menke, J.W. (1993) Grazing in the Sierra Nevada: home-range and space use patterns of mule deer as influenced by cattle. *California Fish and Game*, **79**, 145–166.

Madsen, J. (1991) Status and trends of goose populations in the Western Palearctic in the 1980s. *Ardea*, **79**, 113–122.

McNaughton, S.J. (1979) Grazing as an optimization process: Grass–ungulate relationships in the Serengeti. *American Naturalist*, **113**, 691–703.

McNaughton, S.J. (1984) Grazing lawns: animals in herds, plant form and coevolution. *American Naturalist*, **124**, 863–886.

McNaughton, S.J. (1985) Ecology of a grazing system: The Serengeti. *Ecological Monographs*, **55**, 259–294.

Meffe, G.K., Carroll, C.R. and Pimm, S.L. (1994) Community level conservation: species interactions, disturbance regimes, and invading species, in *Principles of Conservation Biology*, (eds G.K. Meffe and C.R. Carroll), pp. 209–236, Sinauer Associates, Sunderland, Massachusetts.

Meininger, P.L., Schekkerman H. and Van Roomen, M.W.J. (1995) Populatieschattingen en 1%-normen van in Nederland voorkomende watervogelsoorten: voorstellen voor standaardisatie. *Limosa*, **68**, 41–48.

Mills, L.S., Soulé, M.E. and Doak, D.F. (1993) The keystone-species concept in ecology and conservation. *BioScience*, **43**, 219–224.

Noss, R.F. (1992) The Wildlands Project; land conservation strategy. *Wild Earth* **Special Issue**, 10–25.

Oksanen, L. (1990) Predation, herbivory, and plant strategies along gradients of primary productivity, in *Perspectives on Plant Competition*, (eds J.B. Grace and D. Tilman), pp. 445–474, Academic Press, Toronto.

Paine, R.T. (1969) A note on trophic complexity and community stability. *American Naturalist*, **103**, 91–93.

Pickett, S.T.W., Parker, V.T. and Fiedler, P.L. (1993) The new paradigm in ecology: implications for conservation biology above the species level, in *Conservation Biology; the Theory and Practice of Nature Conservation Preservation and Management*, (eds P.L. Fiedler and S.K. Jain), pp. 65–88, Chapman & Hall, New York.

Polman, G.K.R. and Schmidt-ter Neuzen, S. (eds) (1987) *Ontwikkelingsvisie Oostvaardersplassen*, Flevobericht nr 282, Ministerie van Verkeer en Waterstaat, Rijksdienst voor de IJsselmeerpolders, Lelystad.

Prins, H.H.T. (1996) *Ecology and Behaviour of the African Buffalo: social inequality and decision-making*, Chapman & Hall, London.

Putman, R.J., Langbein, J., Hewison, A.J.M. and Sharma, S.K. (1996) Relative roles of density-dependent and density-independent factors in population dynamics of British deer. *Mammal Review*, **26**, 81–101.

Rietkerk, M. and Van de Koppel, J. (1997) Alternate stable states and threshold effects in semi-arid grazing systems. *Oikos*, **79**, 69–76.

Rietkerk, M., Ketner, P. Stroosnijder, L. and Prins, H.H.T. (1996) Sahelian rangeland development; a catastrophe? *Journal of Range Management*, **49**, 512–519.

Rose, P.M. and Scott, D.A. (1994) *Waterfowl Population Estimates*, IWRB Publication 29, Slimbridge, UK.

Schultz, E. (1992) *Waterbeheersing van de Nederlandse Droogmakerijen*, Van Zee tot Land 58, Rijkswaterstaat, Directie Flevoland, Lelystad.

Sinclair, A.R.E. (1985) Population regulation of the Serengeti wildebeest: a test of the food limitation hypothesis. *Oecologia*, **65**, 266–268.

Smith, L.M. and Kadlec, J.A. (1983) Seed banks and their role during drawdown of a North American marsh. *Journal of Applied Ecology*, **20**, 673–84.

Soikkeli, M. and Salo, J. (1979) The bird fauna on abandoned shore pastures. *Ornis Fennica*, **56**, 124–132.

Soulé, M.E. (1987) Where do we go from here? in *Viable Populations for Conservation*, (ed. M.E. Soulé), pp. 175–183, Cambridge University Press, Cambridge.

Stanier R.Y., Adelberg, E.A. and Ingraham, J.L. (1976) *General Microbiology*, Macmillan Press, London.

Summers, R.W. and Critchley, C.N.R. (1990) Use of grassland and field selection by Brent geese *Branta bernicla*. *Journal of Applied Ecology*, **27**, 834–836.

Ter Heerdt, G.N.J. and Drost, H.J. (1994) Potential for the development of marsh vegetation from the seed bank after a drawdown. *Biological Conservation*, **67**, 1–11.

Thalen, D.C.P. (1984) Large mammals as tools in the conservation of diverse habitats. *Acta Zoologica Fennica*, **172**, 159–163.

Van de Koppel, J., Huisman, J., Van der Wal, R. and Olff, H. (1996) Pattern of herbivory along a gradient of primary productivity: an empirical and theoretical investigation. *Ecology*, **77**, 736–745.

Van den Tempel, R. and Osieck, E.R. (1994) *Areas Important for Birds in the Netherlands*, Technisch Rapport Vogelbescherming Nederland 13E, Vogelbescherming Nederland, Zeist.

Van der Meijden, R., Weeda, E.J., Holverda, W.J. and Hovenkamp, P.H. (1990) *Heukels' Flora van Nederland*, 21st edn, Wolters-Noordhoff, Groningen.

Van der Toorn, J. and Hemminga, M.A. (1994) Use and management of common reed *Phragmites australis* for land reclamation, The Netherlands, in *Wetlands and Shallow Continental Waterbodies*, Vol. 2, (eds B.C. Patten, S.E. Jorgensen and H. Dumont), pp. 363–371, Academic Publishing, The Hague.

Van der Valk, A.G. (1994) Effects of prolonged flooding on the distribution and biomass of emergent species along freshwater wetland coenocline. *Vegetatio*, **110**, 185–196.

Van der Valk, A.G. and Davies, C.B. (1976) The seedbanks of prairie glacial marshes. *Canadian Journal of Botany*, **54**, 1832–1838.

Van der Valk, A.G. and Davies, C.B. (1978) The role of seed banks in the vegetation dynamics of prairie glacial marshes. *Ecology*, **59**, 322–335.

Van Deursen, A.M. (1994) *Modellering van de Vegetatieontwikkeling in het Binnenkaadse Gebied van de Oostvaardersplassen bij Verschillende Waterpeilscenario's*, Flevobericht nr 355, Ministerie van Verkeer en Waterstaat, Rijkswaterstaat Directie Flevoland, Lelystad.

Van Eerden, M.R., Vulink, J.T., Polman, G.K.R. *et al.* (1995) Oostvaardersplassen: 25 jaar pionieren op weke bodem. *Landschap*, **12**, 23–39.

Van Wieren, S.E. (1991) The management of populations of large mammals, in *The Scientific Management of Temperate Communities for Conservation*, (eds I.F. Spellerberg, F.B. Goldsmith and M.G. Morris), pp. 103–127, Blackwell Oxford.

Vera, F.W.M. (1988) *De Oostvaardersplassen: van spontane natuuruitbarsting tot gerichte natuurontwikkeling*, IVN and Grasduinen Oberon, Amsterdam.

Vitousek, P.M. and White, P.S. (1981) Process studies in succession, in *Forest Succession: Concepts and Applications*. (eds D.C. West, H.H. Shugart and S.D.B. Botkin), pp. 267–275, Springer-Verlag, New York.

Voslamber, B. (1994) *Ecologisch Onderzoek in het Proefgebied 'De Waterlanden'; Deel E: Dierecologische aspecten*, Flevobericht nr 364, Ministerie van Verkeer en Waterstaat, Rijkswaterstaat Directie IJsselmeergebied, Lelystad.

Voslamber, B. and Buijse, A.D. (1996) *Lepelaars in de Oostvaardersplassen: de beschikbaarheid van prooivis en het foerageersucces van Lepelaars (Platalea leucorodia)*, Flevobericht nr 411, Ministerie van Verkeer en Waterstaat, Rijkswaterstaat Directie Flevoland, Lelystad.

Vulink, J.T. and Drost, H.J. (1991) Nutritional characteristics of cattle forage plants in the eutrophic nature reserve Oostvaardersplassen, The Netherlands. *Netherlands Journal of Agricultural Science*, **39**, 263–271.

WallisDeVries, M.F. (1995) Large herbivores and the design of large-scale nature reserves in Western Europe. *Conservation Biology*, **9**, 25–33.

WallisDeVries, M.F and Daleboudt, C. (1994) Foraging strategy of cattle in patchy grassland. *Oecologia*, **100**, 98–106.

Westoby, M., Walker, B. and Noy-Meir, I. (1989) Opportunistic management for rangelands not at equilibrium. *Journal of Range Management*, **42**, 266–274.

Ydenberg, R.C. and Prins, H.H.T. (1981) Spring grazing and the manipulation of quality by barnacle geese. *Journal of Applied Ecology*, **183**, 443–453.

Zijlstra, M., Loonen, M.J.J.E., Van Eerden, M.R. and Dubbeldam, W. (1991) The Oostvaardersplassen as a key moulting site for greylag geese (*Anser Anser*) in Western Europe. *Wildfowl*, **42**, 42–52.

8

The practical use of grazing in nature reserves in The Netherlands

Harm Piek
Vereniging Natuurmonumenten, PO Box 9955, 1243 ZS 's Graveland, The Netherlands

8.1 INTRODUCTION

In considering the concept of grazing in nature reserves, a clear distinction should be made between grazing by wild herbivores and grazing by domestic herbivores which is deployed by the ranger in order to control vegetation development. Wild herbivores, such as roe deer (*Capreolus capreolus*), red deer (*Cervus elaphus*), hare (*Lepus europeus*), rabbit (*Oryctolagus cuniculus*), herbivorous birds and insects, are not included in this chapter. Although such herbivores may have a considerable impact on the vegetation, they are, as a rule, not used as a steering mechanism to achieve management objectives. The domestic herbivores which have been introduced in nature reserves, such as cattle, horses, sheep and goats, are being used predominantly as a kind of steering mechanism (Chapter 2).

The following chapter gives an overview of experiences with the deployment of domestic grazers in conservation management. During the past decades grazing has become a popular management measure amongst park rangers. Prior to 1970 grazing was applied as a management measure in only 10 Dutch nature reserves; in 1995 there were at least 400 nature reserves (covering a total of about 45 000 ha) in which large herbivores had been introduced. On the one hand, this popularity is a result of the positive effects of grazing, but on the other hand it cannot be denied that grazing has become 'fashionable' as a modern and innovative measure. After 25 years of experience in grazing, it is important that the knowledge and insights which have been gained from practice are presented. It should be emphasized that these observations are the result of practical management. They are, there-

Grazing and Conservation Management. Edited by M.F. WallisDeVries, J.P. Bakker and S.E. Van Wieren. Published in 1998 by Kluwer Academic Publishers, Dordrecht. ISBN 0 412 47520 0.

fore, frequently not supported by solid evidence from scientific investigations. Lacking this support, some of the conclusions should be viewed as working hypotheses rather than ascertained facts.

The Dutch society for nature conservation, Vereniging Natuurmonumenten, applies grazing by cattle, horses and sheep on a large scale. First of all it uses large grazers in areas where nature and landscape have often developed under the influence of traditional grazing by livestock. Heathland, chalk grassland, dune grassland, riverine grassland, dike slopes, salt marshes, wet grassland in valleys and peatland have been grazed for many centuries by way of agricultural exploitation. A continuation of this form of land use in the light of nature management offers good possibilities for the conservation and management of the natural value of these ecosystems. Another case concerns recently formed territory, such as in the southwestern part of The Netherlands, where land outside the dikes has been reclaimed by the implementation of the Delta project; grazing is used there on a large scale in the development of new conservation areas (Slim and Oosterveld, 1985). In recent decades grazing has been particularly applied in 'nature development' projects (Chapter 2). Particularly in efforts to convert former arable land and fertilized agricultural grassland into nutrient-poor grassland, heaths, rough grassland with tall herbs, scrub and woodland, grazing has proved to be a quite suitable measure.

In this chapter the most important management objectives and the effects of grazing are described, with emphasis on areas managed by Vereniging Natuurmonumenten. The aspects of grazing will be evaluated insofar as they concern the type of herbivore, herbivore density, grazing in relation to the available area and the food situation, the duration and season of grazing, as well as practical factors such as the well-being and health of the animals and their relation to the public.

8.2 OBJECTIVES OF GRAZING MANAGEMENT

To determine a management objective for a given site, it is essential to consider the conservation management strategy. Natuurmonumenten operates on the following nature management strategies or principles for the conservation, restoration and/or development of landscapes (Natuurmonumenten, 1993; see also Chapters 2 and 11):

- conservation, restoration and development of near-natural landscapes;
- conservation, restoration and development of semi-natural landscapes;
- conservation and restoration of species-rich cultivated landscapes.

Grazing is applied in all three strategies as a means of achieving any of the above landscape types.

In **near-natural landscapes** the landscape patterns and plant and animal communities are largely the result of natural and spontaneous processes without human intervention. In practice, this means that there is no longer any management. Measures that stimulate major natural key processes are, however, unavoidable. In this respect, introduction of grazing is an important key process (WallisdeVries, 1994; Vera, 1997; Chapter 1). Grazing constitutes one of the important key processes in landscape formation. In principle, the grazers, once introduced, are not regulated with regard to number, sex ratio, age, location or duration of grazing. In this type of landscape the grazers are, in fact, an integrated natural part of the entire ecosystem just as trees, fungi, predators and abiotic factors (e.g. climate, geological and pedological processes) and grazing is not a regulation method for the conservation or development of certain landscape patterns or types of vegetation. In this case cattle and horses should be regarded as more or less wild herbivores and as substitutes for the now extinct aurochs (*Bos primigenius*) and tarpan (*Equus ferus ferus*). Within this strategy their function is comparable to that of non-regulated wild herbivores such as deer and red deer. At the moment this type of grazing is only applied in a very limited number of nature reserves (Oostvaarderplassen, Chapter 7; and Veluwezoom National Park).

In **semi-natural landscapes** such as moors, dune grassland and riverine grassland, grazing is, contrary to the management of near-natural landscapes, an explicitly selected management method. As a matter of fact the strategy for semi-natural landscapes does not aim for a largely natural and spontaneous landscape formation, but, rather, the major key processes are geared towards a particular landscape structure or vegetation and its characteristic target species. As regards grazing, this implies that the number and type of herbivore, the duration and season are determined by the ranger in order to conserve, restore or develop a certain landscape structure (e.g. rough grassland or wood-pasture), a particular vegetation type, or a selected set of target species. Another important feature of this management strategy is the emergence of spatial heterogeneity in the vegetation structure, which comes about by variation in grazing intensity in the area. This type of management is the most widespread in grazed nature reserves.

In a strategy for species-rich **cultivated landscapes** the landscape and vegetation structure are regulated to a large extent by the ranger, who may carry out a variety of management measures. The objective is to preserve or restore a certain plant and animal community or a 'target nature type' with its characteristic target species by active intervention. It applies to old cultivated landscapes such as can be found in brook valleys, peat polders, wooded parks surrounding manor houses and country estates. It concerns mostly the management of enclosed land, parcel by parcel, by mowing, cutting trees and sods and burning as well as grazing. Management by grazing is often linked with traditional agricultural grassland management. In many pasture areas with a management focused on waders, geese or herb-rich meadows, cattle grazing is a common management measure. For the conservation of historically valuable cultivated landscapes (e.g. country estates) grazing by endangered and rare indigenous breeds of cattle, horses and sheep is considered a good option.

8.3 EVALUATION OF THE EFFECTS OF GRAZING

In the following section the several methods of grazing and the effects of grazing are described with respect to the major landscape types. Evaluation of grazing management is only possible when an objective for the planned management has been set. Such an objective is part of the management plan for each area. Which objective is chosen depends on the actual nature conservation interest (particularly the presence of target species and communities) as well as on the prospects for recovery and restoration of lost values or of values to be newly developed, respectively. In this respect it is important to have insight into the key ecological processes that are operative in the area or can be restored. It is also important to know the influence of any disturbing factors from the surrounding territory (e.g. water drainage, inputs from water and atmospheric pollution). The deployment of grazing is a derivative of the selected objective. The objective indicates which type of terrain and target communities can be expected, given the local circumstances and operating processes. Following this analysis, it is determined whether grazing would constitute a key process. If so, it is then determined which grazers and what grazing intensity are required to obtain a desired level of impact. Usually it is a matter of determining the balance between plant productivity and its utilization by the grazers.

The evaluation of grazing is described per terrain type rather than per management strategy for each landscape type. The goal is conservation, restoration and development of nature conservation interest irrespective of the chosen management strategy per landscape: a particular management strategy may occur in different landscapes. The evaluation is based on monitoring data. It mostly concerns data on developments in vegetation structure, species groups and, in particular, indicator species. Both the indicator species of relevant processes (e.g. development of woodland, tall herb vegetation, nutrient status) and the indicators of nature conservation interest (e.g. endangered species) are being monitored. Monitoring takes place mainly by descriptions of permanent test sections and permanent plots, surveys of indicator species over entire nature reserves, and vegetation mapping with or without remote sensing techniques.

To determine a certain development it is essential that monitoring is carried out during a period of at least five years. Many grazing projects have been running for more than 10 years and have been followed at various degrees of intensity.

8.3.1 Heathlands

The objective of grazing heathlands is the conservation or improvement of this dwarf shrub vegetation with the corresponding flora and fauna. This can be achieved by steering the most important processes: impeding the encroachment of trees and shrubs, preventing the spreading of grasses and thickets and arresting or delaying the ageing process of heather. Apart from ensuring the prevalence of heather species, grazing is also a means of conserving or developing a greater structural variation of the heathland. The preservation of species-rich grass-heaths and the establishment of gradual transitions to woodland are considered particularly important.

Until about 1985 grazing in most heathland reserves was by herds of roaming heathland sheep (especially Drenthe Heath sheep). Later cattle and ponies were deployed to an increasing extent, either in combination with sheep or as substitutes for heathland sheep. Cattle are often used only in seasonal grazing, especially when modern breeds such as Limousin, Charolais, Hereford and young Holstein-Friesian cattle are used. For year-round grazing cattle breeds such as Scottish Highland or Galloway and pony breeds such as Iceland, Shetland or New Forest are favoured (Figure 8.1).

Grazing of heathland with flocks of heathland sheep, as was the practice in past centuries, appears to be no longer adequate to influence sufficiently the processes described above. As it turned out, sheep were no longer capable of curbing the spreading of grasses such as *Molinia caerulea* and *Deschampsia flexuosa,* which resulted from increased atmospheric deposition of nitrogen as well as mineralization of organic soil due to desiccation. Especially on humid and wet heathland, the increase in *Molinia* could not

Figure 8.1 Galloway cattle are often used for year-round grazing in areas with large numbers of visitors.

be contained by sheep grazing (not even at densities of more than 2 sheep/ha). Subsequently, many rangers decided to increase the number of sheep per hectare, but they often discovered that the remaining heath (both old shrubs and young seedlings) deteriorated rather than improved. Supplementary management measures such as mechanical cutting of sods and mowing were applied ever more frequently. The use of cattle turned out to be more successful when it came to stopping the growth of grasses and increasing the amount of heather (*Calluna vulgaris*) and bell-heather (*Erica tetralix*). In sites such as, for instance, Wapserveld, Wolfhezerheide, Kampina and Posbank, the coverage of heather increased by 20–40% (in places by even 80%). Characteristic heathland plant species such as *Gentiana pneumonanthe*, *Rynchospora alba*, *R. fusca* and *Carex panicea* and animal species such as sand lizard (*Lacerta agilis*) and wheatear (*Oenanthe oenanthe*) have either increased in number or have re-established. Species of rich grass-heaths (*Violion caninae*) also increased in places and on sand dune soils pioneer species of the *Thero-Airion* and *Spergulo-Corynephoretum* communities developed in well-trodden spots. In some places rare species such as grayling butterfly (*Hipparchia semele*), the spider *Dictyna latens* and common lizard (*Lacerta vivipara*) have even been spotted.

In many grazed heathlands there has been an increase in small grazers like rabbits and species that live on dung. The increase of short and young grass has clearly been an improvement in the food supply for small grazers (Chapter 6). As a result, food conditions for predators, especially foxes (*Vulpes vulpes*), have also become more favourable. Heathland grazing has also accelerated the turnover of organic matter. Thanks to the presence of dung, dung fungi (such as *Hygrophorus* species), dung beetles (Scarabaeidae) and badgers (*Meles meles*) have increased. However, in many heathlands the introduction of sheep, cattle or horses has resulted in a decline in the number of roe deer. The reason for this may be found in food competition during winter, disturbance, and disappearance of sufficient cover or shelter (Chapter 6).

Another motive for changing to cattle grazing on heathland was that historical research (Bieleman, 1987) has shown that cattle grazing in Dutch heathlands occurred on a large scale before the eighteenth century. Knowledge acquired from successful cattle grazing on heathland abroad (e.g. the New Forest, England; Putman, 1986) has also played a role in the conversion from sheep to cattle.

To an increasing extent heathlands are being grazed together with grassland (mostly former fertilized agricultural land) and woodlands. As a bonus it has been found that the former sharp boundaries between the different types of terrain are fading and are actually forming interesting transition zones.

When grazing heathland, the availability of food is important for determining the type and density of grazing animals, as well as the season of grazing. Based on historical data a stocking rate of about one heathland sheep per hectare was initially planned, irrespective of heathland type, the cover of grasses, and food conditions during winter. It soon became clear that this general directive was inadequate. The following problems occurred.

- The encroachment of grasses and trees were not curbed, or these even increased.
- In winter the food availability or quality was insufficient, which led to starvation (Chapter 9), or food supplements had to be provided at very high costs.
- In the absence of favourable effects, excessive grazer densities were adopted (up to sometimes two sheep/ha).

This resulted in negative effects on the herpetofauna, entomofauna and ground-breeding birds, especially nightjar (*Caprimulgus europaeus*). Some rare and threatened plant species such as *Gentiana pneumonanthe* and *Arnica montana*, and some lichens, consequently decreased.

In the event of year-round grazing of heathland the availability of food during winter is of great importance. In many wet and humid heathlands, food quality in winter is very low on account of the poor digestibility and low protein content of, especially, *Molinia caerulea* (Bokdam and WallisDeVries, 1992). Year-round grazing often turns out to be a fiasco there, particularly in situations where humid and wet heathlands have a limited area of dry sand ridges on which *Calluna* and *Deschampsia flexuosa* form the main food source during winter. At a density of one sheep/ha over the entire heathland acreage, the grazing of these drier ridges was much too intense. Grazing with cattle or ponies at more than one head/6 ha has proved to have negative effects and often means a sharp decline in the condition of the animals. In view of this, year-round grazing of wet and humid heathland should generally be rejected, unless it is carried out with a drastically smaller number of grazers during winter or in the presence of an alternative feeding area, such as woodland (which is usually rich in *Deschampsia*), former agricultural grassland or dry heathland.

In retrospect we can now conclude that, even in drier heathlands, a stocking rate of one sheep/ha or more than one head of cattle or ponies per 3 ha is too intensive, in particular for the herpetofauna and entomofauna. On dry heaths a stocking rate of one head of cattle or ponies per 5–6 ha is feasible and year-round grazing is also possible. Besides the availability of sufficiently digestible food with adequate protein content (Chapter 9), in order to determine the appropriate stocking rate it is, of course, also important to know what effect one wishes to achieve.

In many situations a temporary form of intensive grazing may be desirable in order to restore the heathland vegetation. Particularly in areas overgrown with grasses, temporary intensive grazing with cattle or horses may be essential despite the fact that it may have a negative influence on vulnerable species. Of course it is essential that the stocking rate be reduced to the desired level once the regeneration of heather has taken place. If temporary intensive grazing is omitted, it can be predicted that characteristic heathland species will disappear in due course. To limit the negative effects of intensive grazing it is advisable to restore a heathland in parts, in a step-by-step process. Additional measures such as cutting sods to curb the grass encroachment and the accumulation of organic matter may also be beneficial.

8.3.2 Wet and humid grasslands

Grazing on humid and wet grasslands takes place on tidal flats in coastal areas, dune valleys, and on former agricultural grasslands in peat polders, river forelands and brook valleys. It mainly involves cattle grazing. The main breeds used for year-round grazing are Heck cattle (Chapter 7), Scottish Highland and Galloway cattle. In some areas Konik horses (Chapter 7), Shetland ponies and other pony breeds are also deployed. For seasonal grazing practically all breeds of beef cattle and young dairy cattle are used; with regard to the choice of breeds there are hardly any limitations because the food supply of these grasslands is plentiful during the summer season. Sheep are used comparatively little because of the risk of liver-fluke, among other reasons.

In wet and humid grasslands grazing management is mostly focused on maintaining wholly or partially open landscapes by preventing bush encroachment and by preserving and restoring nutrient-poor grasslands with high species and structural diversity. In nature reserves for waders and geese, grazing is mainly aimed at obtaining a sufficient food supply and favourable vegetation structure for a good breeding or winter-staging habitat. In river forelands, grazing may also be aimed at limiting the extent of riverine woodland and scrub: excessive flow resistance at flood levels increases the risk of inundation of the hinterland.

In areas with seasonal grazing the stocking rate is usually higher than one head of cattle/ha if encroachment of rough grasses and tall herbs is not desired. If this is possible or desirable, a density of one head of cattle/3 ha can be applied for year-round grazing.

The effectiveness of grazing on this type of terrain has varied a great deal and its success strongly depends on abiotic environmental factors. On calcareous embankments, valuable humid and nutrient-poor grassland vegetation often develops. Many rare and endangered plant species settle there – species that are normally found in humid dune slacks, such as *Orchis majalis*, *Parnassia palustris*, *Blackstonia perfoliata* and various species of sedges and rushes. This development can be perceived particularly in the enclosed inlets of the sea and in the Lauwersmeer (on the northern coast of The Netherlands). Plant communities with nature conservation interest such as the *Lolio-Potentillion anserinae* develop in situations with unstable groundwater, as on marine and alluvial clay soils. On embankments rich in lutum and organic matter the results are clearly less favourable (Figure 8.2). The grazed grassland vegetation there consists in considerable part of species from fertilized grasslands. Besides, there is often a substantial growth of tall grasses and herbs, i.e. common species such as *Cirsium arvense*, *Urtica dioica*, *Epilobium hirsutum*, *Deschampsia caespitosa* and *Festuca arundinacea*.

In the drier parts of peat polders and brook valleys, grazing often represents a successful form of management to develop former fertilized grasslands into valuable vegetation. Plant communities such as the *Lolio-Potentillion anserinae*, *Lolio-Cynosuretum* and sometimes even nutrient-poor grassland vegetation may develop here. In contrast, grazing on wet peat grassland and wet brook valley grasslands has not been successful, since development towards *Calthion palustris*, *Caricion curto-nigrae* and quagmire communities (*Caricion davallianae*) hardly occurs. These communities developed in the past as a result of haymaking. In places where these communities already exist, grazing may even create a disturbance and lead to the impoverishment of these vegetation types (Chapter 5). Due to the physically low carrying capacity of these peat soils, trampling may enhance the dominance of species such as *Juncus effusus*, *Rumex obtusifolius*, *Alopecurus geniculatus* and *Agrostis stolonifera*. This effect occurs quite often, especially in case of high stocking rates and grazing outside the growing season, when the topsoil no longer dries up. A positive asset is that such soil disturbance

Figure 8.2 In wet grassland, grazing frequently does not lead to the desired short grassland vegetation.

may sometimes prove beneficial to rare pioneer communities (*Nanocyperion* and *Montion*).

In wet grasslands where the objective is to favour waders or geese, grazing appears only to be a profitable management measure if it is combined with mowing (but see Chapter 7). In many reserves with this objective, management by mowing and then pasturing, or by grazing and then baring (Dutch *bloten*, a form of mowing in which part of the soil is laid bare) the ungrazed rough grassland and tall herbs parts has been carried out with success. On account of the encroachment of tall grasses and herbs, merely grazing will generally not lead to a dense wader population and the available food will be insufficient for most geese species. If a high stocking rate is applied to suppress the growth of rough grasses and tall herbs, too many wader nests are trampled. To get around this problem, the grazers are often put to pasture only after the breeding season has passed. As a result of the relatively high productivity of the grassland in early summer (sometimes increased by fertilization) the excessive growth of rough grasses necessitates mowing or baring.

Finally, there also is a positive side to the grazing of wet grasslands. The occurrence of rough grasses, rushes and tall herbs (e.g. *Cirsium palustre*) on grazed humid grasslands is often highly beneficial for the entomofauna and small mammals. The rough grasslands also form a good habitat for grass snakes, and birds of prey frequently forage there on the abundance of mice. Bird species feeding on insects, such as the red-backed shrike (*Lanius collurio*), also prefer these grazed grasslands.

8.3.3 Dry grasslands

Grazing is a frequently applied measure in dry grasslands such as river dunes, sandy and dry alluvial clay soils, dike slopes, calcareous grassland, grass heaths and dry former tidal sand flats of former sea inlets. In nature reserves the acreage of this type of terrain is rather limited and in many places the remnants are too small for grazing. Grazing takes place with both cattle and horses, but a considerable part of the dry grassland is also grazed with sheep. Seasonal grazing as well as year-round grazing (if the area is larger than 5 ha) is applied. River dune grasslands (including *Juniperus communis* scrub) and grasslands on dry strips of alluvial clay are often grazed, together with humid types of terrain in river forelands. Grass heaths on dry sands are mostly grazed in combination with heather vegetation. Calcareous grasslands as well as dike slopes are grazed with sheep. Both free-ranging flocks and shepherded flocks are used (Figure 8.3).

In the case of year-round grazing, breeds such as Galloway and Scottish Highland cattle, Icelandic pony, Konik horse and Shetland pony are used. Besides these, young cattle and beef cattle of more productive stock are used for seasonal grazing. Heathland sheep (Drenthe Heath, Veluwe Heath,

Figure 8.3 Small areas such as dikes can be grazed in rotation by shepherded flocks of sheep.

Kempen Heath and Mergelland) are used as well as Zeeland Milk sheep and the modern Texel. Stocking rates vary greatly. On extremely nutrient-poor sites (river dunes, inland sand dunes) densities of one head of cattle or pony per 5–15 ha may be found. If adjacent nutrient-rich or humid terrains are grazed together with dry areas, the stocking rate on the drier parts is often substantially higher and it may lead to overgrazing of the dry grasslands. On nutrient-rich soils (river basins, calcareous and loess grasslands) the stocking rate can be considerably higher. A density of one head of cattle/pony per 1–3 ha is applied in many places. Sheep on calcareous grasslands and dike slopes are kept at densities of one sheep per 0.1–1 ha. Dry soils that have been fertilized for agricultural purposes are generally grazed much more intensively (up to 20 sheep/ha) in restoration management. In addition to domestic grazers, dry grasslands are often also grazed intensively by rabbits. On river dunes this kind of grazing may be so intensive that they can outcompete horses and cattle (Bokdam, 1987).

Grazing management is usually aimed at conserving, restoring or developing nutrient-poor grasslands with high species and structural diversity.

Target plant communities consist of *Festuco-Sedetalia* and *Trifolion medii*, and communities of calcareous grassland, river basin grassland, scrub and fringes (Chapter 4). Scrub communities frequently develop. Besides rare and endangered plant communities, management by grazing also focuses on animal groups such as entomofauna, small mammals and songbirds.

The effects of management by grazing in dry grasslands may as a rule be characterized as very positive. The dominance of rough grasses, tall herbs and woods has been reduced by the grazers to acceptable levels and numerous characteristic target species have reacted positively. The most remarkable results have been achieved on calcareous grasslands, river dunes and grasslands on old alluvial clay soils. On dry, acid and oligotrophic sandy soils the results may not always be termed positive in view of acidification and eutrophication through atmospheric deposition. In areas with high grazing densities there is a negative impact on the development of scrub and tall herb vegetations and consequently on the entomofauna. Grazing in very small plots also leads to a marginal development of tall herb, scrub and woodland vegetation; in particular, the effect of grazing may turn out to be negative in cases of year-round grazing in small areas.

8.3.4 Tidal grassland and salt marshes

Grazing has been taking place for centuries on salt marshes and other coastal regions that are influenced by the sea (Bakker, 1985). In The Netherlands few areas of this type are left ungrazed. The important motive underlying grazing is to avoid the spreading of rough grasses in the drier parts of the terrain and thereby to preserve the characteristic salt marsh vegetation and communities of desalinating environments, such as the *Lolio-Potentillion anserinae*. Controlling the extent of rough grassland is also important for conserving a favourable habitat for waders and terns and good foraging areas for herbivorous birds such as geese and wigeon (*Anas penelope*).

Grazing is mainly carried out with cattle and sheep. Horses are rarely used for this purpose on account of problems with hooves and suspected risks of getting stuck in creeks. Due to the limited food supply in winter, the main system is seasonal grazing during the summer. This often concerns young Holstein-Friesian cattle and Texel sheep. Generally speaking, relatively high grazing densities of more than 1.5 head of cattle/ha or 10 sheep/ha are applied. This results in a reduction of tall vegetation with *Elymus athericus*, *Artemisia* and the like. Certain species of the *Puccinellion maritimae*, too, decrease with intensive grazing (e.g. *Plantago maritima* and *Limonium vulgare*). On the other hand, species such as *Juncus gerardi*, *Armeria maritima* and *Juncus maritimus* and other types of trampling-resistant species may benefit. The variety of vegetation structure generally increases, which in turn increases the richness of species at a small scale. In desalinating environments, keeping down the amount of rough and dense grass vegetation by grazing has a positive influence on preserving salt marsh vegetation for a longer period (Chapter 5).

Due to intensive grazing, relatively large numbers of birds' eggs are trampled. Intensive grazing may also promote embankment erosion on salt marshes. Decreasing the stocking rate on many of the areas beyond the dikes is advisable. Salt marshes that have not been grazed until now should remain ungrazed or else should be grazed at very low stocking rates to maintain habitat for species that are vulnerable to grazing.

8.3.5 Dunes

Although practically all coastal dunes were grazed until the twentieth century, it is only comparatively recently that a restricted number of dunes have been designated for management by grazing again. In the past two decades grazing has been reintroduced in a large number of dune regions, mainly with cattle and horses, and to a lesser extent with sheep. Good examples of grazed dunes are Zwanenwater and Duin en Kruidberg in Noord-Holland, Zeepeduinen and Westduinen in Zeeland, and Moksloot on the Isle of Texel.

The absence of grazing by large grazers during a long period has resulted in a sharp increase of rough areas with *Calamagrostis epigejos*, scrub and woodland. Species as *Carex arenaria* and *Rubus caesius* also have increased strongly. These increases were at the expense of the acreage of dune grasslands and communities which depend on open sand (e.g. *Tortulo-Phleetum arenarii*). Most humid dune slacks are subject to mowing management.

Both year-round and seasonal grazing are now applied in dunes. Breeds used for this purpose are mainly Shetland, Icelandic and Fjord ponies and Scottish Highland, Blonde d'Aquitaine and young Holstein-Friesian cattle. The stocking rate may vary between one head of cattle per 3–4 ha to one head of cattle/pony per 20 ha. The density is mainly determined by the presence or absence of humid dune valleys and the acreage of dense scrub. If a relatively large area of humid dune valleys is available, seasonal grazing is practised. In winter the food supply in dry dunes is low on account of the relatively poor nutritive content of species such as *Calamagrostis epigejos* and *Ammophila arenaria*. For year-round grazing, ponies are generally used at low density, although cattle sometimes perform at least as well as horses.

Despite the still limited experience with grazing in dunes, it can already be established that this management measure has a positive effect on the vegetation. The encroachment of trees and shrubs is kept low on the one hand, but on the other hand grazing of dense rough grassland vegetation may stimulate bush encroachment because of better conditions for germination and establishment of (for instance) birches. Grazing creates a greater diversity in vegetation structure and thereby promotes species richness. There is a positive development with respect to dune slack vegetation, dune heathland, dune grassland, grass-heaths and dune scrub. Pioneer vegetation (*Nanocyperion* and *Tortulo-Phleetum arenarii*) may reappear or is also enhanced. Due to better light conditions in short vegetation, vulnerable mosses and lichens develop favourably despite trampling. Vegetation dominated by *Rosa pimpinellifolia* is also strongly increasing. Notwithstanding the fact that the soil is broken up by trampling in places, grazing has not (yet) led to local sand-drifts in the dunes. Rabbit density also appears to increase under grazing.

8.4 DILEMMAS AND PROBLEMS IN THE USE OF GRAZING

Besides the effects of grazing management described above, a number of dilemmas also occur, as explained in the following sections.

8.4.1 Inadequate stocking rates in relation to the objective

In many reserves, particularly on grassland, intensive grazing is applied. As a consequence there is hardly any development of tall herbs, scrub and woodland with their associated fauna (particularly entomofauna) in these areas. This vegetation development is still given insufficient room, even in grasslands where there is no habitat objective for waders or geese. In many situations the development of extensive stands of *Cirsium arvense* or *Juncus effusus* is viewed as a problem for surrounding agricultural land. Another point is that quite often the grazing of heathlands is too intense. Especially in situations where the stocking rate is tuned to counter the spreading of grasses, the heath vegetation is grazed relatively intensively, at the expense of characteristic target species.

In contrast, former fertilized grasslands are often grazed insufficiently in proportion to their productivity and the future vegetation development that is aimed for. A low stocking rate on highly productive former agricultural land results in extensive stands of *Urtica dioica, Rumex obtusifolius* and *Cirsium* species, which may be quite persistent. Implementing a high initial stocking rate that is gradually reduced with decreasing plant production rates in time offers better opportunities for developing valuable rough grassland vegetation.

Grazing management without additional mowing measures and with a (too) low stocking rate in wader and geese territories may result in a decreasing number of breeding birds and foraging geese because of substantial senescence and encroachment of rough grasses. Extensive grazing in wader and geese sanctuaries – with the exception of greylag goose (*Anser anser*) and bean goose (*Anser fabalis*) – must be advised against.

8.4.2 Problems of rank order

Plant species and vegetation types representing a rare minority in number or acreage, compared with other species and vegetation types, may be grazed excessively despite a low overall grazing intensity. Inevitable consequences are trampling damage, overgrazing or excessive dunging, which often lead to a marked decrease of the rarer species or communities. This phenomenon can be seen in (for example) the establishment of trees or *Empetrum nigrum* in open vegetation, mesotrophic vegetation (e.g. fen vegetation) in oligotrophic heathland, or humid dune slacks in dry dunes. Rare plant species such as orchids, *Arnica montana* and *Juniperus communis* in nutrient-poor grassland and heathland are more damaged than would be expected from their relatively small numbers.

These ranking order problems generally occur in sensitive parts of grazed territory, which by their ranking order belong to the minority. It generally concerns a combination of, for instance, wet/few versus dry/many or oligotrophic/few versus eutrophic/many. The latter occurs quite often in situations where small heathlands are grazed together with a large acreage of former cultivated grounds. Ranking order problems can be solved by strengthening the weaker component in the terrain or by excluding it from grazing. Expansion of the grazed area or decreasing the stocking rate is usually not found to be adequate.

8.4.3 Grass growth and grazing

The spreading of grasses such as *Molinia caerulea* and *Deschampsia flexuosa*, and sometimes also the growth of *Rubus*, does not always show the expected decrease despite grazing. The grasses do become shorter, and so other species can establish, but the grasses are still only disappearing partially. Grass encroachment increases rather than decreases if grazing takes place with sheep.

The influence of nitrogen deposition as a result of atmospheric deposition is not removed by grazing, so that grass growth remains a threat. The balance of the competitive relationship between grasses and dwarf shrubs is only tipped in favour of the latter under more or less continuous grazing. Besides, it looks as if a number of target heathland species do not reappear despite grass reduction, because of excessive nitrogen deposition and a decrease in the acid : base ratio caused by acidification. Supplementary management measures such as cutting sods or mowing are often necessary in order to regenerate the original heathland (see also Chapter 5).

8.4.4 Transitions between open vegetation and woodland disturbed by grazing

When mowing or grazing is discontinued, a succession to woodland will be the result in most places in The Netherlands. In the intermediate stage to forest, rare and valuable communities are found. Well known examples are scrub vegetations of *Cytisus scoparius*, *Juniperus communis*, *Rubus* spp., *Salix* spp., *Prunus spinosa* and *Crataegus monogyna* and semi-open fringe communities between woodland and open vegetation. Their structural diversity is typically high.

When grazing is introduced, preservation of these transition stages is often a set objective, but despite grazing these grounds develop into woodland or they disappear. Perhaps the problem is even more general. When previously ungrazed vegetation becomes grazed, grazing initially always alters the earlier vegetation structure and thereby constitutes a disturbance (even at low stocking rate). This is expressed in the dying off of dwarf shrubs, *Molinia* tussocks, and the breakdown of accumulated litter, top soil and moss layers. This transformation is often accompanied by the occurrence of disturbance indicators (e.g. *Rumex acetosella*, *Ceratocapnos claviculata*, *Agrostis* spp., *Juncus effusus*). This period of so-called degeneration has a temporary nature and after five years of grazing the symptoms of degeneration clearly decrease. When the population of grazers continues to increase from a low initial level to the ultimately desired number, the degeneration symptoms seem to last longer. It is therefore recommended that the desired number of grazers (provided it is known) is used as soon as possible after introduction rather than gradually increasing the number of grazers.

Besides these negative impacts it must be noted that temporary degenerative processes may create favourable conditions for pioneer stages that may develop if the number of grazers is temporarily decreased. As a matter of fact, little or no research has been done so far about the effects if the numbers of grazers are fluctuating instead of constant.

8.4.5 Inability of grazing to prevent succession to woodland

Grazing with cattle and horses and, to a lesser extent, with sheep is not always successful in stopping the encroachment of birch (*Betula pendula*) and Scots pine (*Pinus sylvestris*) in nature reserves that carry the objective of keeping an open landscape. Birch and pine are not usually eaten by cattle and horses. If there is a strong regeneration of these trees the landscape will be closed or semi-open despite its being grazed. Supplementary management measures such as cutting and mowing offer a solution. The introduction of browsers such as goats and in large nature areas perhaps also red deer is possible.

8.4.6 Overestimation of the effects of grazing

Reserve managers show a tendency to overestimate the effects of grazing in comparison with the influence of other factors. Abiotic processes such as eutrophication, acidification, water drainage, changes in ecohydrological dynamics and, in the case of restoration, measures such as changing the groundwater regime, decreasing nutrient availability and cutting of sods, must generally be considered to be of greater influence on the species composition of communities than the herbivores.

Quite often, biotic processes such as the spreading of grasses in heathland are caused by desiccation and atmospheric deposition, and can only partially be stopped by grazing. In many grazed nature reserves, measures are taken with regard to the groundwater regime, cutting sods and trees and mowing. In this respect it becomes difficult to assign positive and negative effects of grazing unequivocally to the grazing that has taken place. Leaving out these supplementary measures may be a consideration from a scientific point of view, but in management they are usually necessary and cannot be omitted.

8.4.7 Insufficient variation in terrain types

In many sites with year-round grazing there is a lack of sufficiently nutritious winter food (Figure 8.4). The chief cause of this deficiency lies in the fact that there is not enough variation in terrain types in most grazing areas (WallisDeVries, 1994; Chapter 9). There is a particular shortage of terrain types that also offer sufficient and qualitatively good food during winter (e.g. peat polders, wet heathland, river forelands, wet brook valley grasslands).

This leads to additional feeding measures or temporary removal of livestock to stables. High labour costs (feeding, removing dung and production of stable fodder) and the costs of buildings often constitute an impediment to stabling. There are generally few possibilities to transfer grazers to winter grazing areas because these are too distant or too small to accommodate the entire herd. Suitable winter areas are mainly found in the drier areas

Figure 8.4 In small and homogeneous areas, grazing can be a matter of dining out ...

with sufficient evergreen grass and dwarf shrubs. For year-round grazing it is advisable to include sufficient terrain types that can be used by the grazers during the winter. Transferring grazers to winter grazing areas (a kind of transhumance) is a possibility but it should be noted that the vegetation growth is substantially less during winter. As a rule of thumb, 10 ha of nutrient-poor winter grazing land should be required for 1 ha of nutritious summer grazing ground.

8.4.8 Extra feeding, rearing and care of grazers

The health and well-being of grazers must be considered in reaching the decision for grazing management (Figure 8.5). There is a legal responsibility to care for the animals one keeps. From the perspective of effective and efficient grazing management, good health and well-being of the grazers is important because of (for example) lower costs and fewer fluctuations in stocking rate. With a good understanding of what grazers need and of the potential of a certain terrain, this responsibility can be met adequately. Regular veterinary check-ups are advisable.

When utilizing rare breeds or when a new herd is being built up, it is sensible to obtain as much young stock as one can get. A consequence of breed-

Figure 8.5 The legal responsibility for the well-being of animals may lead to practices, such as supplementary winter feeding, that are questionable from a conservation point of view.

ing of cattle and horses in order to market them is that one strives to produce marketable animals. In these situations, keeping grazing animals as such becomes an objective in itself. It implies that caring for the animals becomes a more intensive job than would be necessary simply to maintain a herd of grazers to carry out grazing management. Extra feeding and selection on the basis of productive factors may come at the expense of the effectiveness of grazing for nature reserve management.

Due to extra feeding, grazing in the terrain diminishes or the diet selection changes (e.g. excessive feeding on heather and other woody plants). The selection criteria for greater deployability and self-sustainability (hardiness, etc.) of the grazers may lose out if selection takes place with a view to increasing market value.

8.4.9 Fences

Quite often fences present unwanted barriers for wild fauna. In particular, fences to enclose sheep are a great impediment for fleeing mammals, and birds and roe deer sometimes harm themselves on barbed-wire. Fences of smooth wire (say four or five wires), with or without high-powered electric voltage, in combination with entrances and exits at various places, seem to work quite well in practice. Wire-netted fences should be provided with hatches for smaller mammals (e.g. badgers).

8.4.10 Relationship between the public and grazing

The presence of large grazers in a nature reserve brings up both positive and negative reactions from the public. To many people, spotting a grazer is an added dimension to their experience of nature, particularly when it concerns spectacular, 'wild' and primitive-looking animals, such as Scottish Highland cattle, Heck cattle or Konik horses. These large grazers form an added attraction. The deployment of flocks of sheep with a shepherd speaks to people's romantic imaginations and brings back an atmosphere of old times. Many visitors prefer a more natural form of management with animals, rather than one that uses machines, and they therefore value grazing more than mowing and cutting trees or sods.

Besides this positive appreciation there are also negative elements. The presence of dung on paths, trampling effects, flowers and trees that have been grazed down – these all sometimes give rise to criticism. Keeping grazers outdoors during winter may give the impression that the animals are unattended and suffering, and starving, which raises public concern. Furthermore, some people fear the possible aggressive behaviour of horses and cattle.

Providing information about the objectives and methods of grazing is therefore essential. Good ways of doing so include placing noticeboards at reserve entrances and frequently organizing excursions. It is important not only to inform visitors from urban areas but also to involve the local rural population in the information projects. Although accidents in confrontations with visitors and grazers very rarely happen, advising visitors about correct behaviour will minimize the chance of such accidents occurring. For instance, extra feeding with bread, walking (unleashed) dogs, approaching a cow or mare who is with calf or foal, riding a gelding or stallion in an area where other stallions are roaming free – these actions carry an extra risk and are best avoided. Grazing of city parks or nature reserves that are close to urban areas and receive large numbers of visitors brings an extra risk.

REFERENCES

Bakker, J.P. (1985) The impact of grazing on plant communities, plant populations and soil conditions on salt marshes. *Vegetatio*, **62**, 391–398.

Bieleman, J. (1987) *Boeren op het Drentse Zand; Een nieuwe visie op de 'oude' landbouw*, Hes Studia Historica, Vol. XV, Utrecht.

Bokdam, J. (1987) Foerageergedrag van jongvee in het Junner Koeland in relatie tot het voedselaanbod, in *Begrazing in de Natuur* (eds S. De Bie, W. Joenje and S.E. Van Wieren), pp. 165-186, Pudoc, Wageningen.

Bokdam, J. and WallisDeVries, M.F. (1992) Forage quality as a limiting factor for cattle grazing in isolated Dutch nature reserves. *Conservation Biology*, **6**, 399–408.

Natuurmonumenten (1993) *The Management of Nature Reserves owned by the Vereniging Natuurmonumenten. Main principles and strategies*, Vereniging Natuurmonumenten, 's Graveland.

Putman, R. J. (1986) *Grazing in Temperate Ecosystems. Large Herbivores and the Ecology of the New Forest*, Croom Helm, London.

Slim, P.A. and Oosterveld, P. (1985) Vegetation development on newly embanked sandflats in the Grevelingen (The Netherlands) under different management practices. *Vegetatio*, **62**, 407–414.

Vera, F.W.M. (1997) Metaforen voor de Wildernis: Eik, hazelaar, rund en paard. Doctoral thesis, Wageningen Agricultural University, Wageningen.

WallisDeVries, M.F. (1994) Foraging in a landscape mosaic: diet selection and performance of free-ranging cattle in heathland and riverine grassland. Doctoral thesis, Wageningen Agricultural University, Wageningen.

Part Four

Perspectives and Limitations

9

Habitat quality and the performance of large herbivores

Michiel F. WallisDeVries
Tropical Nature Conservation and Vertebrate Ecology Group, Department of Environmental Sciences, Wageningen Agricultural University, Bornsesteeg 69, 6708 PD Wageningen, The Netherlands

9.1 INTRODUCTION

The provision of habitat of sufficient quality for wildlife or free-ranging herbivores is an essential step in both wildlife and range management. It is well known that quality differences between areas or habitat types affect the performance of animals. In poor ranges wildlife populations are often said to be limited by environmental conditions. Such a claim necessitates a good definition of the word 'limiting' and of the nature of the limitation. Here, limiting is used as in the definition of limiting factor: 'the environmental factor that is of predominant importance in restricting the size of a population' (*Webster's 3rd New International Dictionary*, 1981). The quality of a certain habitat can now be described as a function of one or several limiting factors acting upon a population.

This chapter will review the most important limiting factors determining habitat quality for large herbivores. In discussing limitations it will be necessary to distinguish between individuals and a population as a whole. The first symptoms of habitat limitations will always reveal themselves through the performance of the individual. An analysis of factors determining animal condition therefore deserves special emphasis. However, a limitation at the individual level does not automatically have an impact at the population level, as the inferior performance or failure of an individual may be compensated for by other, more successful conspecifics. Only in the absence of this compensation, and when individuals are suffering from lowered reproduction and survival, will a limitation on individual performance affect the dynamics of a population.

Grazing and Conservation Management. Edited by M.F. WallisDeVries, J.P. Bakker and S.E. Van Wieren. Published in 1998 by Kluwer Academic Publishers, Dordrecht. ISBN 0 412 47520 0.

If a limitation is experienced as severe, managers may choose to interfere. Habitat quality may be improved directly by altering habitat structure or by raising forage production. Alternatively, management actions may aim at an amelioration of animal performance by giving food supplements and shelter or by controlling the numbers of competitors. In all cases that involve nature conservation, such practices can have a large impact on the structure and diversity of a conservation area. An overview of management options in various situations commonly encountered will therefore complete these considerations of habitat quality.

9.2 CONDITION CYCLES IN SEASONAL ENVIRONMENTS

Large herbivores in seasonal environments typically experience an annual cycle of body condition in close relation with the changes in forage supply. Periods of plant growth are accompanied by weight gain and accumulation of energy reserves as fat. Conversely, animals lose body reserves and, hence, weight during periods of plant dormancy. In temperate regions weight loss occurs mainly over winter, while in tropical regions the unfavourable period is the dry season.

The effectively available forage is a combination of both quantity and quality. These are a function of various abiotic factors (section 9.3) and plant characteristics, especially cell structure and chemical composition. An overall assessment of forage quality may be given by considering three constituents: cell wall, lignin and crude protein (Van Soest, 1994). Cell wall content and lignification are negatively related to forage digestibility and intake, whereas a positive relation exists for crude protein content; the relation between forage quality and intake only applies to ruminants, not to non-ruminants such as equids (Duncan, 1992). Forage quality varies according to the season and the species concerned, but general distinctions between different forage classes are nonetheless apparent (Table 9.1). Grasses have a relatively high cell wall content but a low degree of lignification. The reverse is true for browse, while herbaceous dicots are intermediate. Crude protein content is generally highest in herbs and grasses. Mosses and lichens are of poor quality in all respects.

9.2.1 Seasonal variation in forage supply

Changes in the forage supply are determined to a large extent by the pattern of plant growth. Grasses, which represent the staple food of most grazers and intermediate feeders, show a typical production sequence of leaf production, formation and elongation of flowering stems, seed-set and senescence (Ulyatt, 1981; Deinum, 1984). Biomass accumulation occurs up to the stage of seed-setting. It is slowed down by litter fall, which starts as soon as the first leaves die off after having reached their full size. The process of

Table 9.1 Average range of three main forage quality determinants in different forage classes from the temperate zone (in % of dry matter) (Chapin *et al.*, 1980; Prins, 1981; Owen-Smith, 1982; Demment and Van Soest, 1985; Bokdam and WallisDeVries, 1992; Prop and Vulink, 1992; Van Wieren, 1996)

Forage class	Cell wall	Lignin	Crude protein
Grasses	49–76	3–8	6.2–32.5
Herbs	35–57	6–15	8.1–35.6
Browse	27–58	7–22	6.5–20.6
Mosses	64–76	5–25	2.5–12.9
Lichens	83	15	1.3–4.4

maturation and senescence is characterized by reallocation of soluble cell contents to young, growing parts, lignification of cell wall and an increase in cell wall content (Norton, 1982). As a result the digestibility of plant material declines substantially from young leaf to mature leaf to flowering stem (Hacker and Minson, 1981; Ulyatt, 1981).

The process of plant maturation is accompanied by a reduction in the concentration of protein and minerals. Consequently, there is a positive overall relationship between the proportion of green plant material, digestibility, and protein and mineral content (Owen-Smith, 1982; Ben-Shahar, 1993) (Table 9.2). It is not surprising, then, that grazing herbivores tend to concentrate on short swards (Ydenberg and Prins, 1981; McNaughton, 1984; Goodson *et al.*, 1991; WallisDeVries and Daleboudt, 1994) and in the more fertile and productive habitats (McNaughton, 1988, 1990; Seagle and McNaughton, 1992; WallisDeVries and Schippers, 1994). By repeated grazing grass maturation is prevented and its regrowth is stimulated, which maintains forage quality at a relatively stable, high level. Yet, as shown in Table 9.2, not all nutrient levels are as closely correlated and, moreover, such correlations are only valid within a limited range of species in a given habitat type. Nutrient concentrations at a certain plant growth stage can vary significantly between species (Norton, 1982; Ben-Shahar, 1993) and across habitat types (Bokdam and WallisDeVries, 1992). Because the animals need to balance the requirements of an array of essential nutrients, this may generate more complex patterns of selectivity by the foraging herbivore than if a single correlation between the different aspects of forage quality were to exist (Belovsky, 1978; Kreulen, 1985; WallisDeVries and Schippers, 1994; Prins, 1996).

With the approach of winter or the dry season and the associated stress through cold and reduced solar irradiation or through drought, grass growth diminishes and senescence of above-ground biomass gains importance. Herbivores then face a depletion of high quality forage and must rely

Table 9.2 Pearson correlation coefficients between different parameters for forage quality in year-round herbage samples from *Agrostis capillaris/Festuca rubra* grassland (N = 100) (after WallisDeVries, 1989)

	OMD	K	P	Mg	Ca	Na	N
Green	.78	.84	.47	.56	.56	.57	.27
OMD	X	.81	.57	.54	.56	.55	.60
K		X	.62	.67	.64	.53	.49
P			X	.59	.50	.26	.53
Mg				X	.76	.48	.24
Ca					X	.46	.25
Na						X	.22

Green = green dry matter fraction.
OMD = organic matter digestibility.
All values are significantly different from zero at $P < 0.05$.

increasingly on lower quality standing hay or switch to other resources. Below-ground storage organs of plants (roots, rhizomes, tubers and bulbs) are generally not exploited to a great extent, except by wild boar (*Sus scrofa*). Herbaceous dicots rarely provide a good alternative, since they usually senesce earlier than grasses and often constitute only a minor proportion of the vegetation. They do, however, represent a significant proportion of the diet of browsers, and occasionally also intermediate feeders and grazers, during the growing season (Van Dyne *et al.*, 1980; Putman, 1986; Campan *et al.*, 1991; Ralphs and Pfister, 1992). At the end of the season of plant growth, seed mast can offer a rich source of food in good years, but mast production is known to fluctuate widely between years. In temperate deciduous forests, cervids and wild boar rely heavily on acorns and beech nuts during autumn and early winter (Putman, 1986; Feldhamer *et al.*, 1989; Maublanc *et al.*, 1991; Groot Bruinderink *et al.*, 1994). In oceanic climates, half-shrubs, such as bramble (*Rubus* spp.) and wintergreen ivy (*Hedera helix*) are significant sources of winter food, especially to browsers (Putman, 1986; Maublanc *et al.*, 1991).

Mosses and lichens present a relatively poor quality food resource (Hyvärinen *et al.*, 1977; Chapin *et al.*, 1980; Larter and Gates, 1991; but see Westermarck and Kurkela, 1980; Prins, 1981). Lichens and, to a lesser degree, mosses (Prins, 1981) are only important to herbivores in arctic and boreal climates, especially reindeer and caribou (*Rangifer tarandus platyrhynchus* resp. *R. t. pearyi*) (e.g. Kelsall, 1968; Skogland, 1985) but sometimes also bison (*Bison bison*) (Larter and Gates, 1991) or white-tailed deer (*Odocoileus virginianus*) (Gray and Servello, 1995). When available, fungi frequently represent a small but possibly underestimated food resource (Kelsall, 1968; Campan *et al.*, 1991; Launchbaugh and Urness, 1992; Strandberg and Knudsen, 1994). In poor environments they offer a rich source of protein and minerals (Launchbaugh and Urness, 1992; WallisDeVries, 1996). Ferns could be another neglected type of forage. Jackson (cited in Putman, 1986) found that they made up 5% of the diet of fallow deer (*Dama dama*) during winter and Hjeljord (1973) reported that the main winter forage of mountain goat (*Oreamnos americanus*) consisted of rhizomes and petioles of *Athyrium filix-femina*.

Woody browse (including dwarf shrubs) is the staple food for most browser species and remains the main alternative food source to grasses for many other large herbivores as well (Van Dyne *et al.*, 1980). Browse production shows a pronounced, relatively short growth peak. In many temperate tree and shrub species a second cohort of new shoots is also produced in the summer. As in most plants, browse quality declines rapidly with the advancing season. However, in contrast to grasses, this is not due to an increase of cell wall contents (these remain at a comparatively low level, Table 9.1), but rather to a rising concentration of lignin and secondary plant compounds such as tannins or resins that act as a defence against herbivory (Bryant *et al.*, 1992; Van Wieren, 1996). Browsing, like grazing, does stimulate regrowth but the regeneration capacity of woody plants is less developed than in grasses. Repeated browsing therefore results in growth reduction, and lowered levels of nutrients and, apparently, defence (Bryant *et al.*, 1992; Haukioja and Lehtilä, 1992). Although the supply of browse is reduced and of lesser quality in the dormant season, twigs of both coniferous and deciduous trees and shrubs remain a significant, though inferior, source of food for many herbivores even then (Van Dyne *et al.*, 1980), especially the browser species such as moose (*Alces alces*) and the smaller deer (Haukioja and Lehtilä, 1992; Gray and Servello, 1995). A factor contributing to the intensive use of browse in regions or periods with heavy snow fall is its relatively good accessibility.

9.2.2 Seasonal cycles in body condition

The seasonal cycle in the body weight of herbivores is closely correlated with the growth cycle in plants. Livestock on rangelands frequently lose 15–25% of their peak body weight, and even up to 30%, which has been termed a critical limit (McDowell, 1985; Holechek and Herbel, 1986). In general the larger species will be able to draw longer on their body reserves (Lindstedt and Boyce, 1985). Under favourable conditions weight losses are rapidly compensated for when plant growth starts anew. In wild herbivores from temperate environments, seasonal weight and condition cycles have been described in detail for a growing number of species, especially cervids (Mitchell *et al.*, 1976; Adamczewski *et al.*, 1987; Tyler, 1987; Leader-Williams, 1988; Cederlund *et al.*, 1991; Parker *et al.*, 1993; Gerhart *et al.*, 1996; Hewison *et al.*, 1996). For free-ranging domestic animals, seasonal changes in body condition have been investigated in horses (*Equus ferus* forma *caballus*) (Duncan, 1992) (Figure 9.1), cattle (*Bos taurus*) (Van Wieren, 1992; WallisDeVries, 1994, 1996) and Soay sheep (*Ovis aries*) (Grubb, 1974).

The condition cycle can be linked to a deterioration in forage digestibility over the dormant season which causes a reduced rate of food passage and thus a decline of metabolizable energy intake rate (Van Soest, 1994; Van Wieren, 1992). Non-ruminants, such as equids, are able to maintain higher passage rates than ruminants, but still show a reduced daily energy intake with declining digestibility (Weston and Poppi, 1987; Duncan, 1992).

Reviews by Young (1987) and Loudon (1994) have shown that there is an important influence of photoperiod on voluntary intake, with a decline in appetite under short days, i.e. during the winter period. It has further been found in a number of studies that voluntary intake is closely related to basal metabolic activity (Vercoe and Frisch, 1982; Webster, 1985; Ketelaars and Tolkamp, 1991). The cycle in appetite could therefore arise from a photoperiodic cycle in basal metabolism (Young, 1987; Ketelaars and Tolkamp, 1991). An annual metabolic cycle explains the observations of low appetite during winter even if rumen fill does not appear to be limiting (Van Wieren, 1992). Maintenance requirements present the largest share of energy costs. It can thus be speculated that herbivores facing a seasonal drop in forage quality below maintenance requirements will gain a decisive advantage in fitness by lowering their metabolic activity (which also reduces foraging costs to satisfy requirements). In this way they can minimize an inevitable loss of condition.

Figure 9.1 Resource limitation is demonstrated by the low condition score of this Konik horse.

The advantage of low metabolic activity under circumstances of shortage in quality food is lost in domestic animals, which are raised in continuously productive environments. The apparently less pronounced photosensitive variation in domestic sheep and cattle in comparison with wild ungulates could result from a different selection pressure for maintaining this cycle (Young, 1987). As a consequence, domestic animals might be less adapted to 'life in the wild' than their wild counterparts. This speculation remains largely unconfirmed by solid data, however. Even the supposedly superior performance of reputed hardy cattle breeds relative to more productive breeds under free-ranging conditions still remains to be established. Under mild climatic conditions the relevance of breed differences for animal performance seems of minor importance (WallisDeVries, 1994).

9.3 HABITAT INFLUENCES ON INDIVIDUAL PERFORMANCE

Superimposed on the annual cycle in plant production and phenology, additional variation in habitat quality is caused by environmental heterogeneity. Availability of food resources is of major importance for the condition of large herbivores. Habitats may differ in energy resources depending on forage productivity and digestibility, but variation in mineral availability according to substrate and topography can also determine habitat quality to a large extent. Furthermore, other habitat factors, unrelated to food resources, such as the presence of shelter and the accessibility of the terrain may be significant.

9.3.1 Energy resources

The energy supply of a certain habitat is determined by the absolute amount and the quality of available plant biomass. Not all biomass is available and not all available biomass can be regarded as adequate food. The notion that the quality of the available standing crop is of crucial importance in limiting energy resources and determining animal survival was first advanced by White (1978). The inverse relation between forage quantity and quality led Hobbs and Swift (1985) to develop a method for assessing the potential availability of adequate forage. They introduced a threshold value for diet quality which any mix of dietary components must meet to satisfy metabolic requirements. When minerals and secondary plant compounds are not limiting, the energy value of forage is mainly determined by the contents of protein and cell wall (Van Soest, 1994). Energy resources become limiting when the quantity of forage above the threshold level falls below a critical level of demand. These limitations will occur at high population densities (section 9.4) or in habitat with low quality forage. Energy shortage through insufficient forage quality typically occurs in the dormant season. It has been shown to cause significant mortality in various wild herbivore populations towards the end of winter (Skogland, 1983, 1985; Leader-Williams, 1988; Mitchell *et al.*, 1973) or the dry season (Sinclair, 1977; Sinclair and Norton-Griffiths, 1979; Owen-Smith, 1990).

Habitat factors are important in determining forage quality and abundance. Energy shortage is more likely to occur in plant communities with die-back during the dormant season (Figure 9.2). Examples from western Europe are reedlands dominated by *Phragmites communis*, tall grass vegetation with *Calamagrostis epigejos* from sandy coastal areas and wet *Molinia caerulea–Erica tetralix* grass-heaths (Drost and Muis, 1988; Vulink and Drost, 1991; Bokdam and WallisDeVries, 1992). In the absence of mast, deciduous forests also present a poor environment during winter (Groot Bruinderink *et al.*, 1994; Feldhamer *et al.*, 1989). Perzanowski (1978) and Drozdz (1979) have argued that roe deer (*Capreolus capreolus*) require wintergreen understorey plants to survive the winter in deciduous forests. Similarly, Gray and Servello (1995) conclude from feeding trials that white-tailed deer have to rely on a source of more digestible forage in addition to a diet of deciduous browse to get through the winter. Even at low densities, the effects of a seasonal shortage in energy will be more severe in the less productive environments where body reserves are depleted more rapidly (WallisDeVries, 1996).

Maintenance requirements for the dietary energy content vary between species, environmental conditions, condition and reproductive status (Robbins, 1993). The higher digestive capacity of large species allows them

Figure 9.2 In productive riverine landscape types, food quality may drop to low levels in winter, resulting in poor foraging conditions for herbivores.

to reach maintenance requirements at a relatively low dietary digestibility (Illius and Gordon, 1991). It can be calculated from energy requirements and observations on food intake that an average 400–500 kg bovine requires about 50–52% digestible organic matter in the diet (dry matter basis) for maintenance (after WallisDeVries and Schippers, 1994). Large non-ruminants, such as equids, may cope with even lower quality forage (Duncan, 1992; Van Wieren, 1992). Smaller-bodied herbivores will always require a higher dietary digestibility. Crude protein content (CP, estimated as 6.25 × the percentage nitrogen) is often used as a measure for the energy value of forage, because protein is essential in the digestion of plant material. Rough estimates of minimal CP concentrations in the diet to ensure microbial nitrogen requirements for cell wall digestion are 7.2% CP for cattle and sheep (after Agricultural Research Council: ARC, 1980) and 7.0% for horses (after Jarrige and Martin-Rosset, 1984; Duncan, 1992). In contrast to energy, the dietary protein concentrations required for maintenance do seem comparable in small and large species (Prins, 1996; Robbins, 1993).

9.3.2 Mineral resources

The importance of minerals in the nutrition of herbivores has been given much attention in livestock production (ARC, 1980; Gartner *et al.*, 1980; Underwood, 1981; McDowell, 1985). In rangelands and unfertilized pastures mineral deficiencies frequently lead to lowered animal production (meat and milk), reduced reproduction and raised mortality. Mortality can be especially high in juvenile animals, which have high requirements and virtually no body reserves.

At least 22 essential minerals have been identified that herbivores need to satisfy their metabolic requirements (Underwood, 1981). The major share of mineral requirements is taken up by seven so-called macro-elements (Ca, P, K, Na, Cl, Mg and S), but numerous trace elements are also required (Fe, I, Zn, Cu, Mn, Co, Mo, Se, Cr, Sn, V, F, Si, Ni and As).

The diagnosis for a certain deficiency is not always unambiguous, as similar symptoms may be related to different deficiencies. General signs of deteriorating condition (loss of appetite, reduced growth and milk production, weight loss, and lowered reproduction rate) are common to many deficiencies (Table 9.3). The identification of the lacking mineral rests on either one or preferably a combination of three approaches: the assessment of the level of mineral body reserves in comparison with a known standard; the biochemical identification of a specific pathological symptom; or measurement of the effect of supplementation with a certain mineral on animal performance (Suttle, 1986).

Although only animal measurements can adequately prove a suspected deficiency, a survey of mineral concentrations in the forage may help to predict possible deficiencies. However, the standards that are said to indicate

Table 9.3 Pathological symptoms for the most common mineral deficiencies in free-ranging large herbivores (after Underwood, 1981; McDowell, 1985)

Deficient element	Pathological symptoms
Sodium (Na)	General loss of condition, craving for salt, pica behaviour, lowered Na/K ration in saliva
Phosphorus/calcium (P/Ca)	General loss of condition, stiffness, pica behaviour, reduced blood serum levels, impaired bone mineralization and bone resorption leading to rickets and osteomalacia (fragile bones)
Magnesium (Mg)	Stiffness, hypersalivation, excitable behaviour, reduced levels in blood serum and urine, grass tetany (hypomagnesaemia)
Copper (Cu)	General loss of condition, loss of hair colour, stiffness, diarrhoea, anaemia, reduced levels in blood and liver, bone resorption leading to rickets and osteomalacia (fragile bones), myocardial fibrosis (heart failure), enzootic ataxia or swayback (neural disease)
Selenium (Se)	Reduced levels in blood serum, liver and kidney, muscular dystrophy (white muscle disease) aggravated by stress

adequate minimum levels vary considerably between authorities (e.g. ARC, 1980; Little, 1982; McDowell, 1985). The more conservative maintenance standards for macro-elements can often be judged as excessive. Aside from the application of a safety margin, this overestimation is largely due to the practice of assuming a standard ration with a relatively high energy concentration and therefore a modest intake at maintenance. Under field conditions, however, food intake will often be higher and minimum forage concentrations of minerals will then be lower. Table 9.4 shows how maintenance levels of Na, P and Ca for cattle are altered by assuming an intake that is more appropriate under field conditions. The new estimates appear more realistic (see also Morris and Gartner, 1971; Little, 1982; McLean and Ternouth, 1994). Maintenance requirements for phosphorus for other species yield similar results. In feeding experiments, Grasman and Hellgren (1993) estimated the dietary phosphorus requirements of white-tailed deer for maintenance at 0.14% P, and this value is also found for horses when assuming daily requirements following Jarrige and Martin-Rosset (1984) at an intake level of 2.2% body weight. In contrast, it should be noted that horses appear to have much higher Na requirements than ruminants (estimated at 0.12% Na for maintenance; after Jarrige and Martin-Rosset, 1984).

Selection at the individual level may increase the efficiency of mineral utilization and lead to a better performance in poor environments. There is evidence for genetic variation in the utilization of minerals (ARC, 1980; Little,

Table 9.4 Dietary quality requirements (in % of dry matter) for sodium, phosphorous and calcium at maintenance and at 0.5 kg gain/day or a 400 kg steer as recommended by ARC (1980) and as calculated for expected dry matter intake (DMI) under free-ranging conditions (intake estimates after WallisdeVries, 1994, using ARC assumptions for total daily requirements)

	Maintenance		0.5 kg gain/day	
	ARC	Free-ranging	ARC	Free-ranging
% Na	0.080	0.038	0.075	0.046
% P	0.22	0.10	0.28	0.17
% Ca	0.25	0.12	0.36	0.22
DMI (g/kg body weight)	9.5	20.0	13.0	21.0

1982; Suttle, 1986; Odenya *et al.*, 1992). The potential for selection thus exists. Yet it remains unknown whether individuals from populations in poor environments are actually selected for improved mineral utilization.

(a) Evidence for mineral deficiencies in free-ranging large herbivores

Under free-ranging conditions large herbivores risk deficiencies for only a minority of the essential minerals. On commercially productive rangelands the potential for deficiency in sodium is the most widespread, followed by phosphorus, copper, cobalt and selenium (in order of decreasing importance) (McDowell, 1985). Iodine deficiency did occur worldwide in domestic animals and humans from regions remote from the sea and geologically young mountain ranges or recently glaciated areas, but, like sodium deficiency, it has been virtually eliminated by the provision of mineral blocks. Deficiencies of these minerals under free-ranging conditions are often area-specific, i.e. related to the geological substrate, soil type and climate. Although it is difficult to define the environmental conditions associated with an increased risk for mineral deficiency, the following generalization may be attempted: most deficiencies are associated with areas of high rainfall and with soils on igneous rocks or their sandy derivatives. Conversely, deficiencies are less likely on the finer sedimentary substrates under climatically drier conditions.

Confirmed reports of mineral deficiencies affecting the condition, reproduction and survival of wild herbivores are much rarer than for livestock. Table 9.5 gives an overview of documented cases, including a number of studies of cattle in Dutch nature reserves. The higher frequency of occurrence of deficiencies in livestock than in wildlife is probably due to a combination of factors. Firstly, the number of studies on this matter has been much greater in livestock than in wildlife. This has probably led to an under-representation of investigations on deficiencies in the latter group.

Unfortunately, even the status of reported deficiencies often remains uncertain, mainly because of unknown interactions with other factors. Secondly, selection pressure for an efficient use of scarce nutrients is probably higher in wildlife than in livestock, which are more selected for high production under relatively optimal conditions. Thirdly, wildlife is not subject to the rigours of commercial production, which puts an extra pressure on livestock performance. Finally, and perhaps most important of all, wild herbivores mostly occur at much lower densities and in more diverse environments than domestic herbivores. This will increase their chance of finding a suitable nutrient resource and thus prevent deficiencies. Opportunities for habitat selection will then lead to the avoidance of unfavourable habitat.

Whatever the relative importance of these factors, mineral deficiencies in wild and free-ranging domestic herbivores appear largely limited to four elements: sodium, phosphorus, copper and selenium (Tables 9.3 and 9.5). Robbins (1993) reported that cobalt deficiency seems absent in wildlife in regions where livestock is susceptible to deficiency. Likewise, iodine deficiency also appears restricted to humans and domestic animals: Robbins (1993) mentions only one confirmed case of goitre in white-tailed deer.

(b) Calcium and magnesium

Deficiencies in calcium and magnesium in wild herbivores can be considered unlikely. A lack of calcium is usually linked to a lack of phosphorus, since both minerals play a prominent role in bone formation. Calcium deficiency is less common, due to its relatively high concentration in plants, but its availability may be reduced by the formation of calcium oxalates (Gartner *et al.*, 1980; McDowell, 1985). The suggestion of calcium deficiency by Freeland and Choquenot (1990) cannot be substantiated from their data because of confounding effects of other dietary changes with respect to the control group (increased fibre content, decreased phosphorus content). Likewise, the observations by Hyvärinen *et al.* (1977) point to multiple deficiencies and general malnutrition, which precludes identification of the true limiting factor.

Magnesium deficiency can be considered a greater hazard than calcium deficiency, because available body reserves are quickly depleted and the resulting grass tetany can be lethal in a matter of hours (McDowell, 1985). However, critical conditions generally arise only in association with the application of chemical fertilizers. These raise the concentrations of crude protein and potassium in young herbage to excessive levels and thus cause a drastic reduction in magnesium absorption. This has led to mortality in Highland cattle grazing on a newly established conservation area on former agricultural pasture (Van der Ouderaa, 1991). The grass tetany reported by Hyvärinen *et al.* (1977) for reindeer was probably a combined effect of severe malnutrition and the appearance of a flush of high quality forage after snow melt. Jones and Hanson (1985) claimed that the use of mineral licks is a remedy against grass tetany in wild North American ungulates, yet their data do not reveal whether natural licks truly are of vital importance to prevent the disorder.

Table 9.5 Studies presenting evidence of mineral deficiencies in wild ungulates and free-ranging livestock

Element	Animal species	Region	Type of evidence	Reference
Sodium (Na)	Cattle/sheep	Worldwide	F C P M R ES/CG	McDowell, 1985
	Cattle	The Netherlands	F C P CG	WallisDeVries, 1996
	Red deer	Scotland	F C ES/CG	Blaxter *et al.*, 1974
Phosphorus (P)	Cattle/sheep	Worldwide	F C P M R ES/CG	McDowell, 1985
	Cattle	The Netherlands	F C P CG	WallisDeVries, 1996
	Cattle	The Netherlands	F C P	Van Wieren, 1988
	Reindeer	South Georgia	C P M	Leader-Williams, 1988
	Reindeer	Finland	C P CG I?	Hyvärinen *et al.*, 1977
	Donkey[1]	Australia	F C M R CG I?	Freeland and Choquenot, 1990
Calcium (Ca)	Reindeer	Finland	C P CG I?	Hyvärinen *et al.*, 1977
	Donkey[1]	Australia	F C M R CG I?	Freeland and Choquenot, 1990
Magnesium (Mg)	Cattle/sheep	Worldwide	F C P M R ES/CG	McDowell, 1985
	Cattle	The Netherlands	F P M ES	Van der Ouderaa, 1991
	Reindeer	Finland	F C P M CG I?	Hyvärinen *et al.*, 1977
Copper (Cu)	Cattle/sheep	Worldwide	F C P M R ES/CG	McDowell, 1985
	Cattle	The Netherlands	F C P ES	Van der Ouderaa, 1991
	Bontebok[2]	South Africa	F C P M	Zumpt and Heine, 1978[3]
	Blesbok[4]	South Africa	C P CG	Bigalke and Van Hensbergen, 1992
	Moose	Sweden	C P M	Frank *et al.*, 1994
	Elk	California	F C P M R ES	Gogan *et al.*, 1989
	Reindeer	Finland	C P CG I?	Hyvärinen *et al.*, 1977
Selenium (Se)	Cattle/sheep	Worldwide	F C P M R ES/CG	McDowell, 1985
	Mountain goat	British Columbia	F C P M	Hebert and Cowan, 1971
	Pronghorn	Idaho	P M?	Stoszek *et al.*, 1980[3]
	Black-tailed deer	California	C M ES/CG	Flueck, 1994
	Reindeer	Finland	F C P M CG	Kurkela, 1980

[1] *Equus asinus*
[2] *Damaliscus d. dorcas*
[3] Cited in Robbins (1993)
[4] *Damaliscus dorcas phillipsi*
Abbreviation used for type of evidence: F, low forage concentration; C, low condition status; P, pathological symptoms; M, raised mortality; R, lowered reproduction; CG, comparison with control group; ES, confirmation by experimental supplementation; I?, doubts because of possible interactions

(c) *Sodium*

Sodium is generally considered to be the most important limiting mineral nutrient for free-ranging herbivores. It is the only mineral that mammals recognize directly by taste and for which they show a distinct craving when it is in short supply (Denton, 1982; McDowell, 1985). The supply of sodium in forage is often limited, perhaps the more so because it does not appear to be an essential nutrient for terrestrial plants. Freeland *et al.* (1985) even suggested that plants may defend themselves against herbivory by causing sodium depletion through the deleterious effect of consumed plant tannins or other allelochemicals. It may therefore sound surprising that no clear evidence of sodium shortage has been reported from wild large herbivores. The only well documented cases concern domestic animals (Table 9.5). Blaxter *et al.* (1974) found a deficient sodium status, particularly during the growing season, in unsupplemented farmed red deer (*Cervus elaphus*) grazing hill pasture and heather moor in Scotland. However, no further consequences for body condition were reported. In indoor experiments Staaland *et al.* (1980) provided support for a possible occurrence of sodium limitation in free-ranging reindeer.

In contrast to the lack of evidence for sodium limiting the performance of wild herbivores, there is a wealth of information showing that large herbivores react and adapt to the low availability of this element (Figure 9.3). The extensive use of salt licks by ungulates worldwide (reviewed by Denton, 1982; Jones and Hanson, 1985; Robbins, 1993) can often be explained as a behaviourial response to obtain the necessary sodium lacking in available forage (Tracy and McNaughton, 1995). Similarly, Belovsky (1978, 1981) and Fraser *et al.* (1984) argued convincingly that North American moose will forage on aquatic plants to satisfy sodium requirements. The feeding by red deer from Rhum on seaweed (Clutton-Brock *et al.*, 1982) and the drinking of seawater and grazing by reindeer on coastal mudflats (Staaland *et al.*, 1980) might serve the same purpose. Belovsky (1981) has suggested that the availability of sodium from aquatic plants acts as one of the factors limiting population size of moose at Isle Royale, Michigan.

Physiological adjustments (increased aldosterone production, enlargement of adrenal zona glomerulosa and, in ruminants, a lowered Na : K ratio in saliva) provide another mechanism to counter sodium shortage (Blair-West *et al.*, 1968; Weeks and Kirkpatrick, 1976; Denton, 1982). Whether these cases present successful adaptations or reveal a sodium stress limiting herbivore populations is still a matter of debate requiring further experiments.

(d) Phosphorus

Impaired bone mineralization, bone breakage and lowered P concentrations in serum are the main indicative signs of P deficiency. Evidence for P deficiency in large herbivores concerns mainly cattle (McDowell, 1985; Van Wieren, 1988; WallisDeVries, 1996). It is mostly found on soils with low fertility, but occasionally also on soils where complexation of P with other minerals (Fe, Al, Ca and Mg) can drastically reduce its availability (Underwood, 1981; McDowell, 1985). P deficiency remains to be clearly established for wild herbivores. As pointed out above, the data presented by Hyvärinen *et al.* (1977) for reindeer and by Freeland and Choquenot (1990) (for donkeys) are inconclusive because of other sources of malnutrition.

Figure 9.3 A steer feeding on *Juncus bulbosus*, an aquatic plant with high sodium levels. Sodium is one of the few minerals for which herbivores possess a specific appetite.

Leader-Williams (1988) studied the introduced population of reindeer on South Georgia and found that mandibular swellings, an infection with an associated increase in frequency of jaw lesions and dental abnormalities, resulted in inferior condition and significant mortality. The disease showed resemblance to actinomycosis ('lumpy jaw') but its precise nature could not be established. The finding of increased bone porosity and a lower dietary ash content in affected animals suggest the possibility of P deficiency. A relatively high incidence of mandibular lesions (up to 7% of individuals killed by hunters) was also recorded in the large Western Arctic caribou herd of Alaska, but a speculated relationship with range quality was not investigated (Doerr and Dieterich, 1979). Similarly, osteoporotic skull lesions found in 32% of recovered moose skulls from Isle Royale, Michigan (Hindelang and Peterson, 1996) also await confirmation as a symptom of P deficiency.

Studies of Scottish red deer, which inhabit an environment notoriously poor in minerals, have pointed to mineral deficiency as a mortality factor (Mitchell *et al.*, 1973). A natural experiment on the Isle of Rhum with red deer foraging on grassland communities fertilized by gull droppings revealed an increased lifetime reproductive success of hinds using these communities (Iason *et al.*, 1986). The droppings increased both P and N availability, but information from comparable habitat would indicate P as the more limiting nutrient (Bokdam and WallisDeVries, 1992; WallisDeVries, 1996). Murray

(1995) has suggested that the long-distance migration by wildebeest (*Connochaetes taurinus*) to the short grasslands in the Serengeti, Tanzania, may be partly driven by the P requirements of lactating females.

One of the striking symptoms of P deficiency is pica behaviour (the ingestion of non-forage items), and osteophagia (the chewing of bones or antlers) in particular. Pica behaviour is also known as a result of Na deficiency (Denton, 1982; McDowell, 1985) and may be directed at a variety of objects, such as soil, stones, wood, plastic, etc. The consumption of soil (geophagia) has also been described in relation to other nutritional disorders, such as acidosis and trace element deficiency (Kreulen, 1985; McDowell, 1985). Chewing on stones, wood and even plastic can be interpreted as a way of stimulating the flow of saliva (which contains high amounts of Na and P) to relieve rumen microbes from mineral shortage and thus maintain their digestive capacity (WallisDeVries, 1996). However, the chewing and eating of bones and antlers should be seen as a specific response to P deficiency (Green, 1925; Denton, 1982; Blair-West *et al.*, 1989; WallisDeVries, 1996). Osteophagia has been recorded worldwide from a variety of ungulates in poor environments (Sutcliffe, 1977; McDowell, 1985; Robbins, 1993). In extreme cases it results in carnivory (Green, 1925; Furness, 1988; WallisDeVries, 1996) (Figure 9.4). The attractant in bone has been identified as an olfactory cue in the form of an organic constituent of ageing bone (Blair-West *et al.*, 1989). The frequent occurrence of osteophagia in the face of P deficiency and the high P reward from bone (8% P in dry matter) strongly suggest that osteophagia is not only a symptom of shortage but also a successful way of countering it. If this is true, carcasses of large mammals can be expected to play an important role in maintaining the phosphorus balance of large herbivores in P-deficient areas.

(e) Copper

In contrast to most other mineral deficiencies occurring in large herbivores, a lack of copper is not directly related to the level of soil fertility. Copper deficiency occurs both on poor acid soils and on richer alkaline soils. An important factor explaining this anomaly is the reduced absorption of copper in the presence of notably molybdenum and sulphur, but also cadmium, iron and ingested soil (ARC, 1980; McDowell, 1985). Copper deficiency may thus either be the result of a simple copper scarcity in the environment (especially on acid soils of granitic parent material) or a lowered availability of copper through complexation with other minerals. As a guideline, McDowell (1985) gives forage concentrations of < 5 ppm Cu at a Cu : Mo ratio > 2 as an indication of simple Cu deficiency, and 5–15 ppm Cu and a Cu : Mo ratio < 2 (or > 20 ppm Mo and normal Cu) as an indication of a secondary Cu deficiency.

The occurrence of copper deficiency has been well documented not only in livestock (McDowell, 1985) but also in wild herbivores (Table 9.5). An apparently simple copper deficiency was found in tule elk (*Cervus elaphus*

Figure 9.4 Consumption of (dead) rabbits by herbivores on poor soils is a clear sign of mineral phosphorus deficiency.

nannodes) in a coastal area with slightly acidic soils overlying granitic bedrock (Gogan *et al.*, 1989). Secondary copper deficiency arising from a combination of marginal copper concentrations in interaction with molybdenum–sulphur appears to be more frequent. Flynn *et al.* (1976) found moose from the Kenai peninsula in Alaska suffering from faulty hoof keratinization and a decrease in reproductive rates as a result of copper deficiency. Although copper levels in browse were marginal, Flynn *et al.* (1977) demonstrated a significant effect of Mo–S levels on Cu concentration in hair samples from moose of seven different sites in Alaska, including the Kenai population. Further evidence of secondary copper deficiency has been found in cattle grazing coastal areas on maritime sand and clay in The Netherlands (Russel *et al.*, 1956; Van der Ouderaa, 1991; Van Deursen *et al.*, 1993). However, Van der Ouderaa (1991) warns that a diagnosis of copper deficiency merely based on forage or blood serum concentrations often fails to match with other, more serious symptoms of copper shortage.

Anthropogenic influences can also increase the risk of copper deficiency. Thus, Barlow *et al.* (1964) and Terlecki *et al.* (1964) recorded symptoms of ataxia (later recognized as the result of copper deficiency) in different park populations of red deer, whereas no evidence of copper deficiency has been reported from wild British deer populations. A similar situation is known from captive and free-ranging red deer herds in New Zealand (Reid *et al.*, 1980). Deficiencies in confined deer populations may have been the result of a depletion of copper resources through overstocking or habitat uniformity. A peculiar case of copper deficiency, apparently due to deleterious human activities, was documented by Frank *et al.* (1994a). They describe an increasing incidence of (sub)clinical symptoms of copper deficiency in Swedish moose since, 1982. Although the soils of the region are acidic and subject to further acidification by aerial pollution, they claim that acidification cannot solely account for the rapid decrease in hepatic Cu by 50% in 10 years. The authors suggest that this dramatic change is rather the result from an increased availability of Mo in the environment as a consequence of intensive liming to counter the ever progressing acidification! Data from this study also signal other possible mineral imbalances (Cr, Fe, Mn and Mg) that could develop with continuing liming activities.

(f) Selenium

The role of selenium in animal nutrition has been reviewed by the National Research Council (NRC, 1983). The typical pathology associated with Se deficiency is muscular dystrophy, also known as white muscle disease or (capture) myopathy. Serious symptoms appear in ruminants at dietary concentrations below 0.05(–0.10) ppm Se (Brady *et al.*, 1978; ARC, 1980). Low Se availability appears specific for areas with low inherent fertility and high rainfall. It appears to be aggravated by soil acidification as well as the application of plant fertilizers. Moreover, Se demands may rise under exposure to heavy metals and sulphur. As in the case of copper, the risk for Se deficiency is therefore likely to have increased under various types of recent anthropogenic disturbance of nutrient cycles, in particular by atmospheric deposition (Flueck, 1994).

Over the last few decades it has become clear that Se deficiency is far from uncommon in domestic animals on rangelands (Kurkela, 1980; McDowell, 1985). In wildlife the symptoms of muscular dystrophy are frequently revealed by additional stress, such as pursuit for marking or capturing, which can lead to mortality within hours (Harthoorn and Young, 1974). This so-called capture myopathy has been observed in many species of ungulates (NRC, 1983) but further data on condition status and range or dietary quality are usually lacking. Cases for which more complete information has been recorded concern mountain goat, pronghorn (*Antilocapra americana*), black-tailed deer (*Odocoileus hemionus*) and reindeer (Table 9.5). In a field experiment, Flueck (1994) found that Se supplements raised pre-weaning fawn survival in black-tailed deer. It is very possible that ungulates in regions with low Se availability are rendered more vulnerable to predation due to their impaired mobility for flight.

9.3.3 Influence of toxic substances

Toxic substances may be available to free-ranging herbivores in a natural form via toxic plants and in an artificial form through exposure to environmental pollution. The toxicity of plants may be due to a range of substances, which can be grouped in three categories (Harborne, 1991): phenolic metabolites (including tannins), nitrogen-containing compounds (mainly alkaloids and cyanogenic glycosides) and terpenoids (including resins). Some plant species accumulate selenium at toxic levels (> 4–5 ppm) (McDowell, 1985; Robbins, 1993). Goitrogenic substances, causing an impaired uptake of iodine from the thyroid gland, are known from a large number of species, particularly legumes and crucifers (ARC, 1980). Wild herbivores confronted with toxic plants usually avoid them or eat them only in small amounts. Physiological adaptation and conditioned food aversion by learning may also be involved in coping with toxic plants (Laycock, 1978; Provenza, 1995). Thus, poisoning by toxic plants is a rare event. It only occurs with some frequency with the introduction of naïve domestic animals or in overstocked areas where regular food resources have been depleted (Laycock, 1978).

Toxicity arising from environmental contaminants may be due to a wide array of substances, of both organic and inorganic origins. Among the best investigated contaminants for herbivores are heavy metals (lead, cadmium, mercury) and fluorine (ARC, 1980; McDowell, 1985). Lethal levels of these substances may be reached in the vicinity of industrial areas. Radioactive caesium became a concern in large parts of Europe following nuclear fall-out from the damaged Chernobyl reactor in 1986 (e.g. Johanson *et al.*, 1994; Strandberg and Knudsen, 1994). The negative impact of these contaminants on free-ranging herbivore populations mostly remains to be established. Cadmium contamination has received particular attention because it tends to accumulate in the internal organs. Raised liver and kidney concentrations of cadmium (and, to a lesser extent, lead) have been found in many polluted areas (e.g. Lusky *et al.*, 1992; Findo *et al.*, 1993; Wolkers *et al.*, 1994). Evidence of high cadmium levels causing a reduction in animal performance or reproduction in such areas is still lacking, although Storm *et al.* (1994) suggested a possible influence of cadmium on the density of white-tailed deer in the vicinity of a former zinc smelter in eastern Pennsylvania. The accumulation of cadmium in kidneys with age has been illustrated with data from red deer and cattle of the moderately contaminated Veluwe, an area of poor sandy soils in The Netherlands (Figure 9.5); levels of environmental contamination have been reported in DHV (1991). A significant relation between age and cadmium concentration in kidneys is apparent but if the rate of accumulation is extrapolated to the maximum acceptable level for ruminants of 200 ppm Cd (Friberg *et al.*, 1979) the conclusion is that this level will never be reached in the lifetime of an individual. Nevertheless, negative effects of cadmium accumulation at these moderate levels through interaction with other metabolic processes, e.g. by reducing copper absorption, should not be ruled out too rashly.

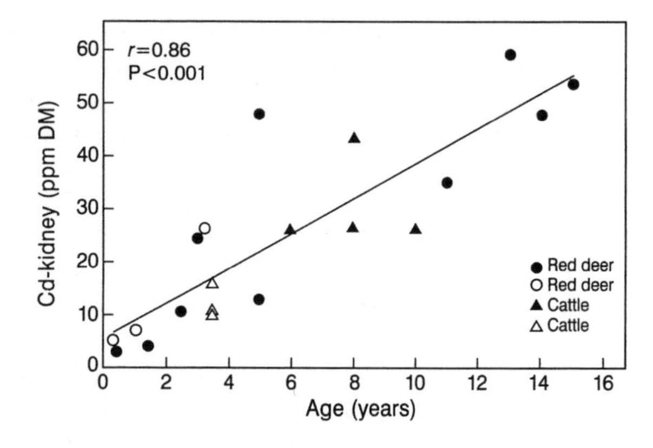

Figure 9.5 Cadmium accumulation in kidneys of red deer and cattle of different ages from the Veluwe area. The correlation coefficient was significantly different from zero ($P < 0.001$). Sources: ●, ▲, Vereniging Natuurmonumenten (unpublished data from 1993–1994); ○, Wolkers *et al.* (1994; data from 1989–1992); △, WallisDeVries (unpublished data from 1991).

9.3.4 Other habitat factors

Limitations on animal performance caused by habitat conditions are not restricted to forage quality and availability, even though the presented evidence shows that forage factors are of paramount importance. Other significant habitat factors can be reduced to two main sources: topography, and hazards of predation and infection by parasites and pathogens.

(a) Topography

Topography is important in various ways besides its indirect impact on forage production and quality. Firstly, it determines the accessibility of a habitat. Moving in steep terrain is costly: the energy required for ascent is 5–20 times that for walking an equal distance on level ground (ARC, 1980; Murray, 1991; Robbins, 1993). Large body size proves a disadvantage for both climbing and descending (Robbins, 1993). The accessibility of food in mountainous areas is restricted by cliffs or by the steepness of slopes. Falls from cliffs have been recorded as an important source of mortality in reindeer seeking adequate winter forage (Leader-Williams, 1988). On the other hand, cliff-dwelling ungulates face a reduced predation risk (see below).

A second aspect of topography is that it determines the drainage pattern of an area and, therefore, the availability of water sources. This is especially important in dry climates or in areas with coarsely textured permeable soils, where the distribution of herbivores is often restricted by the distance to water-holes (Arnold and Dudzinski, 1978; Sinclair, 1983). Water-dependent herbivores will usually stay within a 10 km radius from a water source

(Western, 1975). Periodic floods can also affect animal distribution. Riverine habitat is often unavailable for several weeks a year because of seasonal floods after heavy rains or snow-melt. Exceptional floods sometimes even lead to mass mortality, for example in wood bison (Carbyn *et al.*, 1993).

Finally, topography is the differentiating factor in determining exposure to wind and sun. Exposition evidently has a great impact on the availability and quality of forage by causing variation in vegetation composition. An additional influence is that south-facing slopes and wind-blown ridges often remain relatively free of snow, thus attracting animals by the greater accessibility of forage (Klein, 1965, 1970; Geist, 1971; Hjeljord, 1973). However, whether provided by topography or by vegetation, more sheltered conditions are sought to reduce energy expenditure during resting and rumination (Robbins, 1993). The availability of thermal cover is thought to be a main factor explaining winter dispersion in temperate deer species (Staines, 1974, 1977).

(b) Predation

Predation risk is strongly related to habitat features determining hiding cover, especially for stalking predators. Schaller (1972) attributed the excitement and reluctance of African buffalo (*Syncerus caffer*) when entering thickets to the greater probability of lion (*Panthera leo*) attacks. Prins (1996) showed that mortality of African buffalo in Lake Manyara National Park, Tanzania, concentrated around a grassland–bush ecotone. Yet these observations did not support the contention that buffalo avoided such high-risk areas. It was speculated that, even though predation was the main cause of mortality, the average instantaneous probability of falling victim to a predator was too low to prove relevant for the individual buffalo. Still, a plausible, though not established, hypothesis to explain the aggregation of herbivores in large herds in open habitat is the anti-predator benefit (Fryxell, 1991). Geist (1971) and Schaller (1977) related the relatively insignificant impact of predation on cliff-dwelling ungulates to their ability to retreat to areas inaccessible for predators. Thus, the impact of predation risk on habitat use may be important to explain species differences but remains an open issue at the intraspecific level.

(c) Parasites

The occurrence of external parasites is often related to habitat conditions. Ticks depend on tall vegetation to spread; various species of parasitic flies such as the screw-worm fly and the tsetse fly require the presence of bushes and trees (Sutherst, 1987), in the tundra mosquitos and black flies are prevalent in lower grounds, whereas warble and nostril flies swarm on the drier areas (Kelsall, 1968). Fly harassment significantly affects the behaviour and dispersal of large herbivores. Caribou will congregate and seek relief from warble and nostril flies in ponds, marshes and lake shallows. Feeding will often be inhibited and be restricted to windswept hill tops (Kelsall, 1968; Reimers, 1980; Skogland, 1984). In the Camargue, tabanid flies disrupt horses' feeding activities. Although total feeding time and habitat use appear to be little affected, Camargue horses avoid grazing in areas with high densities of flies during daytime (Duncan, 1992). With regard to contamination by ticks, cattle have been shown to detect and avoid areas infested by cattle tick larvae (*Boophilus microplus*). It has been speculated that migratory shifts of habitat use may occur partly as a response to heavy attack by flies and ticks (Sutherst, 1987).

The link between internal parasites or other pathogens and habitat composition is less evident and its impact on habitat quality for herbivores, if existent, remains largely unknown. An exception may be noted with regard to helminth parasites, concentrations of which occur in the lower horizons of the sward and in the vicinity of dung (Sykes, 1987). Helminth concentrations appear to be selectively avoided by grazing sheep (Sutherst, 1987). However, at low population density infected animals rarely appear to lose condition to a significant extent, with the exception of young or lactating animals (Sykes, 1987). It is therefore not likely that the risk of liver fluke infection will greatly affect habitat use, though it may affect overall habitat quality.

9.4 ANIMAL PERFORMANCE AND POPULATION DYNAMICS

An environmental limitation may be revealed at the level of the individual through a series of symptoms. Among the parameters related to animal performance, Klein (1965, 1970) and Geist (1971, 1978) identified the following signs of environmental limitation: lowered body condition, reduction in growth (body size, horns and antlers), delay of sexual maturity, impaired reproductive success and increased mortality. With a gradual increase in population density, corresponding to a gradual deterioration of habitat quality for the individual, animal growth and condition will be the first parameters affected. Juveniles are relatively vulnerable as they only have small fat reserves (e.g. Mitchell *et al.*, 1976). The reduction in reproduction rate and the rise in mortality are gradual and only slow the population increase, without stopping it (Klein, 1970; Caughley, 1970; Clarke, 1976; Skogland, 1983; Kaji *et al.*, 1988). There may even be a partial compensatory mechanism for the effects on reproduction and mortality, because individual adults raised in poor conditions have a longer life expectancy, apparently as a result of lower maturation rate (Geist, 1978; Seydack and Bigalke, 1992). Once saturation densities have been reached, it appears that the major density-dependent effect on population dynamics is due to increased mortality by starvation, especially in juveniles; lowered fecundity is a much less significant factor (Caughley, 1970; Sinclair, 1977; Skogland, 1985; Clutton-Brock *et al.*, 1985, 1991; Leader-Williams, 1988; Owen-

Smith, 1990; Choquenot, 1991). Mortality is often sex-biased, with a higher mortality among males (Klein, 1970; Clutton-Brock *et al.*, 1982, 1991; Leader-Williams, 1988; Owen-Smith, 1993).

While an inverse relation has thus been identified between animal performance and population density within a certain type of habitat, a positive relation between these variables can be predicted when comparing a range of habitats of varying quality. Three qualitatively different situations may be recognized. In the first situation, inferior habitat will not offer sufficient resources for survival and reproduction, resulting in a total absence or only a transient occurrence of the species concerned. In the second, low quality habitat above the threshold for survival and reproduction allows the establishment of a permanent population, but the average performance of individuals will be suboptimal and the resulting population density will be low. With improving habitat quality, both individual performance and population density will increase until, in the third situation, they reach a maximum under optimal conditions.

However, habitat quality and body condition may not always be positively correlated. Seydack and Bigalke (1992) studied the nutritional ecology and life history of African bushpig (*Potamochoerus porcus*) in two habitats of different quality. They argued that the reproductive investment increases with habitat quality, which leads to a lower condition and life expectancy for individual animals in high quality habitat. Unfortunately, the authors did not report relative animal densities in the different habitat types, which makes it impossible to judge the impact of this factor and thereby weakens their argument.

Population density need not be positively related with habitat quality factors if herbivore numbers are regulated by factors unrelated to habitat quality or by factors that have a stronger negative impact with increasing habitat quality. The occurrence of epidemic disease (Sinclair, 1979; Singer and Norland, 1994; Prins, 1996) and weather calamities such as drought or severe winters (Klein, 1970; Mitchell *et al.*, 1973; Owen-Smith, 1990; NRC, 1992; Dunham, 1994; McRoberts *et al.*, 1995) is often erratic and largely independent of habitat quality, although weather extremes may be mediated by habitat conditions (e.g. drought by water sources, cold by shelter, snow by exposure to wind). Both factors can cause mass mortality and may, ultimately, lead to extinction (DeAngelis and Waterhouse, 1987). It has been argued, however, that wild herbivore populations tend to be in equilibrium with infection by diseases and also by parasites, so that epidemics remain rare and the role of disease and parasites in population control is either absent or constant (Mackintosh and Beatson, 1985; Yuill, 1987; Carbyn *et al.*, 1993). A theoretical analysis by Grenfell (1992) showed that, through their impact on herbivore reproduction, parasites do have the potential to play the role of functional 'top predators'. Prins (1996) found for African buffalo that individuals with a low social status, which graze in the rear of the herd, have a comparatively high parasite load. It was suggested that the poor condition and, hence, survival probability of these individuals could be attributed to the impaired protein utilization caused by parasitism. Of special importance is the close link between the food consumption rate by the herbivores and the transmission rate of the parasites. Parasite influence is therefore likely to be greater for herd-living species and at high population density. It is possible, then, that parasites have a larger impact in high quality habitat, where high herbivore densities may be expected.

The role of predation by large mammal predators in the population regulation of large herbivores continues to be the subject of lively debate between advocates of regulation by food limitation (e.g. Sinclair, 1977; Sinclair *et al.*, 1985; Owen-Smith, 1990; Skogland, 1991; Boutin, 1992) and those claiming a significant role of predators (e.g. Bergerud, 1980; Bergerud *et al.*, 1983; Messier, 1991, 1994; Carbyn *et al.*, 1993). However, some studies have emphasized that the predominance of either factor depends on environmental conditions. Predation is suspected to have a considerable impact on resident (Fryxell *et al.*, 1988; Seip, 1992) and continental populations (Leader-Williams, 1988), in contrast to migratory and island populations. The analysis of moose–wolf (*Canis lupus*) dynamics by Messier (1994) suggests that predation is strongly density-dependent at the lower range of moose density and that wolves limit, but do not regulate, moose numbers at high moose densities. In low quality habitat, predation would therefore be an important regulating factor; whereas in high quality habitat, food resources (and perhaps parasites) would regulate the herbivore population, with predation as a relatively minor but constant source of mortality. Predation may thus further depress population density in low quality habitat.

Finally, the question should be raised as to what degree habitat quality itself is a constant factor. It is clear that 'population quality', as defined by Geist (1971), is not a constant, but this can be the result of factors unrelated to habitat quality, especially stochastic occurrences of diseases and weather extremes. In some cases habitat quality can change in a density-dependent manner. This is true when resources are not renewable and are thus subject to depletion. Such circumstances may arise when herbivores colonize a new environment. Dramatic examples of population eruptions followed by crashes have been recorded from anthropogenic introductions of herbivore species to islands and areas outside their original range (Klein, 1968; Caughley, 1970; Clarke, 1976; Kaji *et al.*, 1988; Leader-Williams, 1988). In these cases the food supply was often overgrazed to the point that both standing crop and productivity of the vegetation declined, so that the population density after the crash was only a fraction of peak density. However, where forage productivity remains relatively unaffected, it appears that the magnitude of the population crash will be considerably reduced and will only result from the depletion of the standing crop. This has been shown theoretically by Caughley (1979), and is supported by evidence from some field studies (e.g. Krefting, 1974; Leader-Williams, 1988; Pastor *et al.*, 1993). Constancy of habitat quality should thus be seen as a function of the productivity of food resources in conjunction with other habitat-related properties reviewed above.

9.5 HABITAT QUALITY AND CARRYING CAPACITY

To evaluate the quality of a certain habitat for an animal population, one would like to assess the maximum density this habitat can sustain, i.e. its carrying capacity. Carrying capacity is an intuitively sensible concept that appears full of pitfalls upon closer examination; some of these have already been touched upon in the previous section. In the field of herbivore ecology carrying capacity has been defined as the maximum animal density at which there is an equilibrium between the herbivores and their food resources in the absence of human interference (Caughley, 1979). In order to include other regulating factors besides the food supply (e.g. territoriality, shelter, water), resources could also be taken in a more general sense.

A critical assumption underlying this definition is that herbivores are regulated by their resources. While the reviewed evidence indicates that this is true in many cases, top-down control by predation and parasites can also play a role. The main difficulty with this notion of carrying capacity, however, is that it assumes stable ecosystem properties and does not, therefore, consider irreversible and dynamic processes. Interactions between herbivores and their food resources, vegetation succession and natural or anthropogenic disturbance may result in altered levels of food production and thus affect potential herbivore density. A further complication is that there may be multiple equilibria in a given ecosystem (DeAngelis and Waterhouse, 1987; Hudson, 1995; Chapter 10). Finally, the whole concept of equilibrium only appears meaningful at a sufficiently large spatial scale and within a restricted time period (DeAngelis and Waterhouse, 1987).

Notwithstanding these pitfalls, the concept of carrying capacity may be useful to give an approximate range of herbivore densities under certain conditions. In order to meet the requirement of ecosystem stability its application should be limited to a distinct type of habitat and a relatively short time scale. Of course this frequently precludes reliable extrapolation of density estimates from one system state to another.

Estimates of herbivore carrying capacity in terms of herbivore biomass/km^2 have been made for different types of African savanna (Table 9.6). The main differentiating factors determining carrying capacity were rainfall, soil nutrient status and herbivore species composition (arid versus moist savanna herbivores). In temperate environments the higher intensity of anthropogenic disturbance and the lower densities of herbivores have resulted in a much more fragmentary insight of potential herbivore density. Herbivore biomass in temperate environments generally ranges between

Table 9.6 Estimates of herbivore carrying capacity (kg live biomass/km^2), based on animal counts, in temperate environments compared with a number of African savannas (with an average annual rainfall higher than 700 mm) with different soil fertility

Region	Area	Vegetation	Herbivore species	Biomass	Reference
Temperate regions					
China	Xanjing	Steppe	Sheep	6700	National Research Council, 1992
Canada	Elk Island N.P.	Mixed forest	Bison, moose, elk	5700–6700	Telfer and Scotter, 1975; Hudson, 1995
Wyoming, USA	Yellowstone N.P.	Forest-steppe	Elk, bison, mule deer, bighorn sheep, pronghorn	5300	Singer and Norland, 1994
Scotland	Isle of Hirta	Grassland	Soay sheep	4300	Clutton-Brock *et al.*, 1991
Oklahoma, USA	?	Prairie	Cattle	3700	Ruth, 1949[1]
South Dakota, USA	?	Prairie	Bison, elk, pronghorn, white-tailed deer	2500–3500	Ruth, 1939; Cahalane, 1952[1]
Japan	Nakanoshima Island	Deciduous forest	Sika deer	2500	Kaji *et al.*, 1988
Poland	Bialowieza N.P.	Mixed forest	Red and roe deer, wild boar	2300	Jedrzejewska *et al.*, 1994
England	New Forest	Parkland mosaic	Ponies	2100	Putman, 1986
Oregon, USA	Coast range	Coniferous forest	Elk, mule deer	950	Smithey *et al.*, 1985
Scotland	Highlands	Heather moor/ coniferous forest	Red deer (roe and sheep omitted)	800–2800	Mitchell *et al.*, 1977; Ratcliffe, 1984
Africa					
High soil fertility	(N = 5)	Savanna	22 species	14 300 (11 800–19 700)	East, 1984[2]
Medium “ “	(N = 4)	Savanna	20 species	8200 (5900–11 000)	East, 1984
Low “ “	(N = 18)	Savanna	21 species	1800 (250–4000)	East, 1984; De Bie, 1991

Body weights (female): bison 475 kg, moose 350 kg, elk 250 kg, mule deer (*Odocoileus hemionus*) 70 kg, pronghorn 35 kg, bighorn sheep (*Ovis canadensis*) 70 kg, sheep 50 kg (China), Soay sheep 25 kg, sika deer (*Cervus nippon*) 70 kg, red deer 20 kg, wild boar 80 kg, pony 300 kg

[1] Cited in Telfer and Scotter, 1975

[2] For the Serengeti region the biomass figure (11 800 kg/km^2) has been derived from Sinclair and Norton-Griffiths (1979); this seems to reflect the carrying capacity more closely than the earlier estimate (5100 kg/m^2) by Schaller (1972), used by East (1984).

1000 and 7000 kg/km^2 (Table 9.6; Figure 9.6), which would compare with a low to moderate herbivore density for African savannas. The data are no more than indicative, however, as only the North American studies represent relatively undisturbed conditions (it should be noted that they also represent more continental climatic conditions, which could introduce a bias).

Yet the range in biomass is higher than usually suspected (Van Dyne *et al.*, 1980; Botkin *et al.*, 1981). European standards used for wildlife range much lower, as exemplified by an estimate of 300–500 kg/km^2 (Anonymous, 1988). This is not surprising, as these estimates have been determined to a large extent by the degree of damage to tree growth considered acceptable by foresters. Furthermore, in the temperate zone herbivore densities are often considered for single species instead of a multispecies assemblage and on the assumption of a closed forest environment. Values for single browsing herbivore species from closed temperate and boreal forests range between 350 and 1000 kg/km^2 (Van Dyne *et al.*, 1980; Botkin *et al.*, 1981; Smithey *et al.*, 1985). The example of the Bialowieza forest (Table 9.6) indicates that these figures may be underestimates when several herbivore species with complementary feeding styles are present. Moreover, the assumption of a closed forest in the presence of a multispecies herbivore assemblage including grazers may be erroneous (Singer and Norland, 1994; Campbell *et al.*, 1994; Vera, 1997). In this respect it is illustrative to note that densities of livestock (often in combination with deer) in Europe's former, half-open rangelands (in the period AD 1600–1900) equalled African situations: 8850–15 300 kg/km^2 in the New Forest, England (Putman,

Figure 9.6 Bison herds in North America illustrate that high herbivore densities may occur in the temperate zone. Did they shape the plant community in the past?

1986); 13 000–21 000 kg/km^2 in Fontainebleau, France (Tendron, 1983); and 2100–10 200 kg/km^2 in Bialowieza, Polish-Russian border (Peters, 1992). It may be wondered whether such densities could exist without human interference.

The carrying capacity estimates in Table 9.6 mostly concern larger areas, which contain a number of habitat types. Environmental heterogeneity will cause essential variation in habitat parameters and thus affects herbivore distribution. De Bie (1991) has speculated that the relatively low herbivore density in West African savannas is due to their great homogeneity. It is therefore important to separate homogeneous from heterogeneous environments when discussing their potential to harbour herbivore populations.

9.5.1 Homogeneous environments

It is evident that truly homogeneous environments do not exist. In as far as the notion can be used, homogeneity is a matter of scale and the amount of variation allowed at a certain scale level. Below, the differentiation between habitats according to soil fertility is considered. Over larger areas, variation in soil nutrient status gives rise to important differences in herbivore abundance (Table 9.6) (East, 1984). Similar differences are found on a small scale for low and high soil fertility areas in The Netherlands. These contrasting habitats are often grazed separately, so that the herbivores face a relatively homogeneous environment.

In the experiments by WallisDeVries (1996), maximum cattle density in fertile riverine grassland was estimated at 0.5 animals/ha (25 000 kg/km^2), which surpasses even the highest herbivore densities in Africa. This estimate is probably somewhat exaggerated when applied to larger areas because it is based on pure grassland without marshes or woodland – and even more so because riverine grassland is not a year-round habitat, due to frequent river floods in winter and spring. Still, the present herbivore biomass in the high-fertility Oostvaardersplassen reserve, which harbours still growing populations of Heck cattle, Konik horses, red deer and roe deer, has reached a level of 10 000–20 000 kg/km^2 at a minor (but perhaps not unimportant) level of supplementary feeding (Chapter 7). A comparable estimate, based on forage availability, for low-fertility grass-dominated heathland on Pleistocene cover-sand would be in the order of 4000 kg/km^2. This is close to the estimate for Soay sheep on poor and intermediate quality grassland on the Isle of Hirta. Maximum red deer density on heather moor and conifer plantations in the Scottish Highlands has been reported at 2800 kg/km^2, not counting roe deer and sheep (Table 9.6). It is worth expanding on the comparison of low and high soil fertility habitat as an illustration of a large difference in habitat quality.

In temperate areas winter mortality as a result of inferior condition is a main factor determining population dynamics. The percentage weight loss over winter can then be used as a parameter for survival probability, given the fact that for non-juvenile herbivores most individuals will die beyond 30% weight loss (Franzmann, 1985; Holechek and Herbel, 1986). In cattle grazing high-fertility grassland in The Netherlands, the digestibility of winter forage appears to be an important factor determining weight loss (Figure 9.7), in the absence of flooding. In this productive environment the die-back of tall herbs and graminoids during autumn and winter is virtually complete. Standing dead material is of extremely low digestibility. Short perennial grasses remain green and productive at temperatures above freezing-point, but spells of frost will also cause these grasses to wither. This results in a low but sufficient availability of digestible forage in mild winters and a sharp decline in the supply of digestible forage in severe winters (Vulink and Drost, 1991; Bokdam and WallisDeVries, 1992; WallisDeVries, 1996). Cattle hardly lose weight in the milder years, but approach a critical weight loss in cold winters. Conditions are far less variable in grass-dominated low-fertility habitat. With a large share of wintergreen plants, forage digestibility tends to remain at a higher level. Compared with high-fertility habitat, cattle in low-fertility habitat show higher weight losses over winter (except in colder years), independent of forage digestibility (Figure 9.7). This different performance between habitats can be explained by the shortage of sodium and phosphorus in the low-fertility environment (WallisDeVries, 1996). In terms of body condition and population density, the quality of an isolated low-fertility habitat can be called low or even inadequate if the impact of mineral deficiency proves as severe as in the experiment by

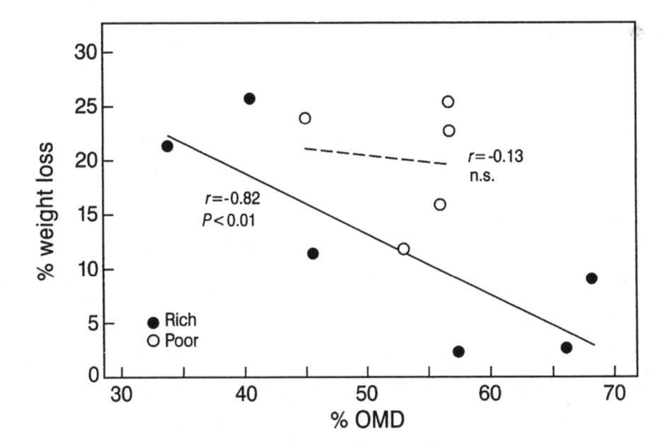

Figure 9.7 The relationship between average organic matter digestibility (OMD) of the winter diet and winter weight loss in cattle from areas with high and low soil fertility (Rich and Poor, respectively). The correlation coefficient was significantly different from zero for the rich areas ($P < 0.01$) but not for the poor areas ($P > 0.10$). (References: Drost, 1986; Vulink and Drost, 1991; Van Deursen *et al.*, 1993; WallisDeVries, 1994; Van Wieren, 1996.)

WallisDeVries (1996). In the absence of flooding, the average habitat quality for high-fertility habitat should, on the contrary, be valued as optimal. However, winter severity would probably lead to periodic population crashes at high densities, as in Soay sheep (Clutton-Brock *et al.*, 1991). Thus, temporal variation in quality would be relatively high in this habitat type. In Chapter 7 it is argued that in areas where recolonization is impossible due to isolation, herbivore population cycles together with vegetation changes leading to a decline in overall forage quality would ultimately lead to local extinction of the herbivores. Large population fluctuations need not be characteristic for high-quality habitats, however. The herbivore assemblage of Lake Manyara National Park, Tanzania, for instance, has been shown to maintain a high and remarkably constant overall biomass, even though fluctuations do occur at the species level (Prins, 1996).

9.5.2 Heterogeneous environments

In a hypothetical heterogeneous environment with patches differing in quality, animals may be thought to be distributed over the landscape so that an equal share of resources is available to each individual. This results in an ideal free distribution (Fretwell and Lucas, 1970), with the highest animal density in the best patch and gradually lower densities in other patches according to the level of resources. Such a pattern has been observed in the distribution of Scottish hill sheep over a range of heathland and grassland communities (Hunter, 1964). Similarly, the density of elephants (*Loxodonta africana*) in Wankie National Park, Zimababwe, was correlated with the availability of salt licks (Weir, 1972). Positive correlations of soil nutrient status and the density of both migrating and resident herbivores have also been found for other species of African herbivores (McNaughton, 1988, 1990; Ben-Shahar and Coe, 1992; Seagle and McNaughton, 1992). Strong support for the occurrence of an ideal-free distribution was found by Wahlström and Kjellander (1995). They compared roe deer populations from a rich and an average habitat. The rich habitat had twice the density of the average one but the performance of does was similar between habitats. This suggests that resources were equally distributed between individuals from both habitats.

A different situation arises when the availability of different essential nutrients is not positively correlated. Habitat heterogeneity may then be a prerequisite for the persistence of a population. Thus, the insufficient sodium supply from terrestrial forage for moose could be compensated for by exploitation of sodium-rich aquatic vegetation from ponds and lakes – at the expense, however, of a relatively high energy cost (Belovsky, 1978; Fraser *et al.*, 1984). In cattle, a comparable trade-off between two habitat types (the riverine and heathland habitats dicussed above) was recorded by WallisDeVries and Schippers (1994). The animals could obtain a higher intake rate of sodium and phosphorus in riverine habitat, whereas in poor heathland habitat the availability of minerals was limited but energy intake rate was higher, especially during winter.

Seasonal variation in food supply and environmental conditions may induce periodic animal movements between habitat types over larger distances, i.e. migration (Fryxell *et al.*, 1988). At a smaller spatial scale, seasonal movements among habitat types in a landscape catena in response to changes in forage quality have been described for African herbivores by Jarman and Sinclair (1979) and Penzhorn and Novellie (1991). Nomadic movement patterns are typical for herbivores that follow the growth flush after local rainfall (Maddock, 1979; Stanley Price, 1989). In mountainous areas migration generally occurs between summer habitat at high altitude and winter habitat in the valleys (Boyce, 1991; Goodson *et al.*, 1991; Albon and Langvatn, 1992). Snow accumulation in autumn triggers the downward movement and the availability of highly nutritious forage during green-up promotes the return to higher altitudes. These factors also appear to determine the long-distance migrations of caribou (Kelsall, 1968), though differential predation risk in summer and winter ranges could present an important additional motive (Seip, 1992; see also McNay and Voller, 1995). Migration of Serengeti wildebeest may well be governed by the spatial segregation of dry season water sources and adequate phosphorus resources in the drier wet season range (Murray, 1995), possibly in conjunction with predator avoidance.

It could be wondered whether the previous examples of resources determining animal distribution patterns indicate a limitation. The animals do not seem to show signs of nutrient shortage, yet they do aggregate where resources are abundant (or predation risk is reduced). Thus, as predicted by Fretwell and Lucas (1970), animals respond to differences in habitat quality even before limitations become manifest through a reduction of animal performance. Limitations would become apparent, however, if environmental heterogeneity were lost by habitat alteration.

9.6 MANAGEMENT CONSIDERATIONS

Considerations of habitat quality are essential for the management of areas with herbivore populations and for projects aiming at the (re)introduction of herbivores. Several basic questions should be addressed in every evaluation of habitat suitability. The following questions pertain to the management of herbivore populations at a minimum degree of human interference.

Does the area contain adequate forage resources?

Forage resources should be surveyed in relation to the feeding style of the herbivore species and with regard to productivity, quality and seasonal variation. Quality aspects of essential importance are forage digestibility (or fibre content), protein concentration and mineral content; they often bear a close relation to soil fertility.

Is the area 'complete' in the sense of ensuring an adequate animal performance on a year-round basis?
In order to answer this question it is necessary to make an inventory of all critical resources, i.e. forage, water, shelter, and potential limitations such as deep snow, river floods and terrain accessibility. For a correct appraisal both seasonal and year-to-year variations should be considered.

What is the maximum population size the area can sustain?
As stated earlier the estimation of carrying capacity is always an approximation assuming relatively stable conditions. De Bie (1991) emphasized the importance of distinguishing favourable and unfavourable years when making an estimate. A carrying capacity estimate should preferably be derived from a comparable existing field situation to reduce the degree of uncertainty (especially regarding seasonal variation in forage supply and the proper use factor of forage production). For grazing herbivores, livestock densities under extensive farming systems may be used as a first approximation. With the exception of African savannas, there is a serious lack of areas with wildlife sufficiently exempt from anthropogenic disturbance to serve as references for estimates of carrying capacity. In temperate regions without supplementary feeding, wildlife densities often appear to be below their potential, whereas livestock densities under extensive land use are usually above carrying capacity because of human interference (seasonal grazing, supplementation, habitat alteration, etc.). In the absence of proper reference sites, calculations of maximum stocking rates on the basis of forage availability can be made to arrive at a preliminary assessment of carrying capacity (Telfer and Scotter, 1975; Holechek, 1988; De Bie, 1991). The calculation is based on the ratio between the maximum proportion of available annual forage production that can be used on a sustainable basis, and the average dry matter intake required for the herbivore (2–3% of body weight on a daily basis). The fraction of forage available for consumption – the proper use factor – has been estimated at 40–60% for both grass and browse species of temperate environments (Telfer and Scotter, 1975; Grant *et al.*, 1982; Holechek, 1988). However, the use of a more conservative estimate of 25% does seem advisable when the calculation is applied for conservation purposes. Excessive grazing and browsing pressure may jeopardize the conservation interests of other species.

Is the area large enough?
The size of an area has not been considered thus far, as it is not a special element of habitat quality. Yet it is essential with respect to the probability that a population is large enough to be persistent. Small populations are vulnerable to extinction in the face of inbreeding depression and stochasticity in environmental factors and genetic constitution (Soulé, 1987;

Gilpin, 1991). How large an area should be depends on its carrying capacity and on the minimum viable population size. A proper threshold for a minimum viable population size is impossible to give as the necessary knowledge is still lacking and as the estimates not only depend on species and environmental characteristics but also on the time scale considered. Nevertheless, for practical purposes a figure of 1000 animals (Thomas, 1990) can be used as a crude and unvalidated estimate of a minimum population size. With animal densities for a single large herbivore species typically ranging between roughly 0.2 and 20/km^2, this would set the hypothetical minimum area size at 50–5000 km^2. This range agrees with the estimate derived from a simple model by Belovsky (1987), yet for some species even larger minimum areas (10 000–100 000 km^2) may be required (Schonewald-Cox, 1983).

What if an area cannot harbour a desired herbivore population on a sustainable basis? Herbivores are often confined to small reserves or they are unable to make use of environmental variation due to habitat fragmentation and artificial barriers between different habitats (Williamson *et al.*, 1988; Klein, 1991; Boyce, 1991; WallisDeVries, 1995). Depending on the management goals it may then be decided that herbivores should be excluded from such a place. However, the presence of herbivores may still be considered valuable, either because of their role as keystone or umbrella species (WallisDeVries, 1995), or because there is no existing area that is more appropriate. Management options to overcome practical limitations then need to be considered. Below, a series of different options has been ordered in a sequence of increasing degree of human interference, reflecting different approaches to conservation measures.

Resource limitations:

- area enlargement;
- provision of complementary habitat;
- improvement of the food supply (fertilization, game meadows, game crops);
- provision of water points, shelter, high-water refuge;
- provision of mineral licks;
- seasonal grazing regime (e.g. by transport between winter and summer range or by winter removal to stable);
- provision of energy and protein supplements.

Small population size:

- area enlargement;
- creation of corridors to other suitable areas;
- periodic genetic exchange with other populations;
- habitat improvement to raise carrying capacity;
- predator and disease control;
- supplementary feeding to prevent mortality under unfavourable circumstances.

9.7 CONCLUSIONS

Habitat quality remains a difficult concept for both scientists and reserve managers, because it involves assessing the role of such a wide range of potentially important limiting factors. It is also an essential notion, to explain why herbivores thrive better in one area than in another and to justify decisions concerning the management of herbivore populations.

This review has shown that the nature of many significant limiting factors is known at least to some extent. Less evident are the critical thresholds at which certain factors become important and still more obscure are the absolute limits determining the boundary between life and death and, ultimately, persistence and extinction. More insight in both interspecific differences and intraspecific variability in the ability to cope with marginal conditions is necessary. The relative availability of resources in different habitats merits further investigation if a better prediction of potential herbivore densities as a function of habitat characteristics is to be achieved. Furthermore, the roles of plant–herbivore equilibria (Chapters 5 and 10) and interspecific resource partitioning (Chapter 6) are essential in determining habitat quality and thus deserve stronger emphasis.

For the purpose of conservation, such scientific knowledge would provide a firmer basis for management decisions. In the meantime, however, it is not necessary to await the outcome of future investigations. It should be an important task for conservation bodies to initiate large-scale experiments with careful monitoring (Chapter 7) to advance our understanding of habitat quality issues. This overview of limiting factors and the guidelines with respect to management will hopefully facilitate such initiatives.

REFERENCES

Adamczewski, J.Z., Gates, C.C., Hudson, R.J. and Price, M.A. (1987) Seasonal changes in body composition of mature female caribou and calves (*Rangifer tarandus groenlandicus*) on an arctic island with limited winter resources. *Canadian Journal of Zoology*, **65**, 1149–1157.

Albon, S.D. and Langvatn, R. (1992) Plant phenology and the benefits of migration in a temperate ungulate. *Oikos*, **65**, 502–513.

Anonymous (1988) *Grofwildvisie*, Ministerie van Landbouw en Visserij, Den Haag, The Netherlands.

ARC (1980) *The Nutrient Requirements of Ruminant Livestock*, CAB, Farnham Royal, UK.

Arnold, G.W. and Dudzinski, M.L. (1978) *The Ethology of Free-ranging Domestic Animals*, Elsevier, Amsterdam.

Barlow, R.M., Butler, E.J. and Purves, D. (1964) An ataxic condition in red deer (*Cervus elaphus*). *Journal of Comparative Pathology*, **74**, 519–529.

Belovsky, G.E. (1978) Diet optimization in a generalist herbivore: the moose. *Theoretical Population Biology*, **14**, 105–134.

Belovsky, G.E. (1981) A possible response of moose to sodium availability. *Journal of Mammalogy*, **62**, 631–633.

Belovsky, G.E. (1987) Extinction models and mammalian persistence, in *Viable Populations for Conservation*, (ed. M.E. Soulé), pp. 35–57, Cambridge University Press, Cambridge.

Ben-Shahar, R. (1993) Patterns of nutrient contents in grasses of a semi-arid savanna. *African Journal of Ecology*, **31**, 343–347.

Ben-Shahar, R. and Coe, M.J. (1992) The relationships between soil factors, grass nutrients and the foraging behaviour of wildebeest and zebra. *Oecologia*, **90**, 422–428.

Bergerud, A.T. (1980) A review of the population dynamics of caribou and wild reindeer in North America, in *Proceedings 2nd International Reindeer/Caribou Symposium, Röros, Norway, 1979*, (eds E. Reimers, E. Gaare and S. Skjenneberg), pp. 556–581, Trondheim, Norway.

Bergerud, A.T., Wyestt, W. and Snider, B. (1983) The role of wolf predation in limiting a moose population. *Journal of Wildlife Management*, **47**, 977–988.

Bigalke, R.C. and Van Hensbergen, H.J. (1992) Observations on a reproductively isolated population of blesbok (*Damaliscus dorcas phillipsi*) in a mineral deficient environment, in *Ongulés/Ungulates 91*, (eds F. Spitz, G. Janeau, G. Gonzalez and S. Aulagnier), pp. 497–503, SFEPM-IRGM, Paris – Toulouse.

Blair-West, J.R., Coghlan, J.P., Denton, D.A. *et al.* (1968) Physiological, morphological and behavioural adaptation to a sodium deficient environment by wild native Australian and introduced species of animals. *Nature*, **217**, 922–928.

Blair-West, J.R., Denton, D.A., Nelson, J.F. *et al.* (1989) Recent studies of bone appetite in cattle. *Acta Physiologica Scandinavica*, **136, Suppl. 583**, 53–58.

Blaxter, K.L., Kay, R.N.B., Sharman, G.A.M. *et al.* (1974) *Farming the Red Deer*, Department of Agriculture and Fisheries for Scotland report, HMSO, Edinburgh.

Bokdam, J. and WallisDeVries, M.F. (1992) Forage quality as a limiting factor for cattle grazing in isolated Dutch nature reserves. *Conservation Biology*, **6**, 399–408.

Botkin, D.R., Mellilo, J.M. and Wu, L.S-Y. (1981) How ecosystem processes are linked to large mammal population dynamics, in *Dynamics of Large Mammal Populations*, (eds C.W. Fowler and T.D. Smith), pp. 373–387, John Wiley, New York.

Boutin, S. (1992) Predation and moose population dynamics: a critique. *Journal of Wildlife Management*, **56**, 116–127.

Boyce, M.S. (1991) Migratory behavior and management of elk (*Cervus elaphus*). *Applied Animal Behaviour Science*, **29**, 539–550.

Brady, P.S., Brady, L.J., Whether, P.A. *et al.* (1978) The effect of dietary selenium and vitamin E on biochemical parameters and survival of young among white-tailed deer (*Oidocoileus virginianus*). *Journal of Nutrition*, **108**, 1439–1448.

Bryant, J.P., Reichardt, P.B., Clausen, T.P. *et al.* (1992) Woody plant-mammal interactions, in *Herbivores – Their Interactions with Secondary Plant Metabolites. Vol. II. Ecological and evolutionary processes*, (eds G.A. Rosenthal and M.R. Berenbaum), pp. 343–370, Academic Press, San Diego.

Campan, R., Bon, R. and Barre, V. (eds) (1991) The wild ungulates of France: a review of recent field studies. *Revue d'Ecologie (Terre et Vie)*, **Suppl. 6**, 1–282.

Campbell, C., Campbell, I.D., Blyth, C.B. and McAndrews, J.H. (1994) Bison extirpation may have caused aspen expansion in western Canada. *Ecography*, **17**, 360–362.

Carbyn, L.N., Oosenbrug, S.M. and Anions, D.W. (1993) *Wolves, Bison and the Dynamics related to the Peace-Athabasca Delta in Canada's Wood Buffalo National Park*, Canadian Circumpolar Research Series no. 4, University of Alberta, Edmonton.

Caughley, G. (1970) Eruption of ungulate populations, with emphasis on Himalayan Thar in New Zealand. *Ecology*, **51**, 53–72.

Caughley, G. (1979) What is this thing called carrying capacity? in *North American Elk: Ecology, Behavior and Management*, (eds M.S. Boyce and L.D. Hayden-Wing), pp. 2–8, University of Wyoming, Laramie, Wyoming.

Cederlund, G.N., Håkan, K.G.S. and Pehrson, Å. (1991) Body mass dynamics of moose calves in relation to winter severity. *Journal of Wildlife Management*, **55**, 675–681.

Chapin III, F.S., Johnson, D.A. and McKendrick, J.D. (1980) Seasonal movement of nutrients in plants of differing growth form in an Alaskan tundra ecosystem: implications for herbivory. *Journal of Ecology*, **68**, 189–209.

Choquenot, D. (1991) Densisty-dependent growth, body condition, and demography in feral donkeys: testing the food hypothesis. *Ecology*, **72**, 805–813.

Clarke, C.M.H. (1976) Eruption, deterioration and decline of the Nelson red deer herd. *New Zealand Journal of Forestry Science*, **5**, 235–249.

Clutton-Brock, T.H., Guiness, F.E. and Albon, S.D. (1982) *Red Deer: the behaviour and ecology of two sexes*, University of Chicago Press, Chicago.

Clutton-Brock, T.H., Major, M. and Guinness, F.E. (1985) Population regulation in male and female red deer. *Journal of Animal Ecology*, **54**, 831–846.

Clutton-Brock, T.H., Price, O.F., Albon, S.D. and Jewell, P.A. (1991) Persistent instability and population regulation in Soay sheep. *Journal of Animal Ecology*, **60**, 593–608.

DeAngelis, D.L. and Waterhouse, J.C. (1987) Equilibrium and nonequilibrium concepts in ecological models. *Ecological Monographs*, **57**, 1–21.

De Bie, S. (1991) *Wildlife Resources of the West African Savannas*. Dissertation, Wageningen Agricultural University Papers 91-2, Wageningen, The Netherlands.

Deinum, B. (1984) Chemical composition and nutritive value of herbage in relation to climate, in *Proceedings 10th General Meeting of the European Grassland Federation*, pp. 338–350, Aas, Norway.

Demment, M.W. and Van Soest, P.J. (1985) A nutritional explanation for body-size patterns of ruminant and non-ruminant herbivores. *American Naturalist*, **125**, 641–672.

Denton, D. (1982) *The Hunger for Salt: an anthropological, physiological and medical analysis*, Springer Verlag, Berlin.

DHV (1991) *De Chemische Kwaliteit van de Nederlandse Heide*, DHV Milieu and Infrastructuur BV, Amersfoort, The Netherlands.

Doerr, J.G. and Dieterich, R.A. (1979) Mandibular lesions in the Western Arctic caribou herd of Alaska. *Journal of Wildlife Diseases*, **15**, 309–318.

Drost, H.J. (1986) Runderen in het riet: Begrazingsonderzoek in de Oostvaardersplassen. *Landbouwkundig Tijdschrift*, **98**, 25–28.

Drost, H.J. and Muis, A. (1988) Begrazing van Duinriet op 'de Rug' in de Lauwersmeer. *De Levende Natuur*, **89**, 82–88.

Drozdz, A. (1979) Seasonal intake and digestibility of natural foods by roe deer. *Acta Theriologica*, **24**, 137–170.

Duncan, P. (1992) *Horses and Grasses: the nutritional ecology of equids and their impact on the Camargue*, Ecological Studies 87, Springer Verlag, New York.

Dunham, K.M. (1994) The effect of drought on the large mammal populations of Zambezi riverine woodlands. *Journal of Zoology, Lond.*, **234**, 489–526.

East, R. (1984) Rainfall, soil nutrient status and biomass of large African savanna mammals. *African Journal of Ecology*, **22**, 245–270.

Feldhamer, G.A., Kilbane, T.P. and Sharp, D.W. (1989) Cumulative effect of winter acorn yield and deer body weight. *Journal of Wildlife Management*, **53**, 292–295.

Findo, S., Hell, P., Farkas, J. *et al.* (1993) Accumulation of selected heavy metals in red and roe deer in the Central West Carpathian Mountains (Central Slowakei) *Zeitschrift für Jagdwissenschaft*, **39**, 181–189.

Flueck, W.T. (1994) Effect of trace elements on population dynamics: selenium deficiency in free-ranging black-tailed deer. *Ecology*, 75, 807–812.

Flynn, A., Franzmann, A.W. and Arneson, P.D. (1976) Molybdenum–sulfur interactions in the utilization of marginal dietary copper in Alaskan moose (*Alces alces gigas*), in *Molybdenum in the Environment*, (eds W.R. Chappell and K.K. Petersen), Vol. 1, pp. 115–124, Marcel Dekker, New York/Basel.

Flynn, A., Franzmann, A.W., Arneson, P.D. and Oldemeyer, J.L. (1977) Indications of copper deficiency in a subpopulation of Alaskan moose. *Journal of Nutrition*, **107**, 1182–1189.

Frank, A., Galgan, V. and Petersson, L.R. (1994a) Secondary copper deficiency, chromium deficiency and trace element imbalance in the Moose (*Alces alces* L.): effect of anthropogenic activity. *Ambio*, **23**, 315–317.

Frank, D.A., Inouye, R.S., Huntly, N. *et al.* (1994b) The biogeochemistry of a North-Temperate grassland with native ungulates: nitrogen dynamics in Yellowstone National Park. *Biogeochemistry*, **26**, 163–188.

Franzmann, A.W. (1985) Assessment of nutritional status, in *Bioenergetics of Wild Herbivores*, (eds R.J. Hudson and R.G. White), pp. 239–259, CRC Press, Boca Raton, Florida.

Fraser, D., Chavez, E.R. and Paloheimo, J.E. (1984) Aquatic feeding by moose: selection of plant species and feeding areas in relation to plant chemical composition and characteristics of lakes. *Canadian Journal of Zoology*, **62**, 80–87.

Freeland, W.J. and Choquenot, D. (1990) Determinants of herbivore carrying capacity: plants, nutrients and *Equus asinus* in Northern Australia. *Ecology*, **71**, 589–597.

Freeland, W.J., Calcott, P.H. and Geiss, D.P. (1985) Allelochemicals, minerals and herbivore population size. *Biochemical Systematics and Ecology*, **13**, 195–206.

Fretwell, S.D. and Lucas, H.L. (1970) On territorial behaviour and other factors influencing habitat distribution in birds. I. Theoretical development. *Acta Biotheoretica*, **19**, 16–36.

Friberg, L., Nordberg, G.F. and Vouk, V.B. (1979) (eds) *Handbook on the Toxicology of Metals*, Elsevier, Amsterdam.

Fryxell, J.M. (1991) Forage quality and aggregation by large herbivores. *American Naturalist*, **138**, 478–498.

Fryxell, J.M., Greever, J. and Sinclair, A.R.E. (1988) Why are migratory ungulates so abundant? *American Naturalist*, **131**, 781–798.

Furness, R.W. (1988) Predation on ground-nesting seabirds by island populations of red deer *Cervus elaphus* and sheep *Ovis*. *Journal of Zoology, Lond.*, **216**, 565–573.

Gartner, R.J.W., McLean, R.W., Little, D.A. and Winks, L. (1980) Mineral deficiencies limiting production of ruminants grazing tropical pastures in Australia. *Tropical Grasslands* **14**, 266–272.

Geist, V. (1971) *Mountain sheep. A study in behavior and evolution*, University of Chicago Press, Chicago.

Geist, V. (1978) *Life Strategies, Human Evolution, Environmental Design – toward a biological theory of health*, Springer, New York.

Gerhart, K.L., White, R.G., Cameron, R.D. and Russell, D.E. (1996) Body composition and nutrient reserves of arctic caribou. *Canadian Journal of Zoology*, **74**, 136–146.

Gilpin, M. (1991) The genetic effective size of a metapopulation. *Biological Journal of the Linnean Society*, **42**, 165–175.

Gogan, P.J.P., Jessup, D.A. and Akeson, M. (1989) Copper deficiency in tule elk at Point Reyes, California. *Journal of Range Management*, **42**, 233–238.

Goodson, N.J., Stevens, D.R. and Bailey, J.A. (1991) Winter–spring foraging ecology and nutrition of bighorn sheep on montane ranges. *Journal of Wildlife Management*, **55**, 422–433.

Grant, S.A., Milne, J.A., Barthram, G.T. and Souter, W.G. (1982) Effects of season and level of grazing on the utilization of heather by sheep. 3. Longer-term responses and sward recovery. *Grass and Forage Science*, **37**, 311–320.

Grasman, B.T. and Hellgren, E.C. (1993) Phosphorus nutrition in white-tailed deer: mutrient balance, physiological responses, and antler growth. *Ecology*, **74**, 2279–2296.

Gray, P.B. and Servello, F.A. (1995) Energy intake relationships for white-tailed deer on winter browse diets. *Journal of Wildlife Management*, **59**, 147–152.

Green, H.H. (1925) Perverted appetites. *Physiology Reviews*, **5**, 336–348.

Grenfell, B.T. (1992) Parasitism and the dynamics of ungulate grazing systems. *American Naturalist*, **139**, 907–929.

Groot Bruinderink, G.W.T.A., Hazebroek, E. and Van der Voet, H. (1994) Diet and condition of wild boar, *Sus scrofa scrofa*, without supplementary feeding. *Journal of Zoology, Lond.*, **233**, 631–648.

Grubb, P.J. (1974) Population dynamics of Soay sheep, in *Island Survivors: the ecology of the Soay sheep of St Kilda*, (eds P.A. Jewell, C. Milner and J.M. Boyd), pp. 242–272, Athlone Press, London.

Hacker, J.B. and Minson, D.J. (1981) The digestibility of plant parts. *Herbage Abstracts*, **51**, 459–482.

Harborne, J.B. (1991) The chemical basis of plant defense, in *Plant Defenses against Mammalian Herbivory*, (eds R.T. Palo and C.T. Robbins), pp. 45–59, CRC Press, Boca Raton, Florida.

Harthoorn, A.M. and Young, E. (1974) A relationship between acid–base balance and capture myopathy in zebra (*Equus burchelli*) and an apparent therapy. *Veterinary Record*, **95**, 337–342.

Haukioja, E. and Lehtilä, K. (1992) Moose and birch: how to live on low-quality diets. *Trends in Ecology and Evolution*, **7**, 19–22.

Hebert, D.M. and Cowan, I.McT. (1971) White muscle disease in the mountain goat. *Journal of Wildlife Management*, 35, 752–756.

Hewison, A.J.M., Angibault, J.M., Boutin, J. *et al.* (1996) Annual variation in body composition of roe deer (*Capreolus capreolus*) in moderate environmental conditions. *Canadian Journal of Zoology*, 74, 245–253.

Hindelang, M. and Peterson, R.O. (1996) Osteoporotic skull lesions in moose at Isle Royale National Park. *Journal of Wildlife Diseases*, 32, 105–108.

Hjeljord, O. (1973) Mountain goat forage and habitat preference in Alaska. *Journal of Wildlife Management*, 37, 353–362.

Hobbs, N.T. and Swift, D.M. (1985) Estimates of habitat carrying capacity incorporating explicit nutritional constraints. *Journal of Wildlife Management*, **49**, 814–822.

Holechek, J.L. (1988) An approach for setting the stocking rate. *Rangelands*, **10**, 10–14.

Holechek, J.L. and Herbel, C.H. (1986) Supplementing range livestock. *Rangelands*, 8, 29–33.

Hudson, R.J. (1995) Temporal and spatial dynamics of natural grazing systems, in *Wild and Domestic Ruminants in Extensive Land Use Systems*, (eds H.J. Schwarz and R.R. Hofmann), pp. 88–105, Proceedings International Symposium 3–4 Oct., 1994, Humboldt-Univ. zu Berlin. Ökologische Hefte der Landwirtschaftlich-Gärtnerischen Fakultät Berlin, Heft 2.

Hunter, R.F. (1964) Home range behaviour in hill sheep, in *Grazing in Terrestrial and Marine Environments*, (ed. D.J. Crisp), pp. 155–171, Blackwell, Oxford.

Hyvärinen, H., Helle, I., Nieminen, M., *et al.* (1977) The influence of nutrition and seasonal conditions on mineral status in the reindeer. *Canadian Journal of Zoology*, 55, 648–655.

Iason, G.R., Duck, C.D. and Clutton-Brock, T.H. (1986) Grazing and reproductive success of red deer: the effect of local enrichment by gull colonies. *Journal of Animal Ecology*, 55, 507–515.

Illius, A.W. and Gordon, I.J. (1991) Prediction of intake and digestion in ruminants by a model of rumen kinetics integrating animal size and plant characteristics. *Journal of Agricultural Science, Camb.*, **116**, 145–157.

Jarman, P.J. and Sinclair, A.R.E. (1979) Feeding strategy and the pattern of resource-partitioning in Ungulates. *Serengeti – Dynamics of an Ecosystem*, (eds A.R.E. Sinclair and M. Norton-Griffiths), pp. 130–163, Chicago University Press, Chicago.

Jarrige, R. and Martin-Rosset, W. (eds) (1984) *Le Cheval: Reproduction, Sélection, Alimentation, Exploitation*, INRA, Paris.

Jedrzejewska, B., Okarma, H., Jedrzejewski, W. and Milkowksi, L. (1994) Effects of exploitation and protection on forest structure, ungulate density and wolf predation in Bialowieza Primeval Forest, Poland. *Journal of Applied Ecology*, 31, 664–676.

Johanson, K.J., Bergström, R., Eriksson, O. and Erixon, A. (1994) Activitiy concentrations of Cs-137 in moose and their forage plants in Mid-Sweden. *Journal of Environmental Radioactivity*, 22, 251–267.

Jones, R.L. and Hanson, H.C. (1985) *Mineral Licks, Geophagy, and Biogeochemistry of North American Ungulates*, The Iowa State University Press, Ames, Iowa.

Kaji, K., Koizumi, T. and Ohtaishi, N. (1988) Effects of resource limitation on the physical and reproductive condition of Sika deer on Nakanoshima Island, Hokkaido. *Acta Theriologica*, 33, 187–208.

Kelsall, J.P. (1968) *The Migratory Barren-Ground Caribou of Canada*, Canadian Wildlife Service, Ottawa.

Ketelaars, J.J.M.H. and Tolkamp, B.J. (1991) Toward a new theory of feed intake regulation in ruminants. Doctoral thesis, Wageningen Agricultural University, The Netherlands.

Klein, D.R. (1965) Ecology of deer range in Alaska. *Ecological Monographs*, **35**, 259–294.

Klein, D.R. (1968) The introduction, increase, and crash of reindeer on St Matthew island. *Journal of Wildlife Management*, **32**, 350–367.

Klein, D.R. (1970) Food selection by North American deer and their response to over-utilization of preferred plant species, in *Animal Populations in Relation to their Food Resources*, (ed. A. Watson), pp. 25–44, Blackwell, Oxford.

Klein, D.R. (1991) Caribou in the changing North. *Applied Animal Behaviour Science*, **29**, 279–291.

Krefting, L.W. (1974) *The Ecology of the Isle Royale Moose with Special Reference to the Habitat*, University of Minnesota Agricultural Experimental Station Technical Bulletin 297-1974, Forestry Series 15.

Kreulen, D.A. (1985) Lick use by large herbivores: a review of benefits and banes of soil consumption. *Mammal Review*, **15**, 107–123.

Kurkela, P. (1980) Green plant feeding of reindeer with reference to selenium, in *Proceedings 2nd International Reindeer/Caribou Symposium, Röros, Norway, 1979*, (eds E. Reimers, E. Gaare and S. Skjenneberg), pp. 207–212, Trondheim, Norway.

Larter, N.C. and Gates, C.C. (1991) Diet and habitat selection of wood bison in relation to seasonal changes in forage quantitiy and quality. *Canadian Journal of Zoology*, **69**, 2677–2685.

Launchbaugh, K.L. and Urness, P.J. (1992) Mushroom consumption (mycophagy) by North American cervids. *Great Basin Naturalist*, **52**, 321–327.

Laycock, W.A. (1978) Coevolution of poisonous plants and large herbivores on rangelands. *Journal of Range Management*, **31**, 335–342.

Leader-Williams, N. (1988) *Reindeer on South Georgia: the ecology of an introduced population*, Cambridge University Press, Cambridge.

Lindstedt, S.L. and Boyce, M.S. (1985) Seasonality, fasting endurance, and body size in mammals. *American Naturalist*, **125**, 873–878.

Little, D.A. (1982) Utilization of minerals, in *Nutritional Limits to Animal Production from Pastures*, (ed. J.B. Hacker), pp. 259–283, CAB, Farnham Royal, UK.

Loudon, A.S.I. (1994) Photoperiod and the regulation of annual and circannual cycles of food intake. *Proceedings of the Nutrition Society*, **53**, 495–507.

Lusky, K., Bohm, D., Stoyke, M. *et al.* (1992) Studies in environmental contaminants in wild boars, red deer, roe deer, mouflon, and fallow deer from the Biosphere Reservation Schorfheide-Chorin. *Archiv für Lebensmittel Hygiene*, **43**, 131–136.

Mackintosh, C.G. and Beatson, N.S. (1985) Relationships between diseases of deer and those of other animals. *Biology of Deer Production, Bulletin*, **22**, 77–82.

Maddock, L. (1979) The 'migration' and grazing succession, in *Serengeti: Dynamics of an Ecosystem*, (eds A.R.E. Sinclair and M. Norton-Griffiths), pp. 104–129, University of Chicago Press, Chicago.

Maublanc, M.L., Cibien, C., Gaillard, J.M. *et al.* (1991) Le Chevreuil. *Revue d'Ecologie (La Terre et la Vie)*, **Suppl. 6**, 155–183.

McDowell, L.R. (1985) *Nutrition of Grazing Ruminants in Warm Climates*, Academic Press, Orlando.

McLean, R.W. and Ternouth, J.H. (1994) The growth and phosphorus kinetics of steers grazing a subtropical pasture. *Australian Journal of Agricultural Research*, **45**, 1831–1845.

McNaughton, S.J. (1984) Grazing lawns: animals in herds, plant form and co-evolution. *American Naturalist*, **124**, 863–886.

McNaughton, S.J. (1988) Mineral nutrition and spatial concentrations of African Ungulates. *Nature*, **334**, 343–345.

McNaughton, S.J. (1990) Mineral nutrition and seasonal movements of African migratory Ungulates. *Nature*, **345**, 613–615.

McNay, R.S. and Voller, J.M. (1995) Mortality causes and survival estimates for adult female Columbian black-tailed deer. *Journal of Wildlife Management*, **59**, 138–146.

McRoberts, R.E., Mech, L.D. and Peterson, R.O. (1995) The cumulative effect of consecutive winters' snow depth on moose and deer populations – a defence. *Journal of Animal Ecology*, **64**, 131–135.

Messier, F. (1991) The significance of limiting and regulating factors on the demography of moose and white-tailed deer. *Journal of Animal Ecology*, **60**, 377–393.

Messier, F. (1994) Ungulate population models with predation: a case study with the North American Moose. *Ecology*, **75**, 478–488.

Mitchell, B., McCowan, D. and Parish, T. (1973) Some characteristics of natural mortality among wild Scottish red deer (*Cervus elaphus* L.), in *Union Internationale des Biologistes du Gibier, Actes du Xe Congrès (1971)*, pp. 437–449, Paris.

Mitchell, B., McCowan, D. and Nicholson, I.A. (1976) Annual cycles of body weight and condition in Scottish Red deer, *Cervus elaphus*. *Journal of Zoology, Lond.*, **180**, 107–127.

Mitchell, B., Staines, B.W. and Welch, D. (1977) *Ecology of Red Deer. A research review relevant to their management in Scotland*, Institute of Terrestrial Ecology, Cambridge.

Morris, J.G. and Gartner, R.J.W. (1971) The sodium requirements of growing steers given an all-sorghum grain ration. *British Journal of Nutrition*, **25**, 191–205.

Murray, M.G. (1991) Maximizing energy retention in grazing ruminants. *Journal of Animal Ecology*, **60**, 1029–1045.

Murray, M.G. (1995) Specific nutrient requirements and migration of wildebeest, in *Serengeti II. Dynamics, Management, and Conservation of an Ecosystem*, (eds A.R.E. Sinclair and P. Arcese), pp. 231–256, University of Chicago Press, Chicago.

Norton, B.W. (1982) Differences between species in forage quality, in *Nutritional Limits to Animal Production from Pastures*, (ed. J.B. Hacker), pp. 89–110, CAB, Farnham Royal, UK.

NRC (1983) *Selenium in Nutrition*, revised edn, National Academy Press, Washington, DC.

NRC (1992) *Grassland and Grassland Science in North China*, National Academic Press, Washington, DC.

Odenya, W.O., Elzo, M.A., Manrique, C. *et al.* (1992) Genetic and environmental factors affecting serum macrominerals and weights in an Angus-Brahman multibreed herd. 2. Heritabilities of genetic, environmental, and phenotypic correlations among serum calcium, phosphorus, and magnesium and weight at weaning. *Journal of Animal Science*, **70**, 2072–2077.

Owen-Smith, N. (1982) Factors influencing the consumption of plant products by large herbivores, in *Ecology of Tropical Savannas*, (eds B.I. Huntley and B.H. Walker), pp. 359–404, Ecological Studies 42, Springer Verlag, Berlin.

Owen-Smith, N. (1990) Demography of a large herbivore, the greater kudu *Tragelaphus strepsiceros*, in relation to rainfall. *Journal of Animal Ecology*, **59**, 893–913.

Owen-Smith, N. (1993) Comparative mortality rates of male and female kudus: the costs of sexual size dimorphism. *Journal of Animal Ecology*, **62**, 428–440.

Parker, K.L., Gillingham, M.P., Hanley, T.A. and Robbins, C.T. (1993) Seasonal patterns in body mass, body composition, and water transfer rates of free-ranging and captive black-tailed deer (*Odocoileus hemionus sitkensis*) in Alaska. *Canadian Journal of Zoology*, **71**, 1397–1404.

Pastor, J., Dewey, B., Naiman, R.J. *et al.* (1993) Moose browsing and soil fertility in the boreal forests of Isle Royale National Park. *Ecology*, **74**, 467–480.

Penzhorn, B.L. and P.A. Novellie (1991) Some behavioural traits of Cape Mountain zebras (*Equus zebra zebra*) and their implications for the management of a small conservation area. *Applied Animal Behaviour Science*, **29**, 293–299.

Perzanowski, K. (1978) The effect of winter food composition on Roe deer energy budget. *Acta Theriologica*, **23**, 451–467.

Peters, K. (1992) *Begrazing door runderen gedurende de laatste eeuwen in het woud van Bialowieza (N.O. Polen/W. Wit–Rusland)*. Undergraduate Report no. 3022, Wageningen Agricultural University, The Netherlands.

Prins, H.H.T. (1981) Why are mosses eaten in cold environments only? *Oikos*, **38**, 374–380.

Prins, H.H.T. (1996) *Ecology and Behaviour of the African Buffalo: social inequality and decision making*, Chapman & Hall, London.

Prop, J. and Vulink, T. (1992) Digestion by barnacle geese in the annual cycle: the interplay between retention time and food quality. *Functional Ecology*, **6**, 180–189.

Provenza, F.D. (1995) Postingestive feedback as an elementary determinant of food preference and intake in ruminants. *Journal of Range Management*, **48**, 2–17.

Putman, R.J. (1986) *Grazing in Temperate Ecosystems: large herbivores and the ecology of the New Forest*, Croom Helm, London.

Ralphs, M.H. and Pfister, J.A. (1992) Cattle diets in tall forb communities on mountain rangelands. *Journal of Range Management*, **45**, 534–537.

Ratcliffe, P.R. (1984) Population dynamics of red deer (*Cervus elaphus* L.) in Scottish commercial forests. *Proceedings Royal Society of Edinburgh*, **82B**, 291–302.

Reid, T.C., McAllum, H.J.F. and Johnstone, P.D. (1980) Liver copper concentrations in re deer (*Cervus elaphus*) and wapiti (*C. canadensis*) in New Zealand. *Research in Veterinary Science*, **28**, 261–262.

Reimers, E. (1980) Activity pattern: the major determinant for growth and fattening in Rangifer? in *Proceedings 2nd International Reindeer/Caribou Symposium, Röros, Norway, 1979*, (eds E. Reimers, E. Gaare and S. Skjenneberg), pp. 466–474, Trondheim, Norway.

Robbins, C.T. (1993) *Wildlife Feeding and Nutrition*, 2nd edn, Academic Press, New York.

Russell, F.C., Duncan, D.L. and Greene, H. (1956) *Minerals in Pasture: deficiencies and excesses in relation to animal health*, Technical Communication No. 15, 2nd edn, CAB, Farnham Royal, UK.

Schaller, G.B. (1972) *The Serengeti Lion: a study of predator–prey relationships*, Chicago University Press, Chicago.

Schaller, G.B. (1977) *Mountain Monarchs*, Chicago University Press, Chicago.

Schonewald-Cox, C.M. (1983) Conclusions – Guidelines to management: a beginning attempt, in *Genetics and Conservation: a reference for managing wild animal and plant populations*, (eds C.M. Schonewald-Cox, S.M. Chambers, B. MacBryde and W.L. Thomas), pp. 141–145, Benjamin/Cummings, Menlo Park, California.

Seagle, S.W. and McNaughton, S.J. (1992) Spatial variation in forage nutrient concentrations and the distribution of Serengeti grazing ungulates. *Landscape Ecology*, 7, 229–241.

Seip, D.R. (1992) Factors limiting woodland caribou populations and their interrelationships with wolves and moose in Southeastern British Columbia. *Canadian Journal of Zoology*, **70**, 1494–1503.

Seydack, A.H.W. and Bigalke, R.C. (1992) Nutritional ecology and life history tactics in the bushpig (*Potamochoerus porcus*): development of an interactive model. *Oecologia*, **90**, 102–112.

Sinclair, A.R.E. (1977) *The African Buffalo: a study of resource limitation of populations*, University of Chicago Press, Chicago.

Sinclair, A.R.E. (1979) The eruption of the ruminants. *Serengeti – Dynamics of an Ecosystem*, (eds A.R.E. Sinclair and M. Norton-Griffiths), pp. 82–103, Chicago University Press, Chicago.

Sinclair, A.R.E. (1983) The adpatations of African ungulates and their effects on community function, in *Ecosystems of the World: 13. Tropical Savannas*, (ed. F. Bourlière), pp. 401–426, Elsevier, Amsterdam.

Sinclair, A.R.E. and Norton-Griffiths, M. (eds) (1979) *Serengeti – Dynamics of an Ecosystem*. University of Chicago Press, Chicago.

Sinclair, A.R.E., Dublin, H. and Borner, M. (1985) Population regulation of Serengeti Wildebeest: a test of the food hypothesis. *Oecologia*, **65**, 266–268.

Singer, F.J. and Norland, J.E. (1994) Niche relationships within a guild of ungulate species in Yellowstone National Park, Wyoming, following release from artificial controls. *Canadian Journal of Zoology*, **72**, 1383–1394.

Skogland, T. (1983) The effects of density-dependent resource limitation on size of wild reindeer. *Oecologia*, **60**, 150–168.

Skogland, T. (1984) Wild reindeer foraging–niche organization. *Holarctic Ecology*, 7, 345–379.

Skogland, T. (1985) The effects of density-dependent resource limitations on the demography of wild reindeer. *Journal of Animal Ecology*, **54**, 359–374.

Skogland, T. (1991) What are the effects of predators on large ungulate populations? *Oikos*, **61**, 401–411.

Smithey, D.A., Wisdom, M.J. and Hines, W.W. (1985) Roosevelt elk and black-tailed deer response to habitat changes related to old-growth forest conversion in southwestern Oregon, in *Proceedings, 1984 Western States and Provinces Elk Workshop, April 17–19*, (ed. R.W. Nelson), pp. 41–55, Edmonton, Alberta.

Soulé, M.E. (ed.) (1987) *Viable Populations for Conservation*, Cambridge University Press, Cambridge.

Staaland, H., White, R.G., Luick, J.R. and Holleman, D.F. (1980) Dietary influences on sodium and potassium metabolism of reindeer. *Canadian Journal of Zoology*, **58**, 1728–1734.

Staines, B.W. (1974) A review of factors affecting deer dispersion and their relevance to management. *Mammal Review*, **4**, 79–91.

Staines, B.W. (1977) Factors affecting the seasonal distribution of red deer (*Cervus elaphus*) at Glen Dye, north-east Scotland. *Annals of Applied Biology*, **87**, 495–512.

Stanley Price, M.R. (1989) *Animal Re-introductions: the Arabian oryx in Oman*, Cambridge University Press, Cambridge.

Storm, G.L., Fosmire, G.J. and Bellis, E.D. (1994) Persistence of metals in soil and selected vertebrates in the vicinity of the Palmerton zinc smelters. *Journal of Environmental Quality*, **23**, 508–514.

Strandberg, M. and Knudsen, H. (1994) Mushroom spores and Cs-137 in faeces of roe deer. *Journal of Environmental Radioactivity*, **23**, 189–203.

Sutcliffe, A.J. (1977) Further notes on bones and antlers chewed by deer and other ungulates. *Journal of the British Deer Society*, **4**, 73–82.

Sutherst, R.W. (1987) Ectoparasites and herbivore nutrition, in *The Nutrition of Herbivores*, (eds J.B. Hacker and J.H. Ternouth), pp. 191–209, Academic Press, Sydney.

Suttle, N.F. (1986) Problems in the diagnosis and anticipation of trace element deficiencies in grazing livestock. *Veterinary Record*, **119**, 148–152.

Sykes, A.R. (1987) Endoparasites and herbivore nutrition, in *The Nutrition of Herbivores*, (eds J.B. Hacker and J.H. Ternouth), pp. 211–232, Academic Press, Sydney.

Telfer, E.S. and Scotter, G.W. (1975) Potential for game ranching in boreal aspen forests of western Canada. *Journal of Range Management*, **28**, 172–180.

Tendron, G. (1983) *La Forêt de Fontainebleau: de l'écologie à la sylviculture*, Office National des Forêts, Fontainebleau.

Terlecki, S., Done, J.T. and Clegg, F.G. (1964) Enzootic ataxia of red deer. *British Veterinary Journal*, **120**, 311–321.

Thomas, C.D. (1990) What do real population dynamics tell us about minimum viable population sizes? *Conservation Biology*, **4**, 324–327.

Tracy, B.F. and McNaughton, S.J. (1995) Elemental analysis of mineral lick soils from the Serengeti National Park, the Konza Prairie and Yellowstone National Park. *Ecography*, **18**, 91–94.

Tyler, N.J.C. (1987) Body composition and energy balance of pregnant and non-pregnant Svalbard reindeer during winter. *Symposium Zoological Society London*, **57**, 203–229.

Ulyatt, M.J. (1981) The feeding value of temperate pastures, in *Grazing Animals*, (ed. F.H.W. Morley), pp. 125–141, Elsevier, Amsterdam.

Underwood, E.J. (1981) *The Mineral Nutrition of Livestock*, CAB, Farnham Royal, UK.

Van der Ouderaa, A.P.M. (1991) *Mineralen – De mineralenstatus van Schotse Hooglanders in terreinen van Staatsbosbeheer*, Rapport, 1991-1. Staatsbosbeheer, Driebergen.

Van Deursen, M., Cornelissen, P., Vulink, J.Th. and Esselink, P. (1993) Jaarrondbegrazing in de Lauwersmeer: zelfredzaamheid van grote grazers en effecten op de vegetatie. *De Levende Natuur*, **94**, 196–204.

Van Dyne, G.M., Brockington, N.R., Szocs, Z. *et al.* (1980) Large herbivore subsystem, in *Grasslands, Systems Analysis and Man*, (eds A.I. Breymeyer and G.M. Van Dyne), pp. 269–537, IBP 19, Cambridge University Press, Cambridge.

Van Soest, P.J. (1994) *Nutritional Ecology of the Ruminant*, 2nd edn, Cornell University Press, Ithaca, NY.

Van Wieren, S.E. (1988) *Runderen in het bos*, Instituut voor Milieuvraagstukken, Vrije Universiteit, Amsterdam.

Van Wieren, S.E. (1992) Factors limiting food intake in ruminants and non ruminants in the temperate zone, in *Ongulés/Ungulates 91*, (eds F. Spitz, G. Janeau, G. Gonzalez and S. Aulagnier), pp. 139–145, SFEPM-IRGM, Paris – Toulouse.

Van Wieren, S.E. (1996) Digestive strategies in ruminants and nonruminants. Doctoral thesis, Wageningen Agricultural University, The Netherlands.

Vera, F.W.M. (1997) Metaforen voor de Wildernis: Eik, Hazelaar, Rund en Paard. Doctoral thesis, Wageningen Agricultural University, Wageningen, The Netherlands.

Vercoe, J.E. and Frisch, J.E. (1982) Animal breeding for improved productivity, in *Nutritional Limits to Animal Production from Pastures*, (ed. J.B. Hacker), pp. 327–342, CAB, Farnham Royal, UK.

Vulink, J.T. and Drost, H.J. (1991) Nutritional characteristics of cattle forage plants in the eutrophic nature reserve Oostvaardersplassen. *Netherlands Journal of Agricultural Science*, **39**, 263–272.

Wahlström, D.J. and Kjellander, F. (1995) Ideal free distribution and natal dispersal in female roe deer. *Oecologia*, **103**, 302–308.

WallisDeVries, M.F. (1989) *Beperkende Factoren in het Voedselaanbod voor Runderen en Paarden op Schrale Graslanden en Droge Heide*, Communication no. 252, Wageningen Agricultural University, The Netherlands.

WallisDeVries, M.F. (1994) Foraging in a landscape mosaic: diet selection and performance of free-ranging cattle in heathland and riverine grassland. Doctoral thesis, Wageningen Agricultural University, The Netherlands.

WallisDeVries, M.F. (1995) Large herbivores and the design of large-scale nature reserves in Western Europe. *Conservation Biology*, **9**, 25–33.

WallisDeVries, M.F. (1996) Nutritional limitations of free-ranging cattle: the importance of habitat quality. *Journal of Applied Ecology*, **33**, 688–702.

WallisDeVries, M.F. and Daleboudt, C. (1994) Foraging strategy of cattle in patchy grassland. *Oecologia*, **100**, 98–106.

WallisDeVries, M.F. and Schippers, P. (1994) Foraging in a landscape mosaic: selection for energy and minerals in free-ranging cattle. *Oecologia*, **100**, 107–117.

Webster, A.J.F. (1985) Differences in the energetic efficiency of animal growth. *Journal of Animal Science*, **61, Suppl. 2**, 92–103.

Weeks, H.P. and Kirkpatrick, C.M. (1976) Adaptations of white-tailed deer to naturally occurring sodium deficiencies. *Journal of Wildlife Management*, **40**, 610–625.

Weir, J.S. (1972) Spatial distribution of elephants in an African National Park in relation to environmental sodium. *Oikos*, **23**, 1–13.

Westermarck, H. and Kurkela, P. (1980) Selenium content in lichen in Lapland South Finland and its effect on the selenium values in reindeer, in *Proceedings 2nd*

International Reindeer/Caribou Symposium, Röros, Norway, 1979, (eds E. Reimers, E. Gaare and S. Skjenneberg), pp. 278–285, Trondheim, Norway.

Western, D. (1975) Water availability and its influence on the structure and dynamics of a savannah large mammal community. *East African Wildlife Journal*, **13**, 265–286.

Weston, R.H. and Poppi, D.P. (1987) Comparative aspects of food intake, in *The Nutrition of Herbivores*, (eds J.B. Hacker and J.H. Ternouth), pp. 133–161, Academic Press, Sydney.

White, T.C.R. (1978) The importance of a relative shortage of food in animal ecology. *Oecologia*, **33**, 71–86.

Williamson, D., Williamson, J. and Ngwamotsoko (1988) Wildebeest migration in the Kalahari. *African Journal of Ecology*, **26**, 269–280.

Wolkers, H., Wensing, T. and Groot Bruinderink, G.W.T.A. (1994) Heavy metal contamination in organs of red deer (*Cervus elaphus*) and wild boar (*Sus scrofa*) and the effect on some trace elements. *Science of the Total Environment*, **144**, 191–199.

Ydenberg, R.C. and Prins, H.H.T. (1981) Spring grazing and the manipulation of food quality by barnacle geese. *Journal of Applied Ecology*, **18**, 443–453.

Young, B.A. (1987) The effect of climate upon intake, in *The Nutrition of Herbivores*, (eds J.B. Hacker and J.H. Ternouth), pp. 163–190, Academic Press, Sydney.

Yuill, T.M. (1987) Diseases as components of mammalian ecosystems: mayhem and subtlety. *Canadian Journal of Zoology*, **65**, 1061–1066.

10

The role of scientific models

Michiel F. WallisDeVries[1] and Johan Van de Koppel[2]
[1]Tropical Nature Conservation and Vertebrate Ecology Group, Department of Environmental Sciences, Wageningen Agricultural University, Bornsesteeg 69, 6708 PD Wageningen, The Netherlands
[2]Laboratory of Plant Ecology, University of Groningen, PO Box 14, 9750 AA Haren, The Netherlands

10.1 INTRODUCTION

The management of large herbivores and the application of grazing in conservation management has been guided by a combination of trial-and-error and scientific investigations. This book has provided an overview of the scientific knowledge concerning relevant aspects of grazing systems and their management. In this chapter we attempt to identify scientific achievements with respect to their function in supporting the management of nature reserves. This will be done by considering fields of research that are of direct significance to management practice (Table 10.1).

For each of these fields an overview of important models is given. The choice to review models has been prompted by their capacity to synthesize current knowledge and experience and to arrive at new insights. Models have become a powerful tool for translating the complexity of reality into meaningful concepts. As tools, models have the potential to supply substantial support for management in the form of decision support models and expert systems.

10.2 AN OVERVIEW OF RELEVANT MODELS

Models come in many different kinds, with different goals and approaches, as reviewed by Morrison *et al.* (1992). DeAngelis (1992) distinguished three functional classes of model: descriptive, explanatory and predictive. Descriptive models describe the important components of a system and the way they interrelate. They may be of a conceptual or quantitative form. Quantitative relations are derived by statistical methods involving especially

Grazing and Conservation Management. Edited by M.F. WallisDeVries, J.P. Bakker and S.E. Van Wieren. Published in 1998 by Kluwer Academic Publishers, Dordrecht. ISBN 0 412 47520 0.

Table 10.1 Major topics involved in conservation management decisions in areas with large herbivores and examples of relevant models or reviews

Topic	Examples
Herbivore basics	
Food intake and digestion	Ketelaars and Tolkamp, 1991; Baker *et al.*, 1992; Illus and Gordon, 1991, 1992; Van Soest, 1994; Gordon and Illus, 1996
Animal performance	Swift, 1983; Hudson and White, 1985; Fox *et al.*, 1988
Foraging and intake	Belovsky, 1978; Sibbald *et al.*, 1979; Illus and Gordon, 1987; Blackburn and Kothmann, 1989; Murray, 1991; Spalinger and Hobbs, 1992; Laca and Demment, 1991, 1996; Dove, 1996
Habitat suitability	
Habitat use	Senft *et al.*, 1983, 1987; Cook and Irwin, 1985; Loza *et al.*, 1992; Fryxell, 1991; Seagle and McNaughton, 1992; Morrison *et al.*, 1992; WallisDeVries and Schippers, 1994; Edwards *et al.*, 1996
Spatial distribution	Coughenour, 1991; Hyman *et al.*, 1991; Pickup, 1994; Turner *et al.*, 1993; Bailey *et al.*, 1996; WallisDeVries, 1996; Milne, 1997; Moen *et al.*, 1997
Habitat evaluation	Connoly, 1974; Caughley, 1976; Benson and Laudenslayer, 1985; Raedeke and Lehmkuhl, 1985; Hobbs and Swift, 1985; Biot, 1993; Abel, 1993; De Ridder and Breman, 1993; Behnke and Scoones, 1993; Stafford Smith, 1996
Population dynamics	
Plant–animal and predator–prey interactions	Rosenzweig, 1971; Noy-Meir, 1975; Caughley, 1976; Caughley and Lawton, 1981; Walker *et al.*, 1981; Oksanen *et al.*, 1981; Owen-Smith, 1988; Dublin *et al.*, 1990; Oksanen, 1992; Messier, 1994; Holt *et al.*, 1994; Grover, 1994, 1995; Abrams and Matsuda, 1996; Van de koppel *et al.*, 1996; Rietkerk and Van de Koppel, 1997
Population genetics, demography and stochasticity	Gilpin and Soulé, 1986; Gilpin, 1991; Lande, 1993, 1995
Herbivore impact	
Plant community dynamics	Walker *et al.*, 1981; Thalen *et al.*, 1987; Fresco *et al.*, 1987; Tilman, 1988; Westoby *et al.*,

Table 10.1 (*continued*)

Topic	Examples
	1989; Milchunas *et al.*, 1988; Milchunas and Lauenroth, 1993; Dublin *et al.*, 1990; Morrison *et al.*, 1992; Pastor and Naiman, 1992; Li, 1995; Loehle *et al.*, 1996; Van Deursen and Van Oene, 1996; Jortisma *et al.*, 1997
Nutrient flows	Innis *et al.*, 1981; Berendse, 1985; Turner, 1988; Pastor and Naiman, 1992; DeAngelis, 1992; Lemaire and Chapman, 1996
Animal community dynamics	Van Wijngaarden, 1985; Oksanen, 1992; Matsuda *et al.*, 1996
Conservation management	
Decision support models	Holling, 1978; Kirkman *et al.*, 1985; Rabinovich *et al.*, 1985; Morrison *et al.*, 1992; Hilborn, 1995; Llewyn *et al.*, 1996; Stafford Smith, 1996
Expert systems	Marcot, 1985); Ritchie, 1989

regression and multivariate analysis. Descriptive models do not rely directly on explanatory mechanisms.

The advance of our understanding of mechanisms and processes is the aim of explanatory models. These are typically based on a set of mathematical equations that quantify the important relationships and serve to investigate the effect of varying system parameters. The two main types of mathematical explanatory model are analytical models, whose results are obtained by solving equations purely by mathematics, and simulation models, which arrive at results through the use of computation. The latter type is considered less elegant from a theoretical point of view but has the power to address situations where mathematical solutions are too complex or no longer exist. Both analytical and simulation models yield predictions of the behaviour of a system that may be tested in the field. Yet the predictions are often of a qualitative nature. The practical use of explanatory models is primarily to identify underlying mechanisms.

Precise predictions of changes in ecosystems are rather the domain of predictive models, which build on the insight provided by theory in combination with empirical data. Predictive models are set up to predict the consequences of human interference or management scenarios. Larger models are typically complex simulation models with different compartments for the various components of a system. Smaller, more pragmatic models often incorporate descriptive components that rely on statistical information or on a more qualitative approach. The major limitation of pragmatic predictive models is that they are usually specific to a certain situation, which precludes broader application.

Before they are applied in management, predictive models should be thoroughly validated by testing the accuracy of their predictions against an independent set of data, i.e. data not used to build the model. This validation stage is crucial but unfortunately it is often omitted, as time or resources are limited. It needs to be emphasized that adequate field data are a necessity for the development of any realistic model if it is to be of any practical use.

10.2.1 Herbivore basics

The topics presented in Table 10.1 each cover some element in the field of conservation management in relation to large herbivores. Basic knowledge of herbivore feeding and performance is essential in the management of herbivores. Good data for food intake, digestion and performance are available for many of the large ungulates, especially livestock (Chapter 9). The data have been incorporated in a variety of models for intake regulation and, less frequently, for animal performance. In the absence of independent validation, the success of these models is often difficult to judge. Many controversies remain to be solved to explain existing variation between species (Van Wieren, 1996) as well as within species (Ketelaars and Tolkamp, 1991). Foraging models, which attempt to capture the process of diet selection, are in an even more premature stage of development. There is little quantitative insight into the process of decision making leading to species preferences and selective foraging (but see Wilmshurst *et al.*, 1995). Still, Laca and Demment (1996) reported promising progress through the application of optimal foraging theory. They argued that weaknesses in optimal foraging theory can be overcome by refining constraints, integrating animal state with behaviour other than foraging and by taking spatial heterogeneity into account.

10.2.2 Habitat suitability

When turning to the habitat level, the previous considerations should temper high expectations . A number of models of habitat choice attempt to circumvent the lack of data on decision making by introducing a qualitative index to describe habitat preference. Senft *et al.* (1987) and Bailey *et al.* (1996) provided a conceptual framework for the process of habitat selection. They argued that different scales of time and space should be recognized and interrelated in a hierarchical manner. While the elements of a more solid basis for habitat selection models are thus being addressed, the existing models are still predominantly explorative. However, some predictive habitat selection models take a more pragmatic, descriptive approach by using statistical information on habitat use. The studies by Senft *et al.*

(1983), Cook and Irwin (1985) and Edwards *et al.* (1996) show that this approach can be quite successful in predicting habitat use and species distributions, although evidently limited to specific cases.

Habitat suitability is always a function of specific conditions and is therefore subject to change. This clearly limits any attempt to estimate carrying capacity (Chapter 9). Yet such an estimate may still be useful for the short term. Only when year-to-year variation (especially erratic rainfall in semi-arid climates) becomes so great as to create conditions of permanent disequilibrium does the concept of an average carrying capacity lose its meaning. Modern range science is now developing new approaches to appropriate management in the face of non-equilibrium dynamics (Westoby *et al.*, 1989; Behnke *et al.*, 1993). Over greater scales of time and space it may again become appropriate to think in terms of equilibria (DeAngelis and Waterhouse, 1987). Milne (1997) shows that the application of fractal geometry can be a powerful tool for evaluating habitat availability across spatial scales. Walker *et al.* (1981) and Rietkerk and Van de Koppel (1997) have fruitfully applied equilibrium models to investigate catastrophic vegetation shifts in semi-arid regions.

In a pragmatic approach developed by the US Fish and Wildlife Service, the habitat is evaluated by means of a habitat suitability index, calculated as the geometric mean of those environmental variables (scaled from 0 to 1) considered most influential to species occurrence (Schamberger and O'Neill, 1985; Morrison *et al.*, 1992). Other models rely on species–habitat matrices and data on occurrence probability (Verner *et al.*, 1985; Morrison *et al.* 1992). The impact of given changes in the relative cover of certain habitats in a landscape can thus be evaluated (Benson and Laudenslayer, 1985; Raedeke and Lehmkuhl, 1985). However, this only provides useful information if changes in the landscape are somehow predictable.

An additional problem when evaluating habitat quality is that there is often more than one limiting resource (Chapter 9). Multiple limitations can be analysed by linear programming (Connoly, 1974; Belovsky, 1978; Seagle and McNaughton, 1992), which predicts the optimal resource use by maximizing a certain resource under the constraints of the requirements of others. A perhaps more justified alternative is balanced maximization of all limiting resources (WallisDeVries and Schippers, 1994), as there is no reason why one limitation should be more important than another. A good mechanistic approach to deal with this problem has not yet been developed, mainly because many temporal and spatial aspects influence the trade-off between different resources when these are unevenly distributed (which is usually the case).

10.2.3 Population dynamics

A different group of models deals with herbivore population dynamics. The study of population dynamics is of importance to the management of large herbivores, as changes in the numbers of herbivores may have a large impact on the ecosystem. Firstly, herbivore population changes may alter competitive interactions between plants and between herbivores. Secondly, the persistence of a given population depends partly on its demographic and genetic structure. It is well established that stochastic factors and inbreeding in small populations pose a threat to their survival (Hedrick *et al.*, 1996). The extinction of the aurochs (*Bos primigenius*) in 1627, despite rudimentary efforts to preserve the species, was probably the first experience of conservation management with this problem (Szafer, 1968). Several models have addressed these phenomena, but they are not well suited for predictive purposes. With the increasing fragmentation of modern landscapes and the resulting isolation of reproductive populations, population dynamics deserves more attention. This also applies to introduced populations of domestic herbivores. Although these are often managed to some degree and their population size is usually kept more or less constant, the risk of inbreeding depression is rarely considered in management.

The interrelationship between herbivore population dynamics and their food supply has been dealt with in many explanatory plant–herbivore and predator–prey models (Rosenzweig, 1971; Noy-Meir, 1975; Caughley, 1976; Walker *et al.*, 1981; Oksanen *et al.*, 1981; Oksanen, 1992; Holt *et al.*, 1994; Grover, 1994, 1995; Abrams and Matsuda, 1996; Van de Koppel *et al.*, 1996). The role of carnivore predation in herbivore dynamics is also addressed by some of these models, although their relevance for large herbivores is still debated (White, 1978; Chapter 9; but see Oksanen, 1992; Messier, 1994). Predator–prey models, however, do provide us with a better understanding of the conditions under which plant–herbivore equilibria occur. Also, non-equilibrium dynamics have become a recent focus of attention in this field (e.g. Hastings and Powell, 1991; Behnke *et al.*, 1993; McCann and Yodzis, 1994).

10.2.4 Herbivore impact

Incorporation of empirical information is essential in predicting the impact of herbivores on plant community dynamics. This has been done in a variety of models, ranging from calculations by Dublin *et al.* (1990) and a statistical model by Milchunas and Lauenroth (1993) to a more complex simulation model by Pastor and Naiman (1992). The integration of models of vegetation dynamics or plant succession (e.g. Shugart, 1984; Tilman, 1988; Van Andel *et al.*, 1993; Kobe, 1996; Chapter 5) and herbivore foraging models (Table 10.1) is only now starting to receive the necessary attention (Van Deursen and Van Oene, 1996; Jorritsma *et al.*, 1997). Dynamic percolation models (Loehle *et al.*, 1996) and non-homogeneous Markovian models (Li, 1995) can be used to put these processes in a spatial context.

A major problem is in making a careful trade-off between complexity and applicability. When too many simplifying assumptions are included, the model will fail – just as it will when it enters into too much detail. In the end the model will only be as useful as its weakest component allows. In the face of unknown relationships and parameters, the emphasis should be on integrating processes of primary importance at the proper scale (Holling, 1992). Especially when dealing with the dynamics of multispecies systems, the complexity becomes so enormous that theoretical insight will aid in the construction of predictive models – for instance, by revealing the most significant interactions. In this regard recent explanatory models by Oksanen (1992), Holt *et al.* (1994) and Grover (1994, 1995) constitute a promising development: they arrive at community assembly rules by analysing the consequences of shared resource utilization and herbivory (or predation in general).

Knowledge about the impact of herbivores on nutrient flow is especially valuable when nutrient turnover and availability are significantly affected, or when dealing with the transport of nutrients between different communities. In modelling, nutrient flows can be linked to competition between plant species and can thus be used to assess changes in the dominance of plant species (Berendse, 1985; DeAngelis, 1992; Pastor and Naiman, 1992). This is a potentially powerful mechanistic approach in addressing vegetation dynamics. It remains unclear, however, whether such models can be successfully extended to the ecosystem level, including the dynamics of rare species and those of higher trophic levels. Such a development has been initiated by Grover (1994, 1995) and remains a major challenge for the future.

The least covered topic among those listed in Table 10.1 is the impact that herbivores may have on the animal community (Chapter 6). This is unfortunate, since the management of nature reserves often focuses on the preservation of particular animal species. In one of the few theoretical contributions to this subject, Prins and Olff (1998) discuss the relations that link the members of large herbivore assemblages. They explain the composition of herbivore assemblages in terms of a combination of competitive exclusion (between similar-sized species) and facilitation (of smaller species by larger ones). Along these lines, Van Wijngaarden (1985) has modelled the impact of African elephants on the composition of the ungulate community. It should be worth examining whether this principle could be applied to examination of the role of large herbivores in shaping the composition of the animal community in a more general perspective, including other species groups as well. The work of Oksanen (1992), Holt *et al.* (1994) and Matsuda *et al.* (1996) provides a theoretical framework for such an endeavour. Using a pragmatic approach, Benson and Laudenslayer (1985) modelled the effect of habitat composition on the occurrence of various target animal species. By including the impact of herbivores on their habitat, one could build an applied model of herbivore–animal community relationships.

10.2.5 Conservation management

Many predictive models have been designed to assist the management in making sensible decisions on a future course of action. When this is the explicit aim of a model, it is called a decision support system. Either it indicates a feasible action plan to fit certain management goals and priorities or it predicts the consequences of alternative types of management. In doing so, decision support systems help to set priorities and to choose between possible options (Hilborn, 1995).

An expert system is a computer program that uses facts and rules to solve a management problem (Marcot, 1985). It is a more elaborate type of system than a decision support system in that it includes a judgement by expert opinion. Dealing with management decisions in modelling implies the addition of another level of information processing to the part of the model covering the biological aspects of a given ecosystem. This introduces new potential weaknesses. The biological part of the model may be simplified to a certain extent, since management is focused on the main issues, not the details. The outcome of the biological analysis, however, should still be valid and reliable. Holling (1978) gives guidelines for an adaptive modelling approach to assist in environmental management. In the management part of the model new factors need to be taken into account, such as financial, logistic and socioeconomic considerations, and these need to be weighed against each other. Unless these factors can be expressed in a common currency, such as money, this will inevitably introduce subjectivity. Where multiple expert opinions are the rule rather than the exception, one might consider explicitly including subjective views on controversial issues in the evaluation of different scenarios.

Finally, validation of any model involving management decisions usually presents great methodological obstacles. It is not surprising, therefore, that operative expert systems do not flourish in the field of grazing and conservation management. Nevertheless, a modelling approach to management decisions does have the great advantage of consistency and the potential to take a multitude of factors into account to evaluate different scenarios. The challenge is then to reflect adequately the basic assumptions and the degree of uncertainty in the model's outcome.

10.3 CHALLENGES FOR THE FUTURE

Efforts in constructing models often reflect the status of scientific knowledge in a certain field. The best covered topics with respect to large herbivores concern the nutrition, foraging behaviour and habitat use of herbivores on the one hand, and plant–herbivore dynamics on the other hand, though it should be noted that the latter topic has been investigated mainly from a theoretical viewpoint. Although the subject of herbivore impact on plant community composition and dynamics has received extensive attention in empirical studies (Chapter 5), explanatory (Holt *et al.*, 1994; Grover, 1994, 1995; Huisman *et al.*, 1998) and predictive models (Pastor and Naiman, 1992) are seemingly lagging behind. The evident explanation is that the study of the interaction between large herbivores and vegetation is still a young and emerging field in science, struggling to cope with the complexity of the systems it addresses. However, advancing our understanding of grazing systems will require integration of data and a model framework which puts the data in a meaningful perspective (Illius and Hodgson, 1996).

Two major fields of study deserve closer attention. The first concerns the mechanisms that determine the distribution of herbivores in spatially heterogeneous environments. The choice between different habitat types and also the extent to which the spatial configuration of habitat elements affects that choice are topics that urgently require further study if herbivore distribution and, hence, impact are to be predicted with reliability. Until recently, predictions of habitat use were achieved mainly by qualitative estimates or through descriptive models. Present developments rather emphasize a mechanistic approach that unravels underlying processes and integrates these to the level of the ecosystem. Such models have the potential to deal with more complex situations. This approach will not yield immediate results, and therefore statistical and qualitative models, crude though they may be, will remain useful for a long time to come. Only mechanistic models can provide a key to convert scientific understanding to applicability in management.

The second field meriting a high priority is that of plant–herbivore dynamics. Here the emphasis should be on extending the theoretical basis to address more specific questions, more closely linked to the ecology of large herbivores in different environments. The central question to be resolved is: under what conditions will large herbivores act as keystone species? This question was raised in Chapter 1, and only partly answered. When humans play their usual dominant role, conditions can be created in which large herbivores do have a large and crucial impact on their environment. This finding has been used to develop successful management practices in nature reserves (Chapters 2, 7 and 8). However, when humans retreat and leave natural processes to decide the outcome, as with the latest trend in conservation management, the result is mostly unclear. Several cases have been presented, showing situations in which large herbivores have a high impact, but also situations in which herbivores do not appear to play a significant role at all (Chapters 1 and 3). Without fundamental insight into the dynamic behaviour of plant–herbivore systems, generalizations on the conditions under which large herbivores act as keystone species lack a coherent basis.

10.3.1 Developing the theoretical framework

An important issue in the management of an area is to identify whether the desired equilibrium situations are achievable. A wrong assessment of the situation may have far-reaching consequences for the chances of realizing the management goals in a particular area. This section presents a theoretical framework that may provide insight into the conditions needed for stable equilibria between plants and herbivores. It is not the aim to provide solutions to the challenges identified above. Rather, the section summarizes recent theoretical developments and puts these into a management perspective.

The mechanisms underlying the dynamics of plant–herbivore systems involve production–consumption relations and their consequences for population dynamics (e.g. Rosenzweig, 1971; Yodzis, 1989). Most simple models that address these relationships consist of two differential equations:

$$dP/dt = f(P) - c(P,H)$$
$$dH/dt = g(P,H)$$

Here P and H are plant density and herbivore density, respectively; $f(P)$ describes plant production as a function of plant density; $c(P,H)$ describes the consumption of plants by herbivores; and $g(P,H)$ describes the rate of change of the herbivore population.

In most grasslands, vegetation quality declines as plant density increases (McNaughton, 1985; Bazely and Jefferies, 1986; Chapter 9) because of increasing densities of dead plant material or increasing vigilance for predators (Van der Wal *et al.*, 1998), or because of decreasing plant digestibility as plants mature (Fryxell, 1991). As a result, herbivore foraging efficiency and herbivore growth are limited by quantity at low plant density and by quality at high plant density.

The local abundance of herbivores in natural systems is determined by herbivore population growth and migrations of herbivores between systems. In heterogeneous systems, movements between patches, on both a small and a large scale, will be the most important factor determining local herbivore numbers. In more homogeneous systems, population growth will be more influential (Van de Koppel, 1997). We will consider two extreme cases. The first case considers a homogeneous system in which herbivore dynamics are solely determined by growth and mortality. The second case considers a patchy system where the abundance of herbivores within a certain patch is determined by migration to and from this patch.

The equilibria of simple two-dimensional models are usually analysed by phase plane analysis (Figure 10.1). Phase planes are drawn in which plant and herbivore isoclines are depicted, for six different situations. An isocline is a line on which net population change is zero. The plant isocline ($dP/dt = 0$) is often represented by an arch-shaped curve; density increases below the isocline and decreases above it. The shape of the herbivore isocline ($dH/dt = 0$) varies with the situation considered (for details, see Van de Koppel *et al.*, 1996, 1997). In situations A, C and E, a system is considered where the dynamics of the herbivore population are determined by growth and mortality. The two vertical herbivore isoclines reflect two different plant densities where herbivore growth rate equals zero. Only in between these isoclines do the herbivores obtain a positive growth. In situations B, D and F, herbivore population dynamics are determined by migration. The herbivore isocline is hump-shaped in such systems; herbivore density increases beneath the isocline, and decreases above it.

At the intercepts of plant and herbivore isoclines, equilibria are found. These equilibria may be either stable or unstable. An equilibrium is stable when a system returns to this equilibrium after a disturbance. In contrast, when an equilibrium is unstable, the system will run away from the equilibrium at the slightest disturbance. A system may contain one or more locally stable states, and a number of unstable equilibria (for a detailed analysis of these systems, see Van de Koppel *et al.*, 1996; Van de Koppel, 1997).

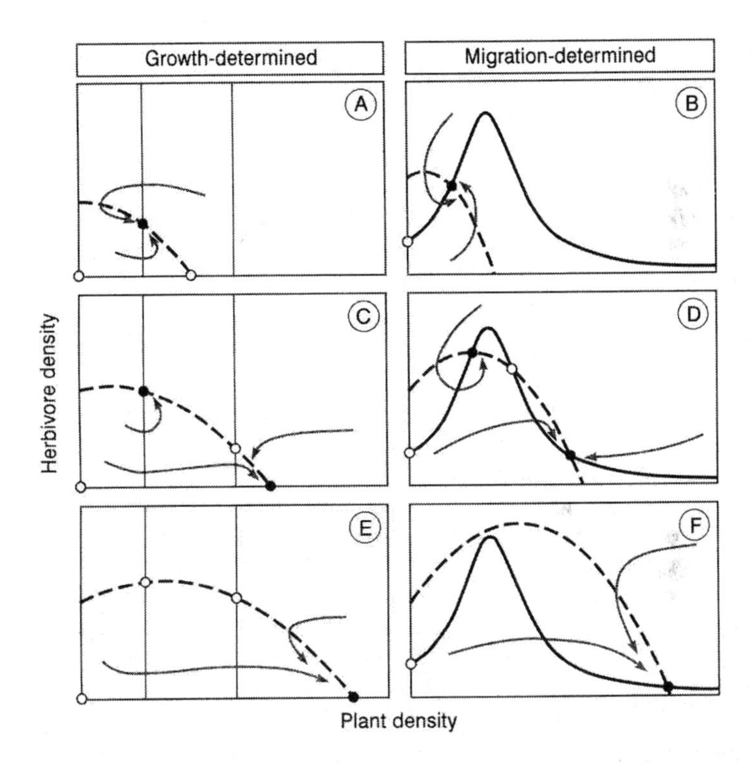

Figure 10.1 Phase plane representations of six basic types of plant–herbivore equilibria (see text for further explanations). Equilibria occur where the plant isocline (dashed line) intersects the herbivore isocline (solid line); ●, stable equilibria; ○, unstable equilibria; arrows indicate trajectories of simulated population development. (A, B) One equilibrium managed by herbivores. (C) Two stable equilibria: one with high herbivore density and open vegetation, one without herbivores and dense vegetation. (D) Two stable equilibria: one with open vegetation and high herbivore density, one with dense vegetation and low herbivore density. (E) One stable equilibrium, at maximum plant biomass without herbivores. (F) One stable equilibrium, near maximum plant biomass and low herbivore density.

In situations A and B in Figure 10.1, the system tends to strong herbivore control on the vegetation because there is no limitation by quality decline when plants reach maximum density (Figure 10.2). This reflects a vegetation of low productivity and high quality, or a vegetation with herbivores that are not limited by low quality vegetation. Examples may be found in alpine meadows, steppe and semi-arid savanna. A similar low plant density equilibrium may be found in systems with mobile herbivores (B). Introduction of herbivores will generally succeed in these systems; there is only one stable equilibrium. When herbivores are absent from the system (for example, in a tree savanna without elephants) and the introduction of herbivores (such as elephants) is considered, then the management should be aware that introduction of herbivores may drastically alter the nature of the system.

Situations C and D represent cases where two stable equilibria may exist, with many, and without (C) or with few (D) herbivores (Figure 10.3). At low plant density, herbivores act as keystone species and control the vegetation. However, beyond a threshold vegetation density, the vegetation may escape herbivore control, which may dramatically reduce herbivore carrying capacity. In most ecosystems situation C, where herbivores risk being totally removed from the system, seems less realistic than D, unless spatial heterogeneity is minimal. Dublin *et al.* (1990) argued that case D may apply to the Serengeti–Mara ecosystem: in the presence of elephants the vegetation would open up, leading to a herbivore-controlled open savanna with high

Figure 10.2 New Forest ponies on preferred stream-side grassland. In less productive systems, large herbivores have greater potential to control or retard vegetation succession.

wildebeest density, but without elephants bush encroachment would lead to a tree savanna with low herbivore numbers. Other cases have been reviewed in Chapter 1.

Vera (1997) claims that case D also represents the temperate deciduous and mixed lowland forests in Europe. Closed forests dominate where the original wild large herbivores have been controlled or eradicated by humans, whereas a more open wood-pasture would have been dominant in the presence of uncontrolled herbivore numbers. At present, with widespread closed forest, herbivores have no potential to open up the vegetation, even if they were to be artificially introduced. A totally different situation would arise if the area were to be sufficiently opened up and herbivores were introduced: this would lead to higher herbivore densities controlling vegetation development. If true, this has far-reaching implications for the management of many nature reserves. Further investigation of this hypothesis is therefore important.

Situations E and F represent productive environments where maximum plant density is high and causes a strong reduction in foraging efficiency. Herbivores are unable to control the vegetation, which grows to reach (E) or approach (F) its maximum plant density independent of initial conditions. At equilibrium, herbivores have been pushed out of the system (E) or their density is low (F). System F can be thought of as a closed equatorial, boreal or temperate montane forest (Figure 10.4) or a dwarf-shrub dominated tundra

Figure 10.3 This exclosure site on poor sandy soil with moderate densities of red deer and roe deer demonstrates the potentially important effects of browsing on forest development.

or heathland. It may also apply to productive temperate wetland areas (Figure 10.5), as argued in Chapter 7. In these systems herbivores can persist at low densities by exploiting spatial heterogeneity, but if spatial heterogeneity is very low, situation F will revert to E. The equilibrium in F has also been found for different underlying assumptions (Tainton *et al.*, 1996).

In the systems described above, the density of herbivores is determined by the availability of plants. However, food availability is not always the most important factor determining herbivore density – other factors, such as water availability, other food sources, or humans, may regulate herbivore numbers. In most managed grazing systems, herbivore numbers are not allowed to fluctuate freely, but instead are kept constant by careful management. In such systems herbivore density is not regulated by the vegetation but is often considered to be constant. This perspective can also be applied to a single vulnerable plant species as part of a plant community: the species may be grazed to extinction when herbivore density is regulated by the bulk of the plant biomass and not by that of the one species. Theoretical analysis suggests that such systems are very susceptible to herbivore impact (Noy-Meir, 1975; May, 1977; Walker *et al.*, 1981; Rietkerk and Van de Koppel, 1997). At low plant density, net plant growth may be severely reduced due to soil degradation (Walker *et al.*, 1981; Rietkerk and Van de Koppel, 1997; Van de Koppel *et al.*, 1997), or due to overgrazing (Noy-Meir, 1975; May, 1977). In such systems, net plant growth is negative at low plant densities,

Figure 10.4 If herbivores cannot act as keystone species in temperate climates, closed forests are expected to develop with a low density of herbivores.

Figure 10.5 In highly productive habitats, even large grazers such as cattle appear unable to control biomass accumulation.

but becomes positive once plant density increases beyond a threshold (Figure 10.6) Two stable states are found in such systems: one dominated by plants, and another devoid of any vegetation. Srivastrava and Jefferies (1996) showed that a positive feedback between reduced plant density and deteriorating soil conditions led to a collapse of the vegetation in salt marshes along the Hudson Bay. The changes resulted from an increase of goose grazing pressure which remained high even after the collapse of the vegetation. Overgrazed situations may persist even after grazing pressure is considerably reduced. To avoid this, herbivore density should be lowered substantially or somehow a regulating effect of the vegetation should be established – for example, by removing an alternative food source elsewhere that maintains high herbivore density over a wider area.

Another situation is where overgrazing leads to the disappearance of both plants and herbivores. This might occur when herbivores are introduced into a very unproductive environment where the plant density allows rapid herbivore growth. In that case food resources are soon depleted, but the regulating effect of the food resource lags behind its availability. The herbivores may then eradicate the plants before they succumb themselves. While this does not seem likely, the population crash of introduced reindeer on St Matthew Island in the Bering Sea after depleting the stock of lichens on the tundra (Klein, 1968) could be an illustration of this situation (see also Caughley and Lawton, 1981).

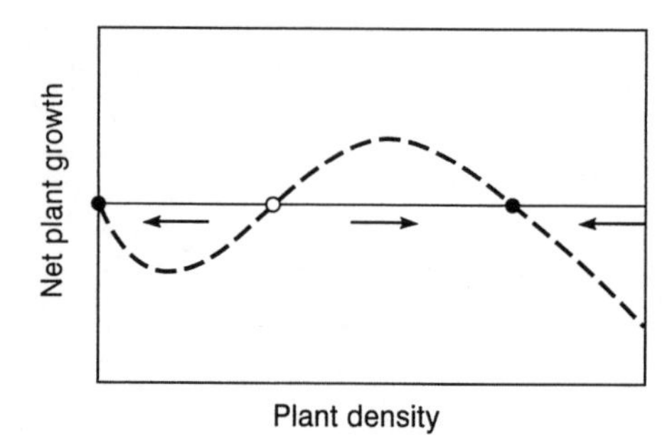

Figure 10.6 Net plant growth rate as a function of plant density. At low plant density, plant growth is negative due to soil degradation or overgrazing. Beyond a threshold density, plant growth rate becomes positive. At high plant density, plant growth is negative again, due to limitations imposed by, for instance, low light levels. Two stable states are found in this system (●), separated by an unstable state (○). Plants persist in one state, while the other state lacks vegetation. Arrows indicate direction of change.

Again, it is important to place these plant–herbivore dynamics in the context of spatial and temporal scales. Different equilibria can apply at different scales. At a sufficiently small spatial and temporal scale, disappearance of both plants and herbivores is quite probable, even though these extinctions are local and temporary when viewed at a larger scale. At a large spatial scale, on the contrary, spatial heterogeneity may cause an overall low herbivore/dense vegetation equilibrium. This occurs when herbivores survive by exploiting temporary gaps of open vegetation (e.g. tree-fall gaps) that cover a relatively small proportion of the total area under consideration. Situation D might arise under conditions of variable spatial heterogeneity over time (e.g. through periodical natural disturbance from fire, storm, floods, disease, etc.). Under these conditions a low herbivore/dense vegetation equilibrium may shift to a high herbivore/open vegetation equilibrium when the proportion of vegetation gaps surpasses a critical threshold value (Figure 10.7).

Let us summarize some practical implications of this excursion into theory. Herbivores do not appear to have a significant role to play where they cannot control vegetation development (situations E and F) even though their presence may be valued as such. If managers still want to use herbivores for vegetation management in these cases, they will have to provide continued artificial extra input (e.g. by supplementary feeding or additional management) to succeed. The management goal, then, is evidently not a natural but a semi-natural landscape. In situations A to E herbivores can clearly function as keystone species (Chapter 1), determining the structure and composition of their environment. Where such situations are identified,

Figure 10.7 At sufficiently high herbivore densities, gaps in old-growth forest remain open for prolonged periods in the New Forest, England. Exploitation of forest gaps may determine whether the forest closes in or more open vegetation develops.

which should be an important task for the future, herbivores definitely have a role to play. However, where multiple stable states exist (C and D), it is of crucial importance to identify the boundary between the two equilibria, in order to prevent the system from shifting from one equilibrium to the other; natural or artificial disturbances may induce such a shift. It is also possible that managers want to preserve a special state at an unstable equilibrium or even at a point of non-equilibrium. In these cases active management will be required continuously. The situation depicted in Figure 10.6 sends a warning against 'petting' herbivores, which may cause a collapse of the vegetation through overgrazing.

We have tried to show here that examining ecosystem dynamics in a theoretical framework offers a meaningful approach in considering management decisions. This also generates questions for future research, centred on the basic questions that determine the occurrence of herbivores as keystone species. What is the shape of the herbivore functional response (the per capita consumption rate) in relation to a combined change in plant density and quality? How does the numerical response (population development) of the herbivores relate to the functional response? When does predation of herbivores play a significant role? How can the analysis of plant–herbivore dynamics be expanded to include spatial heterogeneity or multiple resource limitations? To what extent do plant species characteristics and interactions modify plant–herbivore dynamics? To what extent do different herbivore species interact to shape the structure of plant and animal communities?

10.3.2 Pragmatic predictive modelling

Understandably, present managers cannot await the outcome of the investigations addressing the questions formulated above. Although their insight into the systems they manage could benefit from a better theoretical framework, it will not meet their immediate demands for practical answers. When decisions have to be made, best estimates will be all there is to use. So-called pragmatic predictive modelling should provide an addition to expert opinions in assisting management decisions. The pragmatism should be reflected by the model's ability to address management questions. The modelling technique should preferably be based on a mechanistic approach as far as possible. When mechanistic knowledge is insufficient, statistical or qualitative information can offer short-cuts to achieve the goal of applicability. The extended keystone hypothesis (Holling, 1992), stating that all terrestrial ecosystems are controlled and organized by a small set of key plant, animal and abiotic processes, may provide an organizing principle. This could lead to a simplified model of community dynamics centred on key species and processes. Validation should be sought by testable model predictions. Finally, the model should identify weaknesses that require further scientific development.

An example of a pragmatic predictive model is the PROCOS model, developed over a decade ago and described in Thalen *et al.* (1987). It predicts vegetation dynamics as a result of plant production (as a function of relative growth rate, senescence and decomposition) and consumption. The vegetation is assigned to one of four qualitative macrostructural successional types: short herbaceous, tall herbaceous, scrub and tall woody vegetation. These classes are easy to visualize and to use for habitat evaluation and management purposes. The herbivores are characterized by species-specific forage requirements and qualitative preferences, and by a constant density. Their impact on the vegetation consists of loss of biomass through grazing, browsing and trampling.

The vegetation development predicted by the model was in agreement with general expectations and observed developments in certain areas with a longer history of grazing (Figure 10.8) (Thalen *et al.*, 1987). However, the model was clearly too crude to provide more detailed predictions. Van Deursen and Van Oene (1996) recently developed a different model inspired by PROCOS. They refined the plant production and herbivore modules, added a soil module and included a crude spatial differentiation of the vegetation. The plant module in their model simulated the growth of dominant plant species instead of structural classes. Plant growth was modelled as a function of soil nitrogen availability, temperature, light interception and

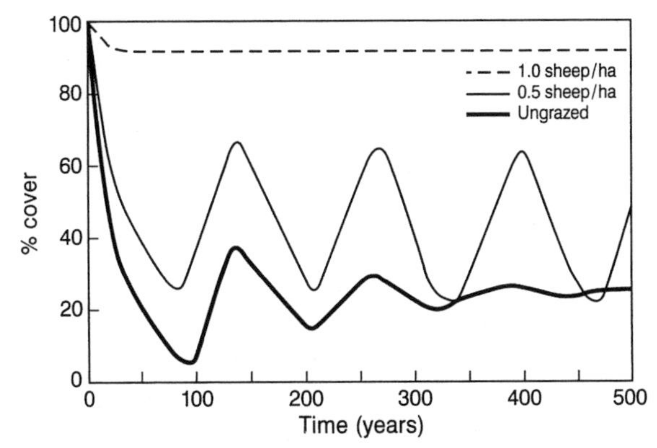

Figure 10.8 Simulated changes in the cover of open herbaceous vegetation over a 500-year period with the PROCOS model in an ungrazed situation, with a constant density of 0.5 sheep/ha and with a constant density of 1.0 sheep/ha. In the ungrazed situation, the proportion of open vegetation stabilizes at a relatively low level. At moderate grazing pressure, open vegetation expands periodically after tree die-off and the prevention of tree establishment by grazing in the remaining open area. At high grazing pressure, only a low cover of shrubs can establish. (After Thalen *et al.*, 1987.)

allocation of nitrogen and carbon to plant parts. Herbivore foraging was determined by the availability of digestible forage in combination with qualitative preferences. The resulting model, called WetDyn, substantially extends the mechanistic basis of the PROCOS model and can therefore be seen as a promising development in pragmatic predictive modelling for grazing management.

The PROCOS and WetDyn models also reveal points of attention for future work. First, extending the mechanistic approach implies the need for more fundamental research into mechanisms that determine ecosystem dynamics. Second, validation of model predictions should receive more attention, as the value of present predictions remains largely unknown. Third, providing the model with a structure of spatial and temporal scales not only appears crucial in keeping the model of manageable size, but is also vital in tackling problems related to spatial heterogeneity and temporal variability. Issues that need to be treated in this context are climatic variation, landscape heterogeneity, gap dynamics in the vegetation, herbivore foraging decisions and herbivore population dynamics. The rapid development of current modelling techniques in landscape ecology should make this next step possible.

10.4 CONCLUDING REMARKS

The preceding overview shows that attempts to integrate information in the form of models are being made in many fields relevant to grazing and conservation management, yet the interaction between science and management is often too weak to provide a scientific basis for management decisions. Science and management are not natural allies, and conservation management is no exception. Managers want immediate answers to practical questions. Scientists want to tackle theoretically interesting questions with perhaps no directly obvious practical relevance. Managers may be reluctant to allow research on their territory, especially when it concerns long-term experimental work that may cause unwanted disturbance. Science, however, often requires long-term studies, experimental manipulation and extensive replication. Each needs the other: the management needs a basis of facts and understanding; science needs field data to generate these. For the relationship to be fruitful, both should be aware of each other's limitations.

The clearest limitation for the usefulness of science to conservation management is that generality and specificity to some extent run contrary rather than parallel to each other. General explanatory models are useful in providing a theoretical framework but they are too broad for application to reality, whereas results of specific case studies cannot be safely extrapolated to other situations. Once sufficient data are available, models become indispensable for integrating information and insights from a variety of disciplines and can be valuable tools for managers. Different types of models may be put to different uses: explanatory models will unravel the mechanisms of processes; predictive models will yield quantitative predictions of future developments after thorough validation. The difficult task for the applied scientist is to find an optimal trade-off between generality and specificity in response to the manager's questions and under the constraints of available knowledge and funding. This requires an open debate between scientists and managers.

ACKNOWLEDGEMENT

We gratefully acknowledge constructive comments by Max Rietkerk.

REFERENCES

Abel, N.O.J. (1993) Reducing cattle numbers on Southern African communal range: is it worth it? in *Range Ecology at Disequilibrium: New Models of Natural Variability and Pastoral Adaptation in African Savannas*, (eds R.H. Behnke, I. Scoones and C. Kerven), pp. 173–195, Overseas Development Institute, London.

Abrams, P.A. and Matsuda, H. (1996) Positive indirect effects between prey species that share predators. *Ecology*, 77, 610–616.

Bailey, D.W., Gross, J.E., Laca, E.A. *et al.* (1996) Mechanisms that result in large herbivore grazing distribution patterns. *Journal of Range Management*, **49**, 386–400.

Baker, B.B., Bourdon, R.M. and Hanson, J.B. (1992) FORAGE: a model of forage intake in beef cattle. *Ecological Modelling*, **60**, 257–279.

Bazely, D.R. and Jefferies, R.L. (1986) Changes in the composition and standing crop of salt marsh communities in response to the removal of a grazer. *Journal of Ecology*, **74**, 693–706.

Behnke, R.H. and Scoones, I. (1993) Rethinking range ecology: Implications for rangeland management in Africa, in *Range Ecology at Disequilibrium: New Models of Natural Variability and Pastoral Adaptation in African Savannas*, (eds R.H. Behnke, I. Scoones and C. Kerven), pp. 1–30, Overseas Development Institute, London.

Behnke, R.H., Scoones, I. and Kerven, C. (eds) (1993) *Range Ecology at Disequilibrium: New Models of Natural Variability and Pastoral Adaptation in African Savannas*, Overseas Development Institute, London.

Belovsky, G.E. (1978) Diet optimization in a generalist herbivore: the moose. *Theoretical Population Biology*, **14**, 105–134.

Benson, G.L. and Laudenslayer, W.F. (1985) DYNAST: Simulating wildlife responses to forest-management strategies, in *Wildlife 2000: Modeling Habitat Relationships of Terrestrial Vertebrates*, (eds J. Verner, M.L. Morrison and C.J. Ralph), pp. 351–356, University of Wisconsin Press, Madison, Wisconsin.

Berendse, F. (1985) The effect of grazing on the outcome of competition between plant species with different nutrient requirements. *Oikos*, **44**, 35–39.

Biot, Y. (1993) How long can high stocking densities be sustained? in *Range Ecology at Disequilibrium: New Models of Natural Variability and Pastoral Adaptation in African Savannas*, (eds R.H. Behnke, I. Scoones and C. Kerven), pp. 153–172, Overseas Development Institute, London.

Blackburn, H.D. and Kothmann, M.M. (1991) Modelling diet selection and intake for grazing herbivores. *Ecological Modelling* **57**, 145–163.

Caughley, G. (1976) Wildlife management and the dynamics of ungulate populations. *Applied Biology*, **1**, 183–246.

Caughley, G. and Lawton, J.H. (1981) Plant–herbivore systems, in *Theoretical Ecology: Principles and Applications*, (ed. R.M. May), pp. 132–166, 2nd edn, Sinauer, Sunderland, Massachusetts.

Connoly, J. (1974) Linear programming and the optimum carrying capacity of range under common use. *Journal of Agricultural Science, Cambridge*, **83**, 259–265.

Cook, J.G. and Irwin, L.L. (1985) Validation and modification of a habitat suitability model for pronghorns. *Wildlife Society Bulletin*, **13**, 440–448.

Coughenour, M.B. (1991) Spatial components of plant–herbivore interactions in pastoral, ranching, and native ungulate ecosystems. *Journal of Rangeland Management*, **44**, 530–542.

DeAngelis, D.L. (1992) *Dynamics of Nutrient Cycling and Food Webs*, Chapman & Hall, London.

DeAngelis, D.L. and Waterhouse, J.C. (1987) Equilibrium and nonequilibrium concepts in ecological models. *Ecological Monographs*, **57**, 1–21.

De Ridder, N. and Breman, H. (1993) A new approach to evaluating rangeland productivity in Sahelian countries, in *Range Ecology at Disequilibrium: New Models of Natural Variability and Pastoral Adaptation in African Savannas*, (eds R.H. Behnke, I. Scoones and C. Kerven), pp. 104–117, Overseas Development Institute, London.

Dove, H. (1996) Constraints to the modelling of diet selection and intake in the grazing ruminant. *Australian Journal of Agricultural Research*, **47**, 257–275.

Dublin, H.T., Sinclair, A.R.E. and McGlade, J. (1990) Elephants and fire as causes of multiple states in the Serengeti–Mara woodlands. *Journal of Animal Ecology*, **59**, 1147–1164.

Edwards, T.C., Deshler, E.T., Foster, D. and Moisen, G.G. (1996) Adequacy of wildlife habitat relation models for estimating spatial distribution of terrestrial vertebrates. *Conservation Biology*, **10**, 263–270.

Fox, D.G., Sniffen, C.J. and O'Connor, J.D. (1988) Adjusting nutrient requirements of beef cattle for animal and environmental variations. *Journal of Animal Science*, **66**, 1475–1495.

Fresco, L.F.M., Laarhoven, H.P.M., Loonen, M.J.J.E. and Moesker, T. (1987) Ecological modelling of short-term plant community dynamics under grazing with and without disturbance, in *Disturbance in Grasslands: Causes, Effects and Processes*, (eds J. van Andel, J.P. Bakker and R.W. Snaydon), pp. 151–167, Junk, Dordrecht, The Netherlands.

Fryxell, J.M. (1991) Forage quality and aggregation by large herbivores. *American Naturalist*, **138**, 478–498.

Gilpin, M. (1991) The genetic effective size of a metapopulation. *Biological Journal of the Linnean Society*, **42**, 165–175.

Gilpin, M. and Soulé, M.E. (1986) Minimum viable populations: processes of species extinction, in *Conservation Biology: The Science of Scarcity and Diversity*, (ed. M.E. Soulé), pp. 19–34, Sinauer, Sunderland, Massachusetts.

Gordon, I.J. and Illius, A.W. (1996) The nutritional ecology of African ruminants: a reinterpretation. *Journal of Animal Ecology*, **65**, 18–28.

Grover, J.P. (1994) Assembly rules for communities of nutrient-limited plants and specialist herbivores. *American Naturalist*, **143**, 258–282.

Grover, J.P. (1995) Competition, herbivory and enrichment: nutrient-based models for edible and inedible plants. *American Naturalist*, **145**, 746–774.

Hastings, A. and Powell, T. (1991) Chaos in a three-species food chain. *Ecology*, **72**, 896–903.

Hedrick, P.W., Lacey, R.C., Allendorf, F.W. and Soulé, M.E. (1996) Directions in conservation biology: comments on Caughley. *Conservation Biology*, **10**, 1312–1320.

Hilborn, R. (1995) A model to evaluate alternative management policies for the Serengeti–Mara ecosystem, in *Serengeti II: Dynamics, Man, and Conservation of an Ecosystem*, (eds A.R.E. Sinclair and P. Arcese), pp. 617–637, University of Chicago Press, Chicago.

Hobbs, N.T. and Swift, D.M. (1985) Estimates of habitat carrying capacity incorporating explicit nutritional constraints. *Journal of Wildlife Management*, **49**, 814–822.

Holling, C.S. (ed.) (1978) *Adaptive Environmental Assessment and Management*, Wiley, Chichester.

Holling, C.S. (1992) Cross-scale morphology, geometry, and dynamics of ecosystems. *Ecological Monographs*, **62**, 447–502.

Holt, R.D., Grover, J. and Tilman, D. (1994) Simple rules for interspecific dominance in systems with exploitative and apparent competition. *American Naturalist*, **144**, 741–771.

Hudson, R.J. and White, R.G. (1985) Computer simulation of energy budgets, in *Bioenergetics of Wild Herbivores*, (eds R.J. Hudson and R.G. White), pp. 261–290, CRC Press, Boca Raton, Florida.

Huisman, J., Grover, J.P., Van der Wal, R. and Van Andel, J. (1998) Competition for light, plant species replacement, and herbivory along productivity gradients, in *Herbivores, Between Plants and Predators*, (eds H. Olff, V.K. Brown and R.H. Drent), pp. 239–269, Blackwell, Oxford.

Hyman, J.B., McAninch, J.B. and DeAngelis, D.L. (1991) An individual-based simulation model of herbivory in a heterogeneous landscape, in *Quantitative Methods in Landscape Ecology*, (eds M.G. Turner and R.H. Gardner), pp. 443–475, Ecological Studies, Vol. 82, Springer, New York.

Illius, A.W. and Gordon, I.J. (1987) The allometry of food intake in grazing ruminants. *Journal of Animal Ecology*, **56**, 989–999.

Illius, A.W. and Gordon, I.J. (1991) Prediction of intake and digestion in ruminants by a model of rumen kinetics integrating animal size and plant characteristics. *Journal of Agricultural Science, Cambridge*, **116**, 145–157.

Illius, A.W. and Gordon, I.J. (1992) Modelling the nutritional ecology of ungulate herbivores: evolution of body size and competitive interactions. *Oecologia*, **89**, 428–434.

Illius, A.W. and Hodgson, J. (1996) Progress in understanding the ecology and management of grazing systems, in *The Ecology and Management of Grazing Systems*, (eds J. Hodgson and A.W. Illius), pp. 429–457, CAB, Wallingford, UK.

Innis, G.S., Noy-Meir, I., Godron, M. and Van Dyne, G.M. (1980) Total-system simulation models, in *Grasslands, Systems Analysis and Man*, (eds A.I. Breymeyer and G.M. Van Dyne), pp. 759–797, IBP 19, Cambridge University Press. Cambridge.

Jorritsma, I., Mohren, G.M.J., Van Wieren, S.E. *et al.* (1997) Bosontwikkeling in aanwezigheid van hoefdieren: een modelbenadering, in *Hoefdieren in het Boslandschap*, (eds S.E. Van Wieren, G.W.T.A. Groot Bruinderink, I.T.M. Jorritsma and A.T. Kuiters), pp. 147–164, Backhuys Publishers, Leiden, The Netherlands.

Ketelaars, J.J.M.H. and Tolkamp, B.J. (1991) Toward a new theory of feed intake regulation in ruminants. Doctoral thesis, Wageningen Agricultural University, Wageningen, The Netherlands.

Kirkman, R.L., Eberly, J.A., Porath, W.R. and Titus, R.R. (1985) A process for integrating wildlife needs into forest management planning, in *Wildlife 2000: Modeling Habitat Relationships of Terrestrial Vertebrates*, (eds J. Verner, M.L. Morrison and C.J. Ralph), pp. 347–350, University of Wisconsin Press. Madison, Wisconsin.

Klein, D.R. (1968) The introduction, increase, and crash of reindeer on St Matthew Island. *Journal of Wildlife Management*, **32**, 350–367.

Kobe, R.K. (1996) Intraspecific variation in sapling mortality and growth predicts geographic variation in forest composition. *Ecological Monographs*, **66**, 181–201.

Laca, E.A. and Demment, M.W. (1991) Herbivory: the dilemma of foraging in a spatially heterogeneous food environment, in *Plant Defenses against Mammalian Herbivory*, (eds R.T. Palo and C.T. Robbins), pp. 29–44, CRC Press, Boca Raton, Florida.

Laca, E.A. and Demment, M.W. (1996) Foraging strategies of grazing animals, in *The Ecology and Management of Grazing Systems*, (eds J. Hodgson and A.W. Illius), pp. 137–158, CAB, Wallingford, UK.

Lande, R. (1993) Risks of population extinction from demographic and environmental stochasticity, and random catastrophes. *American Naturalist*, **142**, 911–927.

Lande, R. (1995) Mutation and conservation. *Conservation Biology*, **9**, 782–791.

Lemaire, G. and Chapman, D. (1996) Tissue flows in grazed plant communities, in *The Ecology and Management of Grazing Systems*, (eds J. Hodgson and A.W. Illius), pp. 3–36, CAB, Wallingford, UK.

Li, B.-L. (1995) Stability analysis of a nonhomogeneous Markovian landscape model. *Ecological Modelling*, **82**, 247–256.

Llewyn, D.W., Shaffer, G.P., Craig, N.J. *et al.* (1996) A decision-support system for prioritizing restoration sites on the Mississippi River alluvial plain. *Conservation Biology*, **10**, 1446–1455.

Loehle, C., Li, B.-L. and Sundell, R.C. (1996) Forest spread and phase transitions at forest-prairie ecotones in Kansas, USA. *Landscape Ecology*, **11**, 225–235.

Loza, H.J., Grant, W.E., Stuth, J.W. and Forbes, T.D.A. (1992) Physiologically based landscape use model for large herbivores. *Ecological Modelling*, **61**, 227–252.

Marcot, B.G. (1985) Use of expert systems in wildlife-habitat modeling, in *Wildlife 2000: Modeling Habitat Relationships of Terrestrial Vertebrates*, (eds J. Verner, M.L. Morrison and C.J. Ralph), pp. 145–150, University of Wisconsin Press, Madison, Wisconsin.

Matsuda, H., Hori, M. and Abrams, P.A. (1996) Effects of predator-specific defence on biodiversity and community complexity in two-trophic-level communities. *Evolutionary Ecology*, **10**, 13–28.

May, R.M. (1977) Thresholds and breakpoints in ecosystems with a muliplicity of stable states. *Nature*, **269**, 471–477.

McCann, U. and Yodzis, P. (1994) Biological conditions for chaos in a three-species food chain. *Ecology*, **75**, 561–564.

McNaughton, S.J. (1985) Ecology of a grazing ecosystem: the Serengeti. *Ecological Monographs*, **55**, 259–294.

Messier, F. (1994) Ungulate population models with predation: a case study with the North American moose. *Ecology*, **75**, 478–488.

Milchunas, D.G. and Lauenroth, W.K. (1993) Quantitative effects of grazing on vegetation and soils over a global range of environments. *Ecological Monographs*, **63**, 327–366.

Milchunas, D.G., Sala, O.E. and Lauenroth, W.K. (1988) A generalized model of the effects of grazing by large herbivores on grassland community structure. *American Naturalist*, **132**, 87–106.

Milne, B.T. (1997) Applications of fractal geometry, in *Wildlife and Landscape Ecology: Concepts of Pattern and Scale for Research and Management*, (ed. J. Bissonette), Springer, New York (in press).

Moen, R.A., Pastor, J. and Cohen, Y. (1997) A spatially explicit model of moose foraging and energetics. *Ecology*, **78**, 505–521.

Morrison, M.L., Marcot, B.G. and Mannan, R.W. (1992) *Wildlife-Habitat Relationships: Concepts and Applications*, University of Wisconsin, Madison, Wisconsin.

Murray, M.G. (1991) Maximising energy retention in grazing ruminants. *Journal of Animal Ecology*, **60**, 1029–1045.

Noy-Meir, I. (1975) Stability of grazing systems: an application of predator–prey graphs. *Journal of Ecology*, **63**, 459–481.

Oksanen, L. (1992) Evolution of exploitation ecosystems: I. Predation, foraging ecology and population dynamics in herbivores. *Evolutionary Ecology*, **6**, 15–33.

Oksanen, L., Fretwell, S.D., Arnuda, J. and Niemelä, P. (1981) Exploitation systems in gradients of primary productivity. *American Naturalist*, **118**, 240–261.

Owen-Smith, R.N. (1988) *Megaherbivores: The Influence of Very Large Body Size on Ecology*, Cambridge University Press, Cambridge.

Pastor, J. and Naiman, R.J. (1992) Selective foraging and ecosystem processes in boreal forests. *American Naturalist*, **139**, 690–705.

Pickup, G. (1994) Modelling patterns of defoliation by grazing animals in rangelands. *Journal of Applied Ecology*, **31**, 231–246.

Prins, H.H.T. and Olff, H. (1998) Species richness of African grazer assemblages: towards a functional explanation, in *Dynamics of Tropical Ecosystems*, (eds D.M. Newberry, H.H.T. Prins and N. Brown), BES Symposium Proceedings, Vol. 37, pp. 449–490, Blackwell, Oxford.

Rabinovich, J.E., Hernandez, M.J. and Cajal, J.L. (1985) A simulation model for the management of vicuna populations. *Ecological Modelling*, **30**, 275–295.

Raedeke, K.J. and Lehmkuhl, J.F. (1985) A simulation procedure for modelling the relationships between wildlife and forest management, in *Wildlife 2000: Modeling Habitat Relationships of Terrestrial Vertebrates*, (eds J. Verner, M.L. Morrison and C.J. Ralph), pp. 377–382, University of Wisconsin Press, Madison, Wisconsin.

Rietkerk, M. and Van de Koppel, J. (1997) Alternative stable states and threshold effects in semi-arid grazing systems. *Oikos*, **78**, 69–76.

Ritchie, J.R. (1989) An expert system for a rangeland simulation model. *Ecological Modelling*, **46**, 91–105.

Rosenzweig, M.L. (1971) Paradox of enrichment: destabilization of exploitation systems in ecological time. *Science*, **171**, 385–387.

Schamberger, M.L. and O'Neill, L.J. (1985) Concepts and constraints of habitat-model testing, in *Wildlife 2000: Modeling Habitat Relationships of Terrestrial Vertebrates*, (eds J. Verner, M.L. Morrison and C.J. Ralph), pp. 5–10, University of Wisconsin Press, Madison, Wisconsin.

Seagle, S.W. and McNaughton, S.J. (1992) Spatial variation in forage nutrient concentrations and the distribution of Serengeti grazing ungulates. *Landscape Ecology*, 7, 229–241.

Senft, R.L., Rittenhouse, L.R. and Woodmansee, R.G. (1983) The use of regression equations to predict spatial patterns of cattle behavior. *Journal of Range Management*, **36**, 553–557.

Senft, R.L., Coughenour, M.B., Bailey, D.W. *et al.* (1987) Large herbivores foraging and ecological hierarchies. *Bioscience*, 37, 789–799.

Shugart, H.H. (1984) *A Theory of Forest Dynamics: The Ecological Implications of Forest Succession Models*, Springer, New York.

Sibbald, A.R., Maxwell, T.J. and Eadie, J. (1979) A conceptual approach to the modelling of herbage intake by hill sheep. *Agricultural Systems*, **4**, 119–134.

Spalinger, D.E. and Hobbs, N.T. (1992) Mechanisms of foraging in mammalian herbivores: new models of functional response. *American Naturalist*, **140**, 325–348.

Srivastava, D.S. and Jefferies, R.L. (1996) A positive feedback: herbivory, plant growth, salinity, and the desertification of Arctic salt-marsh. *Journal of Ecology*, **84**, 31–42.

Stafford Smith, M. (1996) Management of rangelands: paradigms at their limits, in *The Ecology and Management of Grazing Systems*, (eds J. Hodgson and A.W. Illius), pp. 325–357, CAB, Wallingford, UK.

Swift, D.M. (1983) A simulation model of energy and nitrogen balance for free-ranging ruminants. *Journal of Wildlife Management*, **47**, 620–645.

Szafer, W. (1968) The ure-ox, extinct in Europe since the seventeenth century: an early attempt at conservation that failed. *Biological Conservation*, **1**, 45–47.

Tainton, N.M., Morris, C.D. and Hardy, M.B. (1996) Complexity and stability in grazing systems, in *The Ecology and Management of Grazing Systems*, (eds J. Hodgson and A.W. Illius), pp. 275–299, CAB, Wallingford, UK.

Thalen, D.C.P., Poorter, H., Lotz, L.A.P. and Oosterveld, P. (1987) Modelling the structural changes in vegetation under different grazing regimes, in *Disturbance in Grasslands: Causes, Effects and Processes*, (eds J. van Andel, J.P. Bakker and R.W. Snaydon), pp.151–167, Junk, Dordrecht, The Netherlands.

Tilman, D. (1988) *Plant Strategies and the Dynamics and Structure of Plant Communities*, Princeton University Press, Princeton, New Jersey.

Turner, M.G. (1988) Simulation and management implications of feral horse grazing on Cumberland Island, Georgia. *Journal of Range Management*, **41**, 441–447.

Turner, M.G., Wu, Y. Romme, W.H. and Wallace, L.L. (1993) A landscape simulation model of winter foraging by large ungulates. *Ecological Modelling*, **69**, 163–184.

Van Andel, J., Bakker, J.P. and Grootjans, A.P. (1993) Mechanisms of vegetation succession: a review of concepts and perspectives. *Acta Botanica Neerlandica*, **42**, 413–433.

Van de Koppel, J. (1997) Trophic interactions in gradients of primary productivity. Doctoral thesis, Rijksuniversiteit Groningen, Groningen, The Netherlands.

Van de Koppel, J., Huisman, J., Van der Wal, R. and Olff, H. (1996) Patterns of herbivory along a productivity gradient: an empirical and theoretical investigation. *Ecology*, **77**, 736–745.

Van de Koppel, J., Rietkerk, M. and Weissing, F. (1997) Catastrophic vegetation shifts and soil degradation in terrestrial grazing systems. *Trends in Ecology and Evolution*, **12**, 352–356.

Van der Wal, R., Van de Koppel, J. and Sagel, M. (1998) On the relation between herbivore foraging efficiency and plant standing crop. *Oikos*, **82**, 123–130..

Van Deursen, E.J.M. and Van Oene, H. (1996) *Modellering van de Vegetatieontwikkeling in Wetlands onder Begrazing*, Deptartment of Terrestrial Ecology and Nature Conservation, Wageningen Agricultural University, Wageningen, The Netherlands.

Van Soest, P.J. (1994) *Nutritional Ecology of the Ruminant*, 2nd edn, O&B Books, Corvallis, Oregon.

Van Wieren, S.E. (1996) Digestive strategies in ruminants and non-ruminants. Doctoral thesis, Wageningen Agricultural University, Wageningen, The Netherlands.

Van Wijngaarden, W. (1985) *Elephants–Trees–Grass–Grazers: relationships between climate, soil, vegetation, and large herbivores in a semi-arid savanna ecosystem (Tsavo, Kenya)*. Doctoral thesis, Wageningen Agricultural University. ITC Publication No. 4, Enschede, The Netherlands.

Vera, F.W.M. (1997) Metaforen voor de wildernis: Eik, Hazelaar, Rund en Paard. Doctoral thesis, Wageningen Agricultural University, Wageningen.

Verner, J., Morrison, M.L. and Ralph, C.J. (eds) (1985) *Wildlife 2000: Modeling Habitat Relationships of Terrestrial Vertebrates*, University of Wisconsin Press. Madison, Wisconsin.

Walker, B.H., Ludwig, D., Holling, C.S. and Peterman, R.M. (1981) Stability of semi-arid savanna grazing systems. *Journal of Ecology*, **69**, 473–498.

WallisDeVries, M.F. (1996) Effects of resource distribution patterns on ungulate foraging behaviour: a modelling approach. *Forest Ecology and Management*, **88**, 167–177.

WallisDeVries, M.F. and Schippers, P. (1994) Foraging in a landscape mosaic: selection for energy and minerals in free-ranging cattle. *Oecologia*, **100**, 107–117.
Westoby, M., Walker, B. and Noy-Meir, I. (1989) Opportunistic management for rangelands not at equilibrium. *Journal of Range Management*, **42**, 266–274.
White, T.C.R. (1978) The importance of a relative shortage of food in animal ecology. *Oecologia*, **33**, 71–87.
Wilmshurst, J.F., Fryxell, J.M. and Hudson, R.J. (1995) Forage quality and patch choice by wapiti (*Cervus elaphus*). *Behavioral Ecology*, **6**, 209–217.
Yodzis, P. (1989) *Introduction to Theoretical Ecology*, Harper and Row, New York.

11

Grazing for conservation in the twenty-first century

Sipke E. Van Wieren[1] and Jan P. Bakker[2]
[1]Tropical Nature Conservation and Vertebrate Ecology Group, Department of Environmental Sciences, Wageningen Agricultural University, Bornsesteeg 69, 6708 PD Wageningen, The Netherlands
[2]Laboratory of Plant Ecology, University of Groningen, PO Box 14, 9750 AA Haren, The Netherlands

11.1 INTRODUCTION

This chapter evaluates the use of large grazing herbivores, mainly as a tool for conservation. What have we learnt from the past 25 years of research and how do we proceed from here? As can be noted from the previous chapters, grazing has been applied to meet various management objectives. Have these objectives been met, and if not, what are the probable causes? Was grazing the right tool in the first place?

The management objectives that are most frequently encountered can be grouped for convenience in one of the four following classes:

- **Species-oriented objectives.** Management is aimed at maintaining or developing suitable conditions for certain species of (mainly) animals. The species generally share similar habitat and food requirements. Well known examples are the maintenance of grassland by large herbivores to facilitate winter-staging geese and breeding meadow birds.
- **Species-richness within plant communities (species diversity).** Management can be aimed at the maintenance of existing species-rich communities, but is more frequently focused on developments that, for various reasons, threaten species richness. Generally, grazing is used to control a plant species that has gained a dominant position, e.g. *Elymus athericus*, *Calamagrostis epigejos*, *Deschampsia flexuosa*, *Molinia caerulea*, *Pteridium aquilinum* (Figure 11.1). Much of the restoration management using grazing falls in this category.

Grazing and Conservation Management. Edited by M.F. WallisDeVries, J.P. Bakker and S.E. Van Wieren. Published in 1998 by Kluwer Academic Publishers, Dordrecht. ISBN 0 412 47520 0.

Figure 11.1 Competitive plant species can attain dominance in many disturbed systems, especially when they are unpalatable. In this case, ponies were able to reduce the cover of *Calamagrostis epigejos*.

- **Diversity of plant communities (community diversity).** In many other cases management objectives are directed at creating and maintaining a number of more or less well defined plant communities within an area. Although management frequently aims at vegetation patterns and structural diversity in the vegetation, a relationship with the conservation of fauna can often be found – because, of course, so many animal species are closely linked to vegetation structure.
- **Wilderness.** The idea of wilderness is about complete ecosystems that function without human interference. Being a fairly new idea, the wilderness concept as a management option (Noss, 1992) is still much more debated than implemented. Therefore we cannot evaluate it properly, but nevertheless the concept is important because of the (potential) role of large herbivores in shaping community structure (Chapters 1, 10). The outcome of discussions on this topic can shed light on what may be attained with grazing in a more general context.

Within the group of objectives a certain hierarchy can be recognized: they range from object- to more system-oriented. The place in the hierarchy bears a relationship with the extent of knowledge of grazing being a successful instrument or not. The more species-oriented the management aim, the more grazing can be controlled and adjusted. On the other hand, when the objective is more system-oriented, the outcome is much less well defined, and the fine-tuning of grazing is much more difficult and less controlled. Going up through the hierarchy, knowledge and practical experience become less and less, because grazing is superimposed on areas and processes that scale up in space and time (Table 11.1). It is important to bear in mind that this situation will possibly remain so because of the increasing complexity, and related unpredictable behaviour, of the systems aimed for. Below, the main objectives will be evaluated individually.

11.2 SPECIES-ORIENTED OBJECTIVES

The species concerned here are mainly geese and meadow birds. Winter-staging geese need short grassland of good quality, which is found extensively in highly fertilized pasture. Hence, the required forage can easily be provided by a system of summer grazing with large grazers (cattle and horses) at high grazing intensity. If the target bird species are ecologically more or less similar, then the objectives can be met effectively under grazing because the necessary grazing pressure can be relatively easily defined and controlled. In The Netherlands, herbivorous swans, geese and ducks have become dependent on improved grasslands during winter in the course of the twentieth century (Van Eerden *et al.*, 1996). Protein demand proba-

Table 11.1 Components of the landscape and their dynamics as a function of natural processes and human influence in a hierarchical order (after Everts and De Vries, 1991)

Component	Natural processes	Human influences
Climate	Climatic change	Atmospheric pollution, albedo changes
Geological formation	Sedimentation and erosion	Transport of materials
Groundwater	Changes in groundwater table	Groundwater exploitation
	Changes in water flow	Drainage/irrigation
Soil	Soil formation	Soil cultivation/ fertilization
Vegetation	Succession and degeneration, plant interactions	Felling, mowing, burning, sod cutting, grazing, planting, spraying of herbicides
	Plant dispersal	Sowing, species introduction
Fauna	Population dynamics	Hunting, fishing
	Migration	Transhumance, herding
	Dispersal	(Re-)introduction of species
	Herbivore impact	Grazing impact

bly increases with body mass; hence the heavy mute swan (*Cygnus olor*) used the improved grasslands relatively early in the 1940s, whereas the comparatively light wigeon (*Anas penelope*) only started to do so in the 1980s (Figure 11.2). It may therefore be predicted that the heavier species will be the first to decline once fertilizer application decreases.

If the species belong to an ecologically more diverse target group, however, the goal can become more difficult to achieve. The group of meadow birds may serve as an example. Here we can distinguish more critical species such as common snipe (*Gallinago gallinago*), ruff (*Philomachus pugnax*) and common redshank (*Tringa totanus*) and less critical ones such as northern lapwing (*Vanellus vanellus*), oystercatcher (*Haematopus ostralegus*) and Eurasian curlew (*Numenius aquata*). The differences lie in requirements for nest site, feeding habitat, groundwater table, the susceptibility for trampling etc. In practice, it is difficult to provide all the required resources for many species in one grazing system. Agricultural intensification recently coincided with the decrease of black-tailed godwit (*Limosa limosa*) and skylark (*Alauda arvensis*). Problems like this are frequently encountered and the lesson is that not all management aims can be realized at the same time in the same area. If it comes to species, inevitably a choice has to be made (Chapter 6).

It has to be noted that the number of animal species to which the management objectives generally are oriented is very limited. It is likely that more faunal groups or species benefit from low-intensity grazing, but experience in this field is limited.

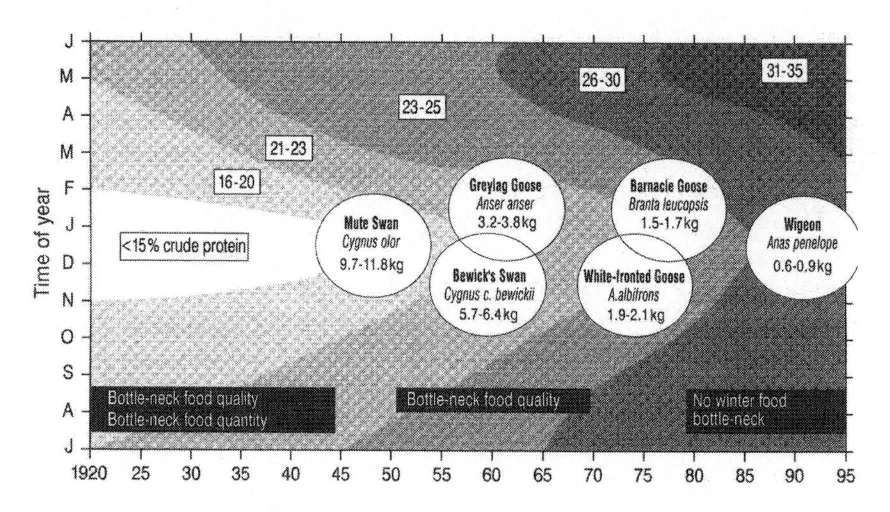

Figure 11.2 Approximated long-term changes in food quality (as crude protein content in dry matter) of grasslands on dairy farms in The Netherlands in the course of the year. Also indicated is the onset of the period of dependence of herbivorous *Anatidae* species on improved grassland during winter. (After Van Eerden *et al.*, 1996.)

11.3 PLANT SPECIES RICHNESS

This objective generally aims at short vegetation on oligotrophic and mesotrophic soils, which harbours many endangered plant species. It has been found that species richness is highest if the vegetation is kept short at a certain level, while no litter accumulation takes place (Chapter 5). Controlled grazing has proved to be an excellent tool in maintaining short vegetation. It is important to establish the proper density of herbivores in order to achieve the utilization factor needed. The required utilization factor depends on soil fertility and, hence, plant production. When the conditions are right, this objective can be met quite effectively. These conditions are still fulfilled in large areas with low-intensity farming systems (Bignal and McCracken, 1996) (Figure 11.3).

The situation changes dramatically if the conditions are not right and this is increasingly the case all over western Europe. Especially when attempts are made to restore species-rich communities from areas that have been altered through reclamation and subsequent agricultural intensification, the following conditions may have changed, sometimes irreversibly: groundwater table and groundwater composition; nutrient availability; and nitrogen input from atmospheric deposition. The result of these changes generally is that one or a few species become dominant. These species are predominantly tall, competitive for light and often clonal. Examples on eutrophic

Figure 11.3 Semi-natural landscape in Extremadura, Spain, extensively grazed for centuries and with a rich flora and fauna, but threatened by agricultural modernization.

soils after agricultural intensification are *Cirsium arvense*, *Urtica dioica* and *Elymus repens*, and on mesotrophic to oligotrophic soils after cessation of former farming systems and/or atmospheric deposition they include *Elymus athericus*, *Pteridium aquilinum*, *Molinia caerulea*, *Deschampsia flexuosa*, *Juncus effusus*, *Calamagrostis epigejos*, *Phragmites australis*, *Brachypodium pinnatum*. So far, the results of tackling these problems with grazing are varied. If the 'problem species' are sufficiently palatable (e.g. *Deschampsia flexuosa*, *Elymus athericus*), then the effect of grazing can be quite positive. If, on the other hand, the 'problem species' are avoided, then the dominant species often persist and reserve managers may become frustrated. Apart from the abiotic conditions, biotic constraints can also be operating. The seedbank of many target species has depleted, and dispersal of diaspores is difficult in the fragmented landscape (Bakker *et al.*, 1996). The message here is that, as long as the abiotic conditions are not right and the target species cannot reach a certain area, grazing cannot solve the problems – but neither can the grazers be blamed for the lack of results.

11.4 DIVERSITY OF PLANT COMMUNITIES

To maintain or to develop macrostructurally diverse vegetation by grazing is the key objective here. Short and tall herbaceous vegetation, scrub and woodland should preferably all be present. The frame of reference generally is the historic wood-pasture (Chapter 4) and examples can still be found in eastern and southern Europe. In western Europe, these systems are scarce and have to be re-established from enclosed grasslands or former arable fields, or a mix of grassland and woodland, and managers try to reach this goal with some kind of grazing management only. Recently, herbivores have been introduced after major changes in abiotic conditions: removal of summer-dikes to give access to water flow and sedimentation as important ecosystem processes, or removal of the topsoil to create relief and oligotrophic conditions as a starting point for restoration. Superimposed on the thus created variation in abiotic conditions, the grazers may enhance diversity in plant communities. The same restrictions with respect to soil seedbank and dispersal of diaspores may apply as for the objectives of increasing species richness. It may be easier for many animal species to reach the target areas, but certainly not all of them if habitats are fragmented.

Scale is important in relation to this objective. Areas have to be large to minimize edge effects, to provide opportunities for herds of large grazers to exert a normal use of home ranges, and also to give room for a more natural social behaviour pattern and social organization.

So far, the results are not clear. To attain variation within grassland is not the problem but to develop a mosaic of the various macrostructural classes appears to be much more difficult. One of the reasons is time: many experiments have only lasted a few decades, at most, and this is far too short a period for ecosystem development. The development of scrub and tall herb fringe communities in particular seems delicate. These are considered important because they contain many endangered species. Furthermore, it is problematic to establish an adequate grazing regime, as the predictability of grazing impact on the development of vegetation patterns is low. Very little is known about the relationship between herbivory and vegetation development, and so far only a few combinations of species and densities have been experimented with. Most experiments have been carried out with true grazers (cattle, sheep, horses), and hardly any attempt has been made to use a more complete herbivore assemblage, including mixed feeders and browsers. The question also arises whether grazing alone can maintain the 'target landscape', with a significant proportion of all macrostructural vegetation classes. Only if large herbivores can act as keystone species can they be expected to accomplish the task. From historic landscapes it is known that additional practices like woodcutting and mowing were the rule, and it is suspected that these will always be needed in some form to reach the management goal.

In summary, very little is known of the potential of large herbivores to create and maintain varied landscapes. Obviously they have a role to play, as they have always done, and it can be questioned whether we will ever know precisely what the contribution of herbivory will be in the making of the landscape as it is superimposed on partly stochastic, abiotic conditions. For the manager this may imply that more flexibility and discontinuity should also be introduced into the management system – for example, to vary herbivore densities. With a clever grazing system most of the community diversity objective can be realized with herbivores, but additional practices in one way or another to ensure structural variation will still be needed.

11.5 WILDERNESS

Within this concept the manager strives towards restoration of ecosystems whose dynamics are exclusively determined by natural processes. For a self-sustaining ecosystem to function properly, it is required – at the least – that species that play a key role are present and can maintain a viable population. This includes the large predators. The condition of 'completeness' means that wilderness areas have to be large and should contain all major habitats typical for the region. As almost all potential wilderness areas are incomplete, reintroduction of important species has to be considered – e.g. beaver (*Castor fiber*) and wolf (*Canis lupus*) – as well as the lifting of barriers to connect target areas which would otherwise be too small.

It will not be surprising that the problems in realizing this objective in western Europe are manifold. It is difficult to get the conditions right: space is limited, the landscape has changed considerably or entirely, and many important species have vanished. As to this last point, a number of serious reintroduction programmes are nevertheless being undertaken or considered at the moment – e.g. lynx (*Lynx lynx*) (Figure 11.4) and European bison (*Bison bonasus*).

To be able to restore a wilderness area, it is important to have a clear idea which key species should be present: a frame of reference is needed. It appears very difficult to reconstruct the potential natural landscape, and some possible options are currently hotly debated. The discussion focuses on the potential role of large herbivores as keystone species in shaping the plant environment (Chapters 1 and 10). The outcome of this discussion is also relevant to the amount of additional management needed for the community-oriented objectives discussed above.

The wilderness concept requires a hands-off approach and this appears particularly difficult, especially when it comes to regulating herbivore populations (Arcese and Sinclair, 1997). When the conditions are met, the outcome of interacting system processes is to a large extent unpredictable. This means that the manager does not exactly know what is going to happen. It may frequently turn out that a hidden management agenda is present and that a variety of sub-objectives are being aimed for, which sometimes conflict with the actual course of events. At this stage the urge to start managing (e.g. herbivore populations) is strong. It has to be realized, though, that through regulating herbivore numbers or the implementation of other powerful management practices, the objective can no longer be marked as 'wilderness' but has fallen into a different category: to increase the diversity of plant communities.

Figure 11.4 Recently the lynx has been successfully reintroduced in various European mountain ranges. Will other large predators follow?

In Europe, wilderness areas are very scarce. The large predators have often been eradicated or are heavily hunted and large herbivore populations are usually controlled. Swedish Lapland is probably the last region where (near) wilderness can be found, in the area comprising the national parks of Stora Sjöfallet, Padjelanta and Sarek. Other potential areas that could be considered are:

- the Swiss National Park, combined with the much larger Stelvio NP in Italy;
- the National Park of the Bavarian Forest (Germany), including a large part of the adjacent Czech Republic and part of the Danube valley;
- Bialowieza National Park (Poland), including a large part of adjoining Belarus;
- Cantabria (Spain);
- Massif Central (France);
- the Isle of Jura (Scotland);
- Abruzzo NP (Italy);
- the French-Spanish Pyrenees (Figure 11.5).

11.6 SUSTAINABILITY OF BIOLOGICAL DIVERSITY

The species-oriented objectives for grazing management can be realized widely in agriculturally exploited systems in many European countries. Only

Figure 11.5 Extensive grazing in the Pyrenees. With increasing abandonment, many mountainous areas have outstanding potential as future wilderness areas.

in countries with very intensive agricultural exploitation need the farmers be compensated for damage by geese, and lower production for taking care of non-critical meadow birds. Reduction of present fertilizer loads on farm grasslands will result in a shift of herbivorous ducks, geese and swans from improved grassland to cereals, because of the superior quality of winter wheat (Van Eerden *et al.*, 1996). The critical meadow birds are incompatible with intensive exploitation, in which case they can only survive in nature reserves.

The objectives for plant species richness can be realized in European countries with low-intensity farming systems, otherwise only in nature reserves. The large-scale objectives for diversity of plant communities and wilderness, respectively, can only be realized in nature reserves.

We know how difficult restoration management is from an ecological point of view. Abiotic conditions, particularly the ecohydrological framework and nutrient levels, are strongly affected from outside the nature reserves. Species loss and habitat fragmentation necessitate (re)introduction of endangered species before they become extinct.

These considerations reveal guidelines for sustainable management of biological diversity. Although existing low-intensity farming systems in Europe should be maintained (Bignal and McCracken, 1996), nature conservation increasingly relies on nature reserves. Nature reserves need to be large for three reasons:

- they should include entire catchment areas in order to become independent from other land use with respect to hydrological conditions and nutrient flows;
- they should be connected with remaining fragments of high biological diversity and new restoration areas in order to enable dispersal (including the spreading of diaspores by a moving ecological infrastructure, be it machinery or large herbivores);
- populations of large herbivores can then display a natural social structure and preserve a higher genetic diversity, which also holds for populations of smaller animals and of plants.

Apart from sustainability, it is important to define the goals of conservation management as exactly as possible. One reason is that it is necessary to decide on the appropriate management practices; the other reason is that it enables the evaluation of the management practised. For example, if the goal is the maintenance of wintering geese or non-critical breeding meadow birds with no botanical conservation interest, the management practices can include high amounts of fertilizer application, drainage and grazing at high stocking rate (Figure 11.6). If the goal is the maintenance of critical meadow birds and colourful grasslands, the management practices can include small amounts of fertilizer application, little drainage, and grazing at low stocking rate. If the goal is the maintenance of plant communities including endangered species, the management should avoid fertilizer application, apply little or no drainage and graze at low to moderate stocking rates, depending on the type of plant community aimed at.

It is evident that in these examples grazing as a tool in conservation is part of a set of other practices. It is also clear that it becomes more difficult to define the goals and practices of conservation management the higher the target is in the series animal species, plant species, plant communities and wilderness, which in fact comprises entire ecosystems.

11.7 FUTURE RESEARCH

After 25 years of probing and experimenting, many questions have been answered but many others remain. What could be the focus of grazing research for conservation in the next 25 years? We will again follow the hierarchy of the objectives discussed earlier and see what is needed.

Grazing for **target animal species** has been successful. We know how to do it for a set of bird species. However, experience with other faunal groups, such as invertebrates, is limited. It is interesting to note that, in contrast to plants, management is very rarely (if ever) aimed at the conservation of animal communities. Comparatively little is known of the effects of extensive grazing at the community level, but it is expected that important insights might be revealed should research focus on this point.

In grazing for **plant species richness**, we have to distinguish between grazing for maintaining species-rich communities and for restoration attempts. By and large we now know quite well how to maintain plant species richness if the abiotic conditions are right. However, it is not always clear which density is needed and which combination of herbivores gives the best

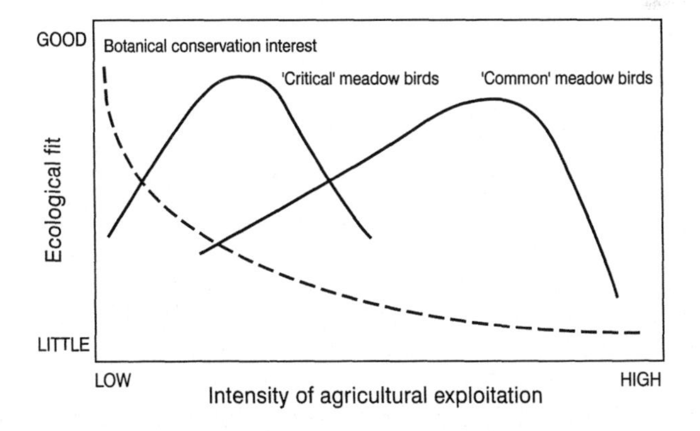

Figure 11.6 Simplified representation of the relationship between the intensity of agricultural exploitation and nature conservation interest (suitability for meadow birds and grassland vegetation). (After Beintema and Müskens, 1987.)

results. The length and timing of the grazing period needs further attention. Research on fine-tuning the grazing management by means of controlled experiments is lacking.

As explained earlier, grazing for restoring species richness frequently fails because of large competitive tall herbs dominating the community. Here research should focus on the control of these dominant species. What are the mechanisms by which these species become dominant? Many of them seem similar in the sense that they monopolize the nutrients available by an internal cycle of nutrient re-allocation between above- and below-ground storage organs. It may be expected that they are more vulnerable to defoliation when having high nutrient levels in their above-ground parts during the growing season. A study of the interaction between nutrient availability and the best timing of defoliation for greatest effect, by means of small-scale laboratory and field experiments, seems called for. The role of herbivores in the control of coarse, unpalatable grasses seems to be limited. In order to induce an effect, experiments could be set up to study the impact of high grazing density for short periods. So, for the dominant species focus is needed on nutrients, including litter decomposition, and regrowth. Most plant species depend on the space left by dominant plant species. These make up the species diversity and include the endangered species. Sometimes expected rare species do not come back despite years of management. Why? Is the seedbank absent? Is dispersal hampered? Introduction experiments may give the desired answers.

We know far too little about the effects of large herbivores on the establishment and maintenance of **diverse plant communities**. Experiments are too brief and research effort has been very limited. There is a particularly large gap in our knowledge on how selective grazing leads to heterogeneity in the landscape. It is poorly understood how certain proportions of macrostructural classes of short and tall herbaceous vegetation, scrub and woodland arise, and it is not known how the spatial arrangement of these classes is determined. Furthermore, we do not know which herbivore densities and herbivore combinations will lead to the desired result. Future research should focus on the following.

- Relevant parameters should be monitored in many areas for a long period. Streamlining of methods is essential. Monitoring should be the responsibility of the management but it is only practical to evaluate management once the goals for conservation have been defined. Scientists should analyse the data, and conduct experiments to study the mechanisms behind the changes observed. This combination of monitoring and research is often lacking.
- Experimental work should include various disciplines, and should focus on a small number of study areas for which many details are already known. This strategy is preferred to spreading out experimental work over many study areas.

- Ideally, controlled experiments with replicates at the landscape level are needed. Although these do not seem to be feasible in western Europe because of practical limitations, it should be realized that generalization of insights from experiments without replication is doubtful.
- The study of dose-effect relationships between key processes in the system, especially between herbivory and vegetation development, requires further attention. Experiments are needed with both constant numbers of herbivores at different stocking rates, and with dynamic herbivore populations. This could be achieved by means of grazing experiments in medium-sized enclosures (> 24 ha), supplemented with intensive small-scale work, e.g. clipping experiments. The data should be used as input in subsequent modelling work (Chapter 10).
- Analysis and study of (large-scale) natural 'experiments' – such as storms, fire, floods and herbivore crashes after diseases or severe winters – may provide valuable evidence to gain insight into ecosystem functioning. Natural experiments deserve more attention as a source of information, especially where the manipulation of ecosystems meets practical limitations.
- The role of herbivores in the dispersal and establishment of plant species is poorly understood. Field data on the amount of diaspores attached to large herbivores have only recently become available (Fischer *et al.*, 1996). It turns out that the shape of diaspores is not a good predictor of the probability of attachment to the herbivores, and so more field work and experiments are necessary. Even so, such data only describe the potential for being transported. Whether diaspores actually reach safe sites for establishment after being released from the animals needs elucidation by experimental evidence. The interplay with the soil seedbank may be important in determining the eventual chance of establishment. An overview of existing data on the seed longevity of many plant species is now available (Thompson *et al.*, 1997) but data for 75% of the endangered species of the northwestern European flora are lacking.

Research with respect to the restoration of **wilderness** areas (Figure 11.7) is still in its infancy. Apart from theoretical work, the following topics require special attention:

- As area is limited, the problem of small population size is relevant. What is the critical minimum size of viable herbivore populations? What are the effects of inbreeding? What is the role of stochasticity in determining population viability?
- As long as systems are incomplete, and large predators (for instance) are absent, it is important to study the mechanisms of population regulation in order to arrive at a well informed strategy on (the necessity of) population control. This demands further research on the relative importance of predation and food limitation in regulating ungulate populations.

Figure 11.7 Yosemite National Park, California, one of the icons in nature conservation. An example for future wilderness areas in Europe?

Long-term experiments are needed in which interspecific effects on population parameters are measured.

- The potential keystone role of herbivores should be studied by the approach discussed earlier combining modelling and small-scale experiments. In addition, key processes and system dynamics can be studied in large-scale field experiments with different herbivore assemblages.
- Our knowledge of the interspecific relationships within herbivore assemblages is limited. More work is needed on herbivore–herbivore interactions. Does competition take place or is facilitation operating? What are the conditions that determine the occurrence of either form of species interaction?

From the above it can be concluded that the way ahead is long. Nevertheless we feel that the questions still to be answered pose a challenge to scientists and managers alike. It is clear that attention is slowly moving towards a systems approach. It is inspiring to see that both large-scale experiments *in situ* and clever computer modelling of complex and chaotic systems are gradually being implemented, and become related. The future lies in cooperation and in an interdisciplinary approach.

REFERENCES

Arcese, P. and Sinclair, A.R.E. (1997) The role of protected areas as ecological baselines. *Journal of Wildlife Management*, **63**, 587–602.

Bakker, J.P., Poschlod, P, Strijkstra, R.J. *et al.* (1996) Seed banks and seed dispersal: important topics in restoration ecology. *Acta Botanica Neerlandica*, **45**, 461–490.

Beintema, A.J. and Müskens, G.J.D.M. (1987) Nesting success of birds in a Dutch agricultural grassland. *Journal of Applied Ecology*, **24**, 743–758.

Bignal, E.M. and McCracken, D.I. (1996) Low-intensity farming systems in the conservation of the countryside. *Journal of Applied Ecology*, **33**, 413–424.

Everts, F.H. and De Vries, N.P.J. (1991) *De Vegetatieontwikkeling van Beekdalsystemen. Een landschapsecologische studie van enkele Drentse beekdalen*, Historische Uitgeverij, Groningen.

Fischer, S., Poschlod, P. and Beinleich, B. (1996) Experimental studies on the dispersal of plants and animals by sheep in calcareous grasslands. *Journal of Applied Ecology*, **33**, 1206–1222.

Noss, R.F. (1992) The Wildlands Project – land conservation strategy. *Wild Earth*, **Special Issue**, 10–25.

Thompson, K., Bakker, J.P. and Bekker, R.M. (1997) *Soil Seed Banks of NW Europe: methodology, density and longevity*, Cambridge University Press, Cambridge.

Van Eerden, M.R., Zijlstra, M., Van Roomen, M. and Timmerman, A. (1996) The response of *Anatidae* to changes in agricultural practice: long-term shifts in the carrying capacity of wintering waterfowl. *Gibier Faune Sauvage*, **13**, 681–706.

Index

Page numbers appearing in **bold** refer to figures; page numbers appearing in *italic* refer to tables.

Printed in Great Britain
by Amazon.co.uk, Ltd.,
Marston Gate.